Human Pathogenic Papillomaviruses

Edited by H. zur Hausen

With 37 Figures and 15 Tables

Springer-Verlag
Berlin Heidelberg New York
London Paris Tokyo
Hong Kong Barcelona
Budapest

Harald zur Hausen
Deutsches Krebsforschungszentrum
Vorsitzender des Stiftungsrates
Im Neuenheimer Feld 280
D-69120 Heidelberg
Germany

Cover illustration: Papillomaviruses inside a cell nucleus, forming dark spots among larger chromatin patches. (By courtesy of Hanswalter Zentgrag, DKFZ, Heidelberg.)

Cover design: Harald Lopka, Jlvesheim

ISBN 978-3-642-78489-7 ISBN 978-3-642-78487-3 (eBook)
DOI 10.1007/978-3-642-78487-3

Softcover reprint of the hardcover 1st edition 1994
Library of Congress Catalog Card Number 15-12910

Typesetting: Thomson Press (India) Ltd., New Delhi;

27/3020 – 5 4 3 2 1 0 – Printed on acid-free paper.

Preface

The hypothesis that cancer of the cervix is infectious was raised more than 150 years ago (Rigoni-Stern 1842). The first cell-free transmissions of papillomas were reported 95 years ago (McFadyan and Hobday 1898). A report of cell-free trans-mission of human warts was published in 1907 (Ciuffo). Thus, since these initial discoveries papillomavirus research has had to go a long way before it became possible to link these infections to cancer of the cervix (Dürst et al. 1983; Boshart et al. 1984). Table 1 lists a selection of the publications that form the basis for our present understanding of the role of human pathogenic papillomaviruses (HPV) in human cancers.

The identification of specific HPV types in cervical cancer in 1983, 1984 and in subsequent years substantially boosted activities in papillomavirus research. In part this is because cancer of the cervix ranks first in cancer incidence in developing countries and is an important cause of cancer death in affluent societies (Parkin et al. 1984; I.A.R.C. 1989). Premalignant cervical lesions, particularly in affluent societies, also contribute considerably to morbidity. Moreover, the availability of cell lines harboring HPV DNA and the identification of HPV genes as oncogenes have permitted in vitro analyses of HPV genome persistence, gene expression, and gene functions.

In addition to this large body of experimental evidence, large scale epidemiological studies have been completed during the past few years that have identified HPV infections as being the major risk factor for cervical cancer (Koutsky et al. 1988; Muñoz et al. 1992; Bosch et al. 1992).

The overwhelming amount of evidence linking specific HPV infections to cancer of the cervix is gradually leading to a shift of interest from questions related to causality towards the understanding of the mechanism by which HPV, the infected host cells, their cellular environment, and other factors contribute to cancer development. Moreover, the roles that members of this group of viruses play in other

Table 1. Selected publications on the history of HPV infections in human cancers (modified from zur Hausen 1991)

Year	Event	References
1842	Incidence of cancer of the cervix depends on sexual contacts and promiscuity	Rigoni-Stern
1898	Cell-free transmission of papillomas in dogs	McFadyan and Hobday
1907	Transmission of human warts by cell-free extracts	Ciuffo
1922	Description of epidermodysplasia verruciformis	Lewandowsky and Lutz
1933	Discovery of the cottontail rabbit papillomavirus (CRPV)	Shope and Hurst
1934–1952	Syncarcinogenesis of CRPV with chemical carcinogens	Rous and Beard 1934 Rous and Kidd 1938 Rous and Friedewald 1944
1946–1966	Experimental transmission of warts from epidermodysplasia verruciformis patients	Lutz 1946 Jablonska and Milewski 1957, Jablonska et al. 1966
1949	Electronmicroscopic demonstration of viral particles in human warts	Strauss et al.
1961	CRPV DNA is sufficient for carcinoma induction in domestic rabbits	Ito and Evans
1965	Physical characterization of HPV DNA	Crawford
1975–1976	Hypothesis that cervical cancer is caused by HPV infections	zur Hausen 1975, 1976
1976	Recognition of koilocytotic atypia of the cervix as HPV-CPE	Meisels and Fortin
1976–1977	Discovery of the plurality of HPV types	Gissmann and zur Hausen 1976, Gissmann et al. 1977, Orth et al. 1977
1978	Discovery of specific HPV types in carcinomas of epidermodysplasia verruciformis patients	Orth et al.
1980–1982	Description of genital wart viruses (HPV 6 and 11)	Gissmann and zur Hausen 1980, Gissmann et al. 1982
1982	Complete DNA sequence of BPV-1	Chen et al.
1983	Identification of HPV 16 in cervical cancer	Dürst et al.
1984	Identification of HPV 18 in cervical cancer and in HeLa cells	Boshart et al.
1985	Identification of specific viral transcripts in cervical cancer cells and of a specific mode of viral DNA integration	Schwarz et al. Yee et al.
1985	Induction of condyloma acuminatum-type lesions in heterografted human keratinocytes by HPV 11	Kreider et al.
1986	Malignant transformation of NIH 3T3 cells by HPV 16 DNA	Yasumoto et al.
1987	Immortalization of human foreskin keratinocytes by HPV 16 DNA	Pirisi et al., Dürst et al.
1988–1991	Different regulation of HPV transcription in non-maligant and malignant HPV positive cells	Rösl et al. 1988, 1991, Bosch et al. 1990, Dürst et al. 1991
1988–1992	Malignant phenotype of human cervical carcinoma cells and malignant transformation of rodent cells depend on HPV E6/E7 gene expression	von Knebel Doeberitz et al. 1988, 1992, Crook et al. 1989
1988–1992	Epidemiological studies revealing high risk HPV infections as major risk factors for cervical cancer development	Koutsky et al. 1988, Muñoz et al. 1992, Bosch et al. 1992
1989	E7 gene product binds the Rb protein	Dyson et al., 1989, Münger et al., 1989
1990	E6 gene product binds the p53 protein	Werness et al.

human cancers and in their prevention, diagnosis, and therapeutic approaches are clearly gaining in importance.

The editor gratefully appreciates the help of the many distinguished colleagues who have contributed to this volume. It hopefully will provide a summary of our present knowledge and experimentation in HPV research. For detailed descriptions of the molecular biology and clinical manifestations of HPV the reader is referred to previous reviews (Salzman and Howley 1987; Bunney et al. 1992).

References

Bosch FX, Schwarz E, Boukamp P, Fusenig NE, Bartsch D, zur Hausen H (1990) Suppression *in vivo* of human papillomavirus type 18 E6–E7 gene expression in non-tumorigenic HeLa-fibroblast hybrid cells. *J Virol* 64:4743

Boshart M, Gissmann L, Ikenberg H, Kleinheinz A, Scheurlen W, zur Hausen H (1984) A new type of papillomavirus DNA, its presence in genital cancer biopsies and in cell lines derived from cervical cancer. *EMBO J* 3:1151

Bunney MH, Benton C, Cubei H (1992) Viral warts: biology and treatment, 2nd edn. *Oxford UP, Oxford.*

Chen EY, Howley PM, Levinson AD, Seeburg PH (1982) The primary structure and genetic organization of the bovine papillomavirus type I genome. *Nature* 299:529

Ciuffo G (1987) Innesto positivo con filtrado di verrucae volgare. *G Ital Mal Venereol* 48:12

Crawford LV (1965) A study of human papillomavirus DNA. *J Mol Biol* 13:362

Crook T, Morgenstern JP, Crawford L, Banks L (1989) Continued expression of HPV 16 E7 protein is required for maintenance of the transformed phenotype of cells cotransformed by HPV 16 plus EJ-ras. *EMBO J* 8:513

Dürst M, Gissmann L, Ikenberg H, zur Hausen H (1983) A papillomavirus DNA from a cervical carcinoma and its prevalence in cancer biopsy samples from different geographic regions. *Proc Natl Acad Sci USA* 80:3812

Düst M, Bosch FX, Glitz D, Schneider A, zur Hausen H (1991) Inverse relationship between HPV 16 early gene expression and cell differentiation in nude mouse epithelial cysts and tumors induced by HPV positive human cell lines *J Virol* 65:796

Dürst M, Dzarlieva-Petrusevska RT, Boukamp P, Fusenig NE, Dürst M (1987) Molecular and cytogenetic analysis of immortalized human primary keratinocytes obtained after transfection with human papillomavirus type 16 DNA. *Oncogene* 1:251

Dyson N, Howley PM, Münger K, Harlow E (1989) the human papillomavirus-16 E7 oncoprotein is able to bind to the retinoblastoma gene product. *Science* 243:934

Gissmann L, zur Hausen H (1976) Human papilloma viruses: Physical mapping and genetic heterogeneity. *Proc Natl Acad Sci USA* 76:569

Gissmann L, zur Hausen H (1980) Partial characterization of viral DNA from human genital warts (condylomata acuminata). *Int J Cancer* 25:605

Gissmann L, Pfister H, zur Hausen H (1977) Human papilloma viruses (HPV): characterization of four different isolates. *Virology* 76: 569
Gissmann L, Diehl V, Schulz-Coulon H-J, zur Hausen H (1982) Molecular cloning and characterization of human papillomavirus DNA derived from a laryngeal papilloma. *J Virol* 44: 393
International Agency for Research on Cancer (IARC) (1989) *Biennial report* 1988/1989. IARC, Lyons
Ito Y, Evans CA (1961) Induction of tumors in domestic rabbits with nucleic acid preparations from partially purified Shope papillomavirus and from extracts of the papillomas of domestic and cottontail rabbits. *J Exp Med* 114: 485
Jablonska S, Milewski B (1957) Zur Kenntnis der Epidermodysplasia verruciformis. *Dermatologica* 115: 1
Jablonska S, Fabjanska L, Formas I (1966) On the viral aetiology of epidermodysplasia verruciformis. *Dermatologica* 132: 369
Koutsky LA, Galloway DA, Holmes KK (1988) Epidemiology of genital papillomavirus infection. *Epidemiol Rev* 10: 122
Kreider JW, Howett MK, Wolfe SA, Bartlett GL, Zaino RJ, Sedlacek TV, Mortel R (1985) Morphological transformation in vivo of human uterine cervix with papillomavirus from condylomata acuminata. *Nature* 317: 639
Lewandowsky T, Lutz W (1922) Ein Fall einer bisher nicht beschriebenen Hauterkrankung (Epidermodysplasia verruciformis). *Arch Dermatol Syphilol* 141: 193
Lutz W (1946) A propos de l'epidermodysplasie verruciforme. *Dermatologica* 92: 30
McFadyean J, Hobday F (1898) Note on the experimental "transmission of warts in the dog." *J Comp Pathol Ther* 11: 341
Meisels A, Fortin R (1976) Condylomatous lesions of the cervix and vagina. I. Cytological patterns. *Acta Cytol* 20: 505
Münger K, Werness B, Dyson N, Phelps W, Harlow E, Howley P (1989) Complex formation of the human papillomavirus E7 proteins with the retinoblastoma tumor suppressor gene product. *EMBO J* 9: 1147
Muñoz N, Bosch FX, de Sanjosé S, Tafur L, Izarzugaza I, Gili M, Viladiu P, Navarro C, Martos C, Ascunce N, Gonzales LC, Kaldor JM, Guerrero E, Lörincz A, Santamaria M, Alonso de Ruiz P, Aristizabal N, Shah K (1992) The causal link between human papillomavirus and invasive cervical cancer, a population based study in Columbia and Spain. *Int J Cancer* 52: 743
Orth G, Favre M, Croissant G (1977) Characterization of a new type of human papillomavirus that causes skin warts. *J Virol* 24: 100
Orth G, Jablonska S, Favre M, Croissant O, Jarzabek-Chorzelska M, Rzesa G (1978) Characterization of two types of human papillomaviruses in lesions of epidermodysplasia verruciformis. *Proc Natl Acad Sci USA* 75: 1537
Parkin DM, Stjernswärd J, Muir CS (1984) Estimates of the worldwide frequency of twelve major cancers. *Bull WHO* 62: 163
Pirisi L, Yasumoto S, Fellery M, Doninger JK, DiPaolo JA (1987) Transformation of human fibroblasts and keratinocytes with human papillomavirus type 16 DNA. *J Virol* 61: 1061
Rigoni-Stern D (1842) Fatti statistici relativi alle malatia cancerose. *G Serv Prog Pathol Therap* 2: 507
Rösl F, Dürst M, zur Hausen H (1988) Selective suppression of human papilomavirus transcription in non-tumorigenic cells by 5-azacytidine. *EMBO J* 7: 1321
Rösl F, Achtstetter T, Hutter K-J, Bauknecht T, Futterman G, zur Hausen H (1991) Extinction of the HPV 18 upstream regulatory region in cervical carcinoma cells after fusion with non-tumorigenic human keratinocytes under non-selective conditions. *EMBO J* 10: 1337

Rous P, Beard JW (1934). The progression to carcinoma of virus-induced rabbit papillomas (Shope). *J Exp Med* 62: 523

Rous P, Friedewald WF (1944) The effect of chemical carcinogens on virus-induced rabbit papillomas. *J Exp Med* 79: 511

Rous P, Kidd JG (1938) The carcinogenic effect of a papilloma virus on the tarred skin of rabbits. I. Description of the phenomenon. *J Exp Med* 67: 399

Schwarz E, Freese UK, Gissmann L, Mayer W, Roggenbuck B, Stremlau A, zur Hausen H (1985) Structure and transcription of human papillomavirus sequences in cervical carcinoma cells. *Nature* 314: 111

Salzman NP, Howley PM (eds) The papovaviridae, vol 2: *the papillomaviruses.* Plenum, New York, 1987

Shope RE, Hurst EW (1933) Infectious papillomatosis of rabbits. With a note on histopathology. *J Exp Med* 58: 607

Strauss MJ, Shaw EW, Bunting H, Melnick JL (1949) "Crystalline" virus-like particles from skin papillomas characterized by intranuclear inclusion bodies. *Proc Soc Exp Biol Med* 72: 46

von Knebel Doeberitz M, Oltersdorf T, Schwarz E, Gissmann L (1988) Correlation to modified human papillomavirus early gene expression with altered growth properties in C4-I cervical carcinoma cells. *Cancer Res* 48: 3780

von Knebel Doeberitz M, Rittmüller C, Geitz D, zur Hausen H, Dürst M (1992) Inhibition of tumorigenicity of cervical cancer cells in nude mice by HPV 18 E6–E7 antisense RNA. *Int J Cancer* 51: 831

Werness BA, Levine AJ, Howley PM (1990) Association of human papillomavirus types 16 and 18 E6 proteins with p53. *Science* 248: 76

Yasumoto S, Burchardt AL, Doniger J, Di Paolo JA (1986) Human papillomavirus type 16 DNA induced malignant transformation of NIH3T3 cells. *J Virol* 57: 572

Yee C, Krishnan-Hewlett Z, Baker CC, Schlegel R, Howley PM (1985) Presence and expression of human papillomavirus sequences in human cervical carcinoma cells lines. *Am J Pathol* 119: 361

zur Hausen H (1975) Oncogenic herpesviruses. *Biochim Biophys* Acta 417: 25

zur Hausen H (1975) Condylomata acuminata and human genital cancer. *Cancer Res* 36: 794

zur Hausen H (1991) Papillomavirus/host cell interactions in the pathogenesis of anogenital cancer. In: Origins of human cancer. *Cold Spring Harbor Laboratory Press, Cold Spring Harbor* pp 685–705

Contents

List of Contributors

(Their addresses can be found at the beginning of their respective chapters.)

Human Pathogenic Papillomavirus Types: An Update

E.-M. DE VILLIERS

In 1917 WAELSCH demonstrated the induction of a lesion differing morphologically from the tumor of origin, when a pigmented flat wart developed on the arm of a volunteer after intradermal inoculation of a cell suspension from a condyloma acuminatum. Subsequently, several other investigators made similar observations (GROSS 1983). These observations can be considered as early hints for the diversity of clinical effects caused by individual types of the human papillomavirus (HPV) group as demonstrated during the past 16 years.

The number of identified and characterized HPV types has increased considerably over the last few years. To date the group consists of 70 distinct types. The initial classification of a type was based on the homology of the DNA genomes utilizing liquid hybridization techniques (reassociation kinetics) (COGGIN and ZUR HAUSEN 1979) (these data were compiled in a review by DE VILLIERS 1989a). A large number of HPV types were isolated from lesions of epidermodysplasia verruciformis (EV) patients and defined as types according to this hybridization technique. Their homology among each other often bordered the cut-off limit for a new type. It therefore happened, after additional data had become available, that HPV-46 (GROSS et al. 1988) had to be reclassified as HPV-20b. To avoid any future confusion, it was decided to leave this number vacant.

This initial definition of an HPV type has recently been modified after sequencing of the DNA genomes became more feasible. At present a new HPV type is defined by a comparison of the DNA sequences of the E6, E7, and L1 open reading frames (ORFs) to the same ORFs of all the known HPV types. If this homology proves to be lower than 90%, the full-length genome constitutes a new type. Homology higher than 90% renders it a subtype of another.

Historically, the HPVs have been grouped according to the location of the lesions from which they were initially isolated, thus the terminology of "cutaneous" or "genital" types. In most instances, a "new type" was used as probe in examining a series of other tumors, mostly though of the same clinical entity as that from which it originated. As more distinct types became

Abteilung Tumorvirus-Charakterisierung, Angewandte Tumorvirologie, Deutsches Krebsforschungszentrum, Im Neuenheimer Feld 242, 69120 Heidelberg, Germany

Current Topics in Microbiology and Immunology, Vol. 186

available, series of tumors of very diverse origin were tested for the presence of any papillomavirus DNA (DE VILLIERS et al. 1986b). The clinical data which have evolved since render it increasingly difficult to divide the HPV types into subgroups. Phylogenetic analyses (CHAN et al. 1992; VAN RANST et al. 1992; Bernard et al., this volume) have been utilized in an attempt to find a suitable classification system for the papillomaviruses. These have been performed with minimal consideration of clinical findings. A subdivision of these viruses, based mainly on sequence information, would certainly present a biased view. The pathogenesis of both benign and malignant tumors, although not completely understood, should be taken into consideration.

In the following, the biological features of these viruses will be weighed against the relationship to each other regarding their genetic composition.

Considering the different clinical types of cutaneous warts, JABLONSKA et al. (1985) proposed a classification according to clinical morphology and the type of infecting HPV. The group of lesions occurring in patients with EV were treated as a separate entity, as were atypical warts such as pigmented plaques in immunosuppressed patients. The cytopathic effects caused by a single infection with, for example, HPV-1, HPV-2, HPV-3, HPV-4, HPV-7 are very distinct, whereas those induced by, for example, an HPV-10 or HPV-28 infection are more difficult to distinguish from each other.

Recently, HPV-63 was identified by recognition of the very distinct type of inclusion body (ICB) seen histologically in multiple punctuated keratotic lesions on the sole of the foot (Egawa, submitted). These were described as a filamentous type of ICB, in contrast with the granular type of ICB seen in an HPV-1-induced lesion. Upon characterization and sequencing of the HPV-63 DNA (EGAWA et al., 1993), the nucleotide homology between the genomes of HPV-1 and HPV-63 was less than 66%.

A third type of ICB, the homogenous type, is induced by infections with HPV-4, HPV-60, and HPV-65 (EGAWA et al., 1993). The genomes of these viruses were demonstrated in common warts and keratotic flat lesions on both the hands and feet, as well as in two biopsies from cutaneous horns (cheek and leg). An interesting aspect of this specific study was that all the lesions containing the HPV DNA were pigmented. The genome sequences of HPV-4 and HPV-65 share 83% nucleotide homology and preliminary sequence data reveal less homology with HPV-60 DNA, but nevertheless sufficiently high to group these three viruses together. HPV-60 was, however, identified in (EGAWA et al. 1990) and isolated from (MATSUKURA et al. 1992) an epidermoid cyst. This points to the fact that one type of HPV can induce different types of clinical lesions. However, not all epidermoid cysts are induced by one type of papillomavirus, but most probably by additional uncharacterized HPV types (MEYER et al. 1991; Egawa, unpublished data).

The intracytoplasmic inclusion bodies are associated with the protein coded for by the E4 ORF of the HPV (DOORBAR et al. 1989, 1991; ROGEL-GAILLARD et al. 1992). The E4 ORFs are very poorly conserved among the different HPV types. Even though they present with the same histological features, the diversity of the E4 ORF sequences of HPV-4, -60, and -65 renders a nucleotide alignment at this stage almost impossible (Delius, personal communication).

An even more remote HPV type is HPV-41. This virus was isolated from facial, foot, and perianal lesions present in the same patient. These lesions differed from each other in their clinical morphology, although the sheet-like appearance of the warts seem to be associated with HPV-41-induced lesions (GRIMMEL et al. 1988; Leigh, personal communication). The histological features are distinct from those previously discussed. HPV-41 has been detected in a number of squamous cell carcinomas of the skin (HIRT et al. 1990; DE VILLIERS 1991). Phylogenetically this virus is being grouped with HPV-1, HPV-63 and the animal viruses canine oral papillomavirus (COPV) and cottontail rabbit papillomavirus (CRPV) (Bernard et al., this volume). The overall nucleotide homology of its genome to other HPVs is below 50%, the closest being HPV-18 with 49% homology. In addition, certain genomic features, for example, E2-binding sites in the long control region, typical for all other papillomaviruses, are not present in the HPV-41 DNA (HIRT et al. 1990).

Epidermodysplasia verruciformis is a rare hereditary disease characterized by the development of multiple cutaneous warts varying from each other, both in morphology as well as the HPV type which they harbor (JABLONSKA 1990). Lesions induced by specific EV HPV types have a predilection for malignant conversion, i.e., HPV-5 and HPV-8. HPV-14, HPV-17, HPV-20, and HPV-47 have each been detected in single squamous cell carcinomas in EV patients.

The HPV types associated with EV are closely related to each other and subgrouped on the basis of percentage nucleotide homology (ORTH 1986). Phylogenetic analyses of the L1 ORF, as well as analysis of the transcription regulatory regions, divided subgroups into even smaller subgroups (Bernard et al., this volume), but these do not seem to correlate with their biological behavior, e.g., their oncogenic potential.

The closely related HPV-3 and HPV-10 are HPV types almost always present in some lesions of EV patients, but are not EV specific. They are associated with plane wart-like lesions which display a distinct histological pattern (JABLONSKA 1990). HPV-28 is less prevalent but shares the above-mentioned characteristics.

Additionally, these viruses are frequently detected in biopsies obtained from cutaneous warts of immunosuppressed patients (OBALEK et al. 1992; de Villiers and Leigh, unpublished results). Although an HPV 10-like cleavage pattern was originally observed in DNA from a vulvar carcinoma of a patient with EV (GREEN et al. 1982), HPV-3 and HPV-10 have been associated with the induction of benign cutaneous lesions. HPV-3 does not seem to induce lesions of mucosal origin, even under conditions of immunosuppression. Phylogeny, however, groups these HPV types with the "mucosal" types.

The presence of more than one HPV type within the same lesion has often been reported. Although the HPVs all have the same genome organization, the cytopathic changes of these lesions seem to be induced by only one HPV type present, for example, lesions harboring HPV-3, HPV-5, as well as HPV-23 display the typical HPV-3-induced cytopathic effect (OBALEK et al. 1992). Utilizing differential in situ hybridization, the DNA of HPV-1 and HPV-63 could be detected within the same cell in a keratotic lesion of the foot. Histologically only the filamentous type of ICB, characteristic for HPV-63-associated lesions, was present (Egawa et al., in press).

HPV 2 has historically been associated with the common hand wart (ORTH et al. 1977). Similar to HPV-3, this HPV type is often prevalent in cutaneous lesions of EV patients, as well as in immunocompromised patients. Different from HPV-3, however, is its association with a number of tumors of the oral mucosa. After the detection of HPV-2 in a tongue carcinoma (DE VILLIERS et al. 1985), studies utilizing in situ hybridization revealed the presence of this HPV type in a number of benign and malignant oral tumors (SHAH 1990). HPV-2 shares an 83% nucleotide homology with HPV-57 (HIRSCH-BEHNAM et al. 1990). This renders the distinction between an HPV-2 and an HPV-57 infection under the conditions of in situ hybridization very unlikely. Additional studies are thus needed to establish the true prevalence of HPV-2 DNA in the mucosal lesions of the oral cavity.

HPV 57, in contrast, was not only isolated and detected in lesions of mucosal origin, but is present in cutaneous lesions as well (DE VILLIERS et al. 1989). HPV-57 was isolated from an inverted papilloma of the maxillary sinus. This lesion later developed into an invasively growing tumor causing the death of the patient. WU et al. (1993) detected this virus in a series of benign squamous papillomas and inverted papillomas with and without carcinoma, all from the nasal cavity.

Interestingly, increasing data are becoming available on the detection of HPV-2 in anogenital warts in children (OBALEK et al. 1990; NUOVO et al. 1991). In our series of 12 anogenital lesions in children aged 1–10 years, only three cases harbored HPV-6/11 DNA, whereas one case contained HPV-2 DNA. Three additional cases harbored HPV-57 DNA, one HPV-57-related sequences, and three cases HPV-27 DNA. HPV 57 has infrequently been detected in cervical intraepithelial neoplasias in adults (de Villiers, unpublished results). HPV 27 was isolated from cutaneous verrucae in an immunosuppressed patient (OSTROW et al. 1989a), but has not been detected in the screening of large series of biopsies of varying origin (de Villiers, unpublished results).

The DNA sequence of HPV-27 shares a very high degree of homology to sequences of HPV-2 (87.9%) and HPV-57 (80.4%) (Delius and Hofmann, this volume). Based on the clinical data already mentioned, it seems justified to group HPV-2, HPV-57, and HPV-27 together. Additional information on the extent to which the anogenital or oral lesions induced by these three HPV types, are localized at the mucosal-cutaneous junction, could be of great interest.

Another HPV type grouped as "mucosal," but whose primary site of infection seems to be the skin, is HPV-7. HPV-7 is very prevalent in the extensive common warts and common wart-like, papillomatous lesions which often develop on the hands of butchers or persons handling meat (JABLONSKA et al. 1987) or fish (RÜDLINGER et al. 1989a). The HPV-7-induced lesions displayed typical histological features and were clinically more hypertrophic and hyperkeratotic (JABLONSKA et al. 1987). The only two patients, other than the mentioned occupationally related group, in whom HPV-7 DNA was detected suffered from extensive cutaneous lesions over prolonged periods of time. Various therapeutic measures proved unsuccessful (DE VILLIERS et al. 1986a). To my knowledge, this HPV type has thus far not been detected in lesions developing in transplant patients. It is, however, often present in oral papillomatous lesions (GREENSPAN et al. 1988), as well as the accompanying facial or neck lesions (DE VILLIERS 1989b) frequently seen in human immunodeficiency virus (HIV)-infected individuals. Contrary to the HPV-7-induced skin lesions, the oral mucosal lesions display a variety of clinical features and are without characteristic histology. The data imply special events occurring in the pathogenesis of HPV-7-associated lesions.

An HPV type which shares a high homology in its DNA sequence to HPV 7 is HPV 40. Relatively little is known about this type. It is infrequently found in condylomata acuminata, as well as in genital intraepithelial neoplasias (DE VILLIERS et al. 1989). The penile intraepithelial neoplasia from which it was originally isolated, was resistant to therapeutic measures (Neumann, personal communication). The only four HPV-40-positive condylomata acuminata found were all situated in the vaginal vault of hysterectomized patients. These two papillomaviruses are phylogenetically grouped with the "high-risk mucosal" types. Clinical data available on HPV-7 certainly does not support this classification, whereas more data are needed in the case of HPV-40.

HPV 34 was isolated from a biopsy of Bowen's disease of the skin (KAWASHIMA et al. 1986). It was subsequently detected in genital intraepithelial neoplasias. Upon characterization of the genome, KAWASHIMA et al. demonstrated the formation of heteroduplexes with the HPV-16 genome in their E1, E2, and L1 ORFs in 50% formamide. Phylogenetically, this HPV type does not fall in the group of HPV-16 viruses (Bernard et al., this volume).

The group of HPV types associated with genital lesions has been treated differentially. Reports on the prevalence of different types dealt with the HPV types available to the authors at the moment the study was conducted. With the advent of the polymerase chain reaction method and additional sequence data it will hopefully soon be possible to fill in the information still lacking. At present 34 of the 70 identified HPV types have been associated with anogenital lesions. Many of the types recently characterized were isolated from genital lesions. These are HPV-61, -62, -64, -66, -67, -68, -69, and -70 (Table 1). HPV types 1–60 have been reviewed in DE VILLIERS (1989a)

Table 1. Recently identified HPV types and their origin

Type	Isolated from	Reference
61	VaIN	T. Matsukura, personal communication
61b	Cervical biopsy	S. Beaudenon, personal communication
62	VaIN	T. Matsukura, personal communication
63	Myrmecia	EGAWA et al., 1993
64	VaIN	T. Matsukura, personal communication
65	Pigmented wart	EGAWA et al., 1993
66	Cervical carcinoma	TAWHEED et al., 1991
67	VaIN	T. Matsukura, personal communication
68	Genital lesion	M. Longuet, personal communication
69	CIN	T. Matsukura, personal communication
70	Vulvar papilloma	S. Beaudenon, M. Longuet, personal communications and BEAUDENON et al., 1987

VaIN, vulvar intraepithelial neoplasia, CIN, cervical intraepithelial neoplasia.

Two large studies screening biopsies from cervical lesions have been conducted LÖRINCZ et al. (1992) used 15 common genital HPV types as probes. The authors grouped the different HPV types according to their prevalence in low- and high-grade squamous intraepithelial leisons, as well as in cancers. The resulting groups were HPV-6, -11, -42, -43, and -44 as low-risk types; HPV-31, -33, -35, -51, -52, and -58 as an intermediate group; and HPV-16, -18, -45, and -56, as high-risk papillomaviruses. In the second study, BERGERON et al. (1992) probed with 18 genital HPV types. HPV-16 was again the most prevalent type detected in 21% low-grade cervical intraepithelial neoplasia (CIN) and 57% high-grade CIN lesions. HPV-30, -39, -45, -51, -52, -56, -58, and -61 were present in 44% low-grade CIN and in only 8% high-grade CIN lesions. It can be noted that, for example, HPV-45 and HPV-52 fall into different categories when comparing the two studies. The prevalence of HPV-52 (YAJIMA et al. 1988; TAKAMI et al. 1991) and HPV-58 (MATSUKURA and SUGASE 1990) in invasive carcinomas of the cervix appears to be higher in Asia. HPV-18 is obviously more prevalent in Africa than in other countries studied (TER MEULEN et al. 1992).

Several of these HPV types have also been demonstrated in cutaneous lesions. HPV-16 has been demonstrated in Bowen's disease of the foot (STONE et al. 1987), an epidermal naevus of the foot and a case of arsenic keratosis (DE VILLIERS 1988), bowenoid lesions of the hand (KÜHNL-PETZOLD et al. 1988; MOY et al. 1989; OSTROW et al. 1989b; de Villiers, unpublished results), a case of bowenoid papulosis of the face (GROB et al. 1991), squamous cell carcinomas of the finger (ELIEZRI et al. 1990) and of the lip (KAWASHIMA et al. 1990), as well as in a keratoacanthoma of the finger (MOY and QUAN 1991). HPV-35 DNA was detected in a bowenoid lesion of the finger (RÜDLINGER et al. 1989b) and HPV-59 in a papillomatous lesion of the lip (Roy-Berman, Gerhardt, and de Villiers, unpublished results).

Comparing the above-mentioned data with the phylogenetic analyses, two HPV types are grouped with the group related to HPV-16, i.e., HPV-32 and HPV-42. HPV-42 has to date only been found in benign genital lesions (BEAUDENON et al. 1987; LÖRINCZ et al. 1992), thus rendering its classification together with the intermediate and high-grade HPV types as questionable. HPV-32, on the other hand, was isolated from a lesion of focal epithelial hyperplasia of the oral cavity and subsequently found in a number of oral papillomas (BEAUDENON et al. 1987; DE VILLIERS 1989c). Upon Southern blot hybridization, this HPV type cross-hybridizes strongly to HPV-18 and not to HPV-16. The same doubt as mentioned for HPV-42, therefore exists.

Human papillomavirus type 26 was originally isolated from four biopsies taken from an immunosuppressed patient. These included verrucae vulgares from the foot and a finger, as well as condyloma acuminata of the right and left medial thighs (OSTROW et al. 1984). Although the detection of HPV-26 has not been reported subsequently, recent data show the presence of this virus in a number of genital lesions (Manos, personal communication). The DNA sequence of HPV-26 shares the highest nucleotide homology to HPV-51 (58.8%) (Delius and Hofmann, this volume).

According to the phylogenetic tree (BERNARD et al., this volume), HPV-30, HPV-53, and HPV-56 are grouped very closely together. HPV-30 was isolated from a laryngeal carcinoma (KAHN et al. 1986), HPV-53 from a normal cervix (GALLAHAN et al. 1989), and HPV-56 from a CIN (LÖRINCZ et al. 1989). Subsequent analysis indicate the rare occurrence of both HPV-30 and HPV-53 in genital lesions and the association of HPV-56 with the high-risk genital HPVs. Despite their very diverses sources of origin, these three HPV types do show varying degrees of homology in hybridization experiments.

The presence of papillomaviral sequences has been described in the cervical carcinoma cell line ME180 (REUTER et al. 1991). The tumor cells did not, however, comprise a complete HPV genome, precluding classification as a new HPV type. The available sequences show more than 90% homology to the recently characterized HPV-68 (Longuet, personal communication).

The closely related HPV-6, HPV-11, and HPV-13 are grouped together. HPV-55 is also closely related. HPV-6 and HPV-11 are commonly associated with benign and low-grade anogenital lesions, as well as with laryngeal papillomas. The notion has been to regard lesions containing these viruses as benign or harmless. This has proved not to be the case. Although the risk for malignant transformation by these viruses appears to be different from, for example, HPV-16 or HPV-18 (BOSHART and ZUR HAUSEN 1986; ROSEN and AUBORN 1991; MCGLENNEN et al. 1992), HPV-6 and HPV-11 have been found in a number of carcinomas of the vulva (RANDO et al. 1986; KASHER and ROMAN 1988; RUSK et al. 1991; BRANDENBERGER et al. 1992; DALING and SHERMAN 1992) or the anus (PALEFSKY et al. 1991; ZAKI et al. 1991). In addition, these two viruses have been detected in carcinomas of the tongue (Kitasato et al., in press), as well as single cases of squamous cell carcinomas of the lung (BYRNE et al. 1987; HELMUTH and STRATE 1987; GUILLON et al. 1991; DILORENZO et al. 1992). Closer characterization mainly of HPV-6

DNA, but also of HPV-11 DNA in these lesions, revealed considerable rearrangements in the upstream regulatory region, such as duplications, deletions, insertions or mutations, thus deviating considerably not only from the respective "prototype" viruses, but among the different isolates themselves. A fair percentage of benign papillomatous lesions of the oral cavity, as well as the nasal and paranasal sinuses, have been shown to contain either HPV-6 or HPV-11 DNA (DE VILLIERS 1989c; KASHIMA et al. 1992). Only in a few benign lesions has the genomic structure of these two viruses been analyzed (FARR et al. 1991; RÜBBEN et al. 1992; Kitasato et al., in press) and rearrangements been determined. Additional data are urgently needed to determine which isolates represent the true prototypes and which mechanisms determine these rearrangements.

Over the years, the methods used for the detection of HPV DNA in lesions varied substantially in their sensitivity and specificity. Except for a few cases, the HPV genomes present were usually not analyzed in more detail. Possible variations in their genomic constellation, as seen in the case of HPV-6, could therefore have been overlooked. Such analyses of other HPV types might be necessary in future.

The isolation and characterization of yet unknown papillomaviruses is a continuing process. The accumulating data suggest a possible association of papillomaviruses with a broader range of tumors in different organs, e.g., upper aerodigestive tract, bladder, and the eye. The number of biopsies from these sites containing known HPV types still remains small, although histological observations of precursor lesions suggest an involvement of papillomaviruses in the development of such lesions. Many of these results have been conflicting and require further clarification. The identification of new, maybe very distantly related, HPV types may nevertheless be useful in studying the etiology of a larger variety of human tumors.

References

Beaudenon S, Kremsdorf D, Obalek S, Jablonska S, Pehau-Arnaudet G, Croissant O, Orth G (1987) Plurality of genital human papillomaviruses: characterization of two new types with distinct biological properties. Virology 161: 374–384

Bergeron C, Barrasso R, Beaudenon S, Flamant P, Croissant O, Orth G (1992) Human papillomaviruses associated with cervical intraepithelial neoplasia. Great diversity and distinct distribution in low- and high-grade lesions. Am J Surg Pathol 16: 641–649

Boshart M, zur Hausen H (1986) Human papillomavirus in Buschke-Löwenstein tumors: physical state of the DNA and identification of a tandem duplication in the noncoding region of a human papillomavirus 6 subtype. J Virol 58: 963–966

Brandenberger AW, Rüdlinger R, Hänggi W, Bersinger NA, Dreher E (1992) Detection of human papillomavirus in vulvar carcinoma. A study by in situ hybridization. Arch Gynecol Obstet 252: 31–35

Byrne JC, Tsao MS, Fraser RS, Howley PM (1987) Human papillomavirus-11 DNA in a patient with chronic laryngotracheobronchial papillomatosis and metastatic squamous cell carcinoma of the lung. N Engl J Med 317: 873

Chan SY, Bernard HU, Ong CK, Chan SP, Hormann B, Delius H (1992) Phylogenetic analysis of

48 papillomavirus types and 28 subtypes and variants: a showcase for the molecular evolution of DNA viruses. J Virol 66: 5714–5725
Coggln JR Jr, zur Hausen H (1979) Workshop on papillomaviruses and cancer. Cancer Res 39: 545–546
Daling JR, Sherman KJ (1992) Relationship between human papillomavirus infection and tumours of anogenital sites other than the cervix. In: Munoz N, Bosch FX, Shah KV, Meheus A (eds) The epidemiology of cervical cancer and human papillomavirus. International Agency for Research on Cancer, Lyon, pp 223–241 (IARC Publ 119)
de Villiers E-M (1988) Implication of papillomaviruses in nongenital tumours. In: de Palo R, Rilke F, zur Hausen H (eds) Herpes and papillomaviruses. Raven, New York, pp 65–70
de Villiers E-M (1989a) Heterogeneity of the human papillomavirus group. J Virol 63: 4898–4903
de Villiers E-M (1989b) Prevalence of HPV 7 papillomas in the oral mucosa and facial skin of patients with human immunodeficiency virus. Arch Dermatol 125: 1590
de Villiers E-M (1989c) Papillomaviruses in cancers and papillomas of the aerodigestive tract. Biomed Pharmacother 43: 31–36
de Villiers E-M (1991) Human papillomaviruses. In: Schmähl D, Penn I (eds) Cancer in organ transplant recipients. Springer, Berlin Heidelberg New York, pp 106–112
de Villiers E-M, Weidauer H, Otto H, zur Hausen H (1985) Papillomavirus DNA in human tongue carcinomas. Int J Cancer 36: 575–578
de Villiers E-M, Neumann C, Oltersdorf T, Fierlbeck G, zur Hausen H (1986a) Butcher's wart virus (HPV 7) infections in non-butchers. J Invest Dermatol 87: 236–238
de Villiers E-M, Schneider A, Gross G, zur Hausen H (1986b) Analysis of benign and malignant urogenital tumors for human papillomavirus infection by labelling cellular DNA. Med Microbiol Immunol 174: 281–286
de Villiers E-M, Hirsch-Behnam A, von Knebel Doeberitz C, Neumann C, zur Hausen H (1989) Two newly identified human papillomavirus types (HPV 40 and HPV 57) isolated from mucosal lesions. J Virol 171: 248–252
DiLorenzo TP, Tamsen A, Abramson AL, Steinberg BM (1992) Human papillomavirus type 6a DNA in a lung carcinoma of a patient with recurrent laryngeal papillomatosis is characterized by a partial duplication. J Gen Virol 73: 423–428
Doorbar J, Coneron I, Gallimore PH (1989) Sequence divergence yet conserved physical characteristics among the E4 proteins of cutaneous human papillomaviruses. Virology 172: 51–62
Doorbar J, Ely S, Sterling J, McLean C, Crawford L (1991) Specific interaction between HPV-16 E1–E4 and cytokeratins results in collapse of the epithelial cell intermediate filament network. Nature 352: 824–827
Egawa K (1994) New types of human papillomaviruses and intracytoplasmic inclusion bodies: a classification of inclusion warts according to clinical features, histology and associated HPV types. Br. J. Dermatol 130
Egawa K, Inaba Y, Ono T, Arao T (1990) "Cystic papilloma" in humans? Demonstration of human papillomavirus in plantar epidermoid cyst. Arch Dermatol 126: 1599–1603
Egawa K, Delius H, Matsukura T, Kawashima M, de Villiers E-M (1993) Two novel types of human papillomavirus, HPV 63 and HPV 65: comparisons of their clinical and histological features and DNA sequences to other HPV types. Virology 194: 789–799
Egawa K, Shibasaki Y, de Villiers E-M (1993) Double infection with HPV 1 and 63 in single cells of a lesion displaying only an HPV 63-induced cytopathogenic effect. Lab Invest vol. 69 (Nov. 93)
Eliezri YD, Silverstein SJ, Nuovo GJ (1990) Occurrence of human papillomavirus type 16 DNA in cutaneous squamous and basal cell neoplasms. J Am Acad Dermatol 23: 836–842
Farr A, Wang H, Kasher MS, Roman A (1991) Relative enhancer activity and transforming potential of authentic human papillomavirus type 6 genomes from benign and malignant lesions. J Gen Virol 72: 519–526
Gallahan D, Müller M, Schneider A, Delius H, Kahn T, de Villiers E-M, Gissmann L (1989) Human papillomavirus type 53. J Virol 63: 4911–4912
Green M, Brackmann KH, Sanders PR, Löwenstein PM, Freel JH, Eisinger M, Switlyk SA (1982) Isolation of a human papillomavirus from a patient with epidermodysplasia verruciformis: presence of related viral DNA genomes in human urogenital tumors. Proc Natl Acad Sci USA 79: 4437–4441
Greenspan D, de Villiers E-M, Greenspan JS, de Souza YG, zur Hausen H (1988) Unusual HPV types in oral warts in association with HIV infection. J Oral Pathol 17: 482–487

Grimmel M, de Villiers, E-M, Neumann C, Pawlita M, zur Hausen H (1988) Characterization of a new human papillomavirus (HPV 41) from disseminated warts and detection of its DNA in some skin carcinomas. Int J Cancer 41: 5-9

Grob JJ, Zarour H, Jacquemier J, Hassoun J, Bonerandi JJ (1991) Extra-genital HPV 16-related bowenoid papulosis. Genitourin Med 67: 18-20

Gross L (1983) Oncogenic viruses, 3rd edn. Pergamon, Oxford

Gross G, Ellinger K, Roussaki A, Fuchs PG, Peter HH, Pfister H (1988) Epidermodysplasia verruciformis in a patient with Hodgkin's disease: characterization of a new papillomavirus type and interferon treatment. J Invest Dermatol 91: 395-407

Guillon L, Sahli R, Chaubert P, Monnier P, Cuttat JF, Costa J (1991) Squamous cell carcinoma of the lung in a nonsmoking, nonirradiated patient with juvenile laryngotracheal papillomatosis. Evidence of human papillomavirus-11 DNA in both carcinoma and papillomas. Am J Surg Pathol 15: 891-898

Helmuth RA, Strate RW (1987) Squamous carcinoma of the lung in a nonirradiated, nonsmoking patient with juvenile laryngotracheal papillomatosis. Am J Surg Pathol 11: 643

Hirt L, Hirsch-Behnam A, de Villiers E-M (1990) Nucleotide sequence of human papillomavirus (HPV) type 41: an unusual HPV type without a typical E2 binding site consensus sequence. Virus Res 18: 179-190

Hirsch-Behnam A, Delius H, de Villiers E-M (1990) A comparative sequence analysis of two human papillomavirus (HPV) types 2a and 57. Virus Res 18: 81-98

Jablonska S (1990) Human papillomaviruses in skin carcinomas. In: Pfister H (ed) Papillomaviruses and human cancer. CRC Press, Boca Raton, pp 46-71

Jablonska S, Orth G, Obalek S, Croissant O (1985) Cutaneous warts. Clinical, histological and virological correlations. In: Jablonska S, Orth G (eds) Clinics in dermatology, vol 3. Lippincott, Philadelphia, pp 71-82

Jablonska S, Obalek S, Favre M, Golebiowska A, Croissant O, Orth G (1987) The morphology of butchers' warts as related to papillomavirus types. Arch Dermatol Res 279: S66-S72

Kahn T, Schwarz E, zur Hausen H (1986) Molecular cloning and characterization of the DNA of a new human papillomavirus (HPV 30) from a laryngeal carcinoma. Int J Cancer 37: 61-65

Kasher MS, Roman A (1988) Characterization of human papillomavirus type 6b DNA isolated from an invasive squamous carcinoma of the vulva. Virology 165: 225-233

Kashima HK, Kessis T, Hruban RH, Wu TC, Zinreich SJ, Shah KV (1992) Human papillomavirus in sinonasal papillomas and squamous cell carcinoma. Laryngoscope 102: 973-976

Kawashima M, Jablonska S, Favre M, Obalek S, Croissant O, Orth G (1986) Characterization of a new type of human papillomavirus found in a lesion of Bowen's disease of the skin. J Virol 57: 688-692

Kawashima M, Favre M, Obalek S, Jablonska S, Orth G (1990) Premalignant lesions and cancers of the skin in the general population: evaluation of the role of human papillomaviruses. J Invest Dermatol 95: 537-542

Kitasato H, Delius H, zur Hausen H, Sorger K, Rösl F, zur Hausen H, de Villiers E-M. Sequence rearrangements in the HPV 6 URR: Are these involved in malignant transition? J Gen Virol (in press)

Kühnl-Petzold C, Grösser A, de Villiers E-M (1988) HPV 16 assozierter bowenoid-verrucösser Tumor der palmaren Haut. Aktuel Dermatol 14: 848-849

Lörincz AT, Quinn AP, Goldsborough MD, McAllsiter, P, Temple GF (1989) Human papillomavirus type 56: a new virus detected in cervical cancers. J Gen Virol 70: 3099-3104

Lörincz AT, Reid R, Jenson AB, Greenberg MD, Lancaster WD, Kurman RJ (1992) Human papillomavirus infection of the cervix: relative risk associations of 15 common anogenital types. Obstet Gynecol 79: 328-337

Matsukura T, Sugase M (1990) Molecular cloning of a novel papillomavirus (type 58) from an invasive cervical carinoma. Virology 177: 833-836

Matsukura T, Iwasaki T, Kawashima M (1992) Molecular cloning of a novel papillomavirus (type 60) from a plantar cyst with characteristic pathological changes. Virology 190: 561-564

McGlennen RC, Ghai J, Ostrow RS, LaBresh K, Schneider JF, Faras AJ (1992) Cellular transformation by a unique isolate of human papillomavirus type 11. Cancer Res 52: 5872-5878

Meyer LM, Tyring SK, Little WP (1991) Verrucous cyst. Arch Dermatol 127: 1810-1812

Moy RL, Quan MB (1991) The presence of human papillomavirus type 16 in squamous cell carcinoma of the proximal finger and reconstruction with a bilobed transposition flap. J Dermatol Surg Oncol 17: 171-175

Moy RL, Eliezri YD, Nuovo GJ, Zitelli JA, Bennett BG, Silverstein S (1989) Human papillomavirus type 16 DNA in periungual squamous cell carcinomas. JAMA 261: 2669–2673
Nuovo GJ, Lastarria DA, Smith S, Lerner J, Comite SL, Eliezri YD (1991) Human papillomavirus segregation patterns in genital and non-genital warts in prepubertal children and adults. Am J Clin Pathol 95: 467–474
Obalek S, Jablonska S, Favre M, Walczak L, Orth G (1990) Condylomata acuminata in children: frequent association with human papillomaviruses responsible for cutaneous warts. J Am Acad Dermatol 23: 205–213
Obalek S, Favre M, Szymanczyk J, Misiewicz J, Jablonska S, Orth G (1992) Human papillomavirus (HPV) types specific of epidermodysplasia verruciformis detected in warts induced by HPV 3 or HPV 3-related types in immunosuppressed patients. J Invest Dermatol 98: 936–941
Orth G (1986) Epidermodysplasia verruciformis: a model for understanding the oncogenicity of human papillomaviruses. Ciba Found Symp 120: 157–174
Orth G, Favre M, Croissant O (1977) Characterization of a new type of human papillomavirus that causes skin warts. J Virol 24: 108–120
Ostrow R, Zachow KR, Thompson O, Faras AJ (1984) Molecular cloning and characterization of unique type of human papillomavirus from an immunedeficient patient. J Invest Dermatol 82: 362–366
Ostrow RS, Shaver MK, Turnquist S, Visknins A, Bender M, Vance C, Kaye V, Faras AJ (1989a) Human papillomavirus-16 DNA in a cutaneous invasive cancer. Arch Dermatol 125: 666–669
Ostrow RS, Zachow KR, Shaver MK, Faras AJ (1989b) Human papillomavirus type 27: detection of a novel human papillomavirus in common warts of a renal transplant recipient. J Virol 63: 4904
Palefsky JM, Holley EA, Gonzales J, Berline J, Ahn DK, Greenspan JS (1991) Detection of human papillomavirus DNA in anal intraepithelial neuplasia and anal cancer. Cancer Res 51: 1014–1019
Rando RF, Groff DE, Chirikjian JG, Lancaster WD (1986) Isolation and characterization of a novel human papillomavirus type 6 DNA from an invasive vulvar carcinoma. J Virol 57: 353–356
Reuter S, Delius H, Kahn T, Hofmann B, zur Hausen H, Schwarz E (1991) Characterization of a novel papillomavirus DNA in the cervical carcinoma cell line ME180. J Virol 65: 5564–5568
Rogel-Gaillard C, Breitburd F, Orth G (1992) Human papillomavirus type 1 E4 proteins differing by their N-terminal ends have distinct cellular localizations when transiently expressed in vitro. J Virol 66: 816–823
Rosen M, Auborn K (1991) Duplication of the upstream regulatory sequences increases the transformation potential of human papillomavirus type 11. Virology 185: 484–487
Rübben A, Beaudenon S, Favre M, Schmitz W, Spelten B, Grußendorf-Conen E-I (1992) Rearrangements of the upstream regulatory region of human papillomavirus type 6 can be found both in Buschke-Löwenstein tumours and in condylomata acuminata. J Gen Virol 73: 3147–3153
Rüdlinger R, Bunney MH, Grob R, Hunter JAA (1989a) Warts in fish handlers. Br J Dermtol 120: 375–381
Rüdlinger R, Grob, R, Yu YX, Schnyder UW (1989b) Human papillomavirus-35-positive verruca with bowenoid dysplasia of the periungual area. Arch Dermatol 125: 655–659
Rusk D, Sutton G, Look K, Roman A (1991) Analysis of invasive squamous cell carcinoma of the vulva and vulvar intraepithelial neoplasia for the presence of human papillomavirus DNA. Obstet Gynecol 77: 918–922
Shah KV (1990) Papillomavirus infections of the respiratory tract, the conjunctive and the oral cavity. In: Pfister H (ed) Papillomaviruses and human cancer. CRC Press, Boca Raton, pp 74–90.
Stone M-S, Noonan C-A, Tschen J, Bruce S (1987) Bowen's disease of the feet. Presence of human papillomavirus 16 DNA in tumor tissue, Arch Dermtol 123: 1517–1520
Takami Y, Kondoh G, Saito J, Noda K, Sudiro TM, Sjahrurachman A, Warsa UC, Sutsudo M, Hakura A (1991) Cloning and characterization of human papillomavirus type 52 from cervical carcinoma in Indonesia. Int. J Cancer 48: 516–522
Tawheed AR, Banderson S, Favre M, Orth G (1991) Characterization of human papillomavirus type 66 from an invasive carcinoma of the uterine cervix. J Clin Microbiol 29: 2656–2660
Ter Meulen J, Eberhardt HC, Luande J, Mgaya HN, Chang-Claude J, Mtiro H, Mhina M,

Kashaija P, Ockert S, Yu X, Meinhardt G, Gissmann L, Pawlita M (1992) Human papillomavirus (HPV) infection, HIV infection and cervical cancer in Tanzania, East Africa. Int J Cancer 51: 515–521
Van Ranst M, Kaplan JB, Burk RD (1992) Phylogenetic classification of human papillomaviruses: correlation with clinical manifestations. J Gen 73: 2653–2660
Waelsch L (1917) Übertragungsversuche mit spitzem Kondylom. Arch Dermatol Syph 124: 625–646
Wu T-C, Trujillo J, Kashima H, Mounts P (1993) Association of human papillomavirus with nasal neoplasia. Lancet 341: 522–524
Yajima H, Noda T, de Villiers E-M, Yajima A, Yamamoto K, Noda K, Ito Y (1988) Isolation of a new type of human papillomavirus (HPV 52b) with a transforming activity from cervical cancer tissue. Cancer Res 48: 7164–7172
Zaki SR, Judd R, Coffield LM, Greer P, Rolston F, Evatt BL (1992) Human papillomavirus infection and anal carcinoma. Retrospective analysis by in situ hybridization and the polymerase chain reaction. Am J Pathol 140: 1345–1355

Primer-Directed Sequencing of Human Papillomavirus Types

H. DELIUS and B. HOFMANN

1 Introduction

Since the involvement of papillomaviruses in the pathogenesis of human tumors was recognized, a great number of different virus isolates have been collected from a variety of lesions. Especially, the observation of a correlation between the development of malignant cervical cancer and the presence of certain types of human papillomavirus (HPV) has spurred an extensive search for possible new types of these viruses (for a review see ZUR HAUSEN 1991).

At the Referenzzentrum für humanpathogene Papillomviren at the DKFZ (German Cancer Research Center) in Heidelberg 70 different types of human papillomaviruses have been registered (see de Villiers, this volume). Classification into types was initially based on the rather crude classification by liquid hybridization (COGGIN and ZUR HAUSEN 1979). However, during the last few years, with the availability of routine nucleotide-sequencing

Institut für Angewandte Tumorvirologie, Deutsches Krebsforschungszentrum, Im Neuenheimer Feld 242, 69120 Heidelberg, Germany

Current Topics in Microbiology and Immunology, Vol. 186

methods, a refined definition of new types has been adopted by Papillomavirus Nomenclature Committee. This definition draws a somewhat arbitrary line between types and sybtypes, but the basis for these decisions is now well defined: the alignments of the nucleotide sequences of the open reading frames (ORFs) E6, E7 and L1 of a new type with the corresponding sequences of any of the known types should not exceed 90% homology in any one of these ORFs. Any closer relationship will be called a subtype of the type exhibiting the highest homology to the newly isolated viral genome.

To date 23 complete nucleotide sequences of human papillomaviruses have been published and submitted to the GenBank and European Molecular Biology Laboratory (EMBL) databases. These are listed in Table 1. The list comprises HPV-1a, HPV-2a, HPV-4, HPV-5a and -5b, HPV-6b, HPV-8, HPV-11, HPV-13, HPV-16, HPV-18, HPV-31, HPV-33, HPV-35, HPV-39, HPV-41, HPV-42, HPV-47, HPV-51, HPV-57, HPV-58, HPV-63, and HPV-65.

We started our sequencing project in 1989 with the intention of ultimately determining all the sequences of the different virus types registered at the *Referenzzentrum*, and in addition the sequences of possible new isolates. We hope that a detailed analysis of the sequence information may contribute to the development of better diagnostic tools and to an understanding of the functional differences displayed by the different viral isolates. In addition, the sequences will constitute a unique set of related virus genomes for phylogenetic studies (see BERNARD et al., this volume).

Table 1. Published sequences

	Type	Total bp	Reference
1	HPV-1a	7615	DANOS et al. 1980
2	HPV-2a	7860	HIRSCH-BEHNAM et al. 1990
3	HPV-4	7353	EGAWA et al. 1993
4	HPV-5a	7746	ZACHOW et al. 1987
5	HPV-5b	7779	YABE et al. 1991
6	HPV-6b	7902	SCHWARZ et al. 1983
7	HPV-8	7654	FUCHS et al. 1986
8	HPV-11	7931	DARTMANN et al. 1986
9	HPV-13	7880	VAN RANST et al. 1992
10	HPV-16	7904	SEEDORF et al. 1985
11	HPV-18	7857	COLE and DANOS 1987
12	HPV-31	7912	GOLDBOROUGH et al. 1989
13	HPV-33	7909	COLE and STREECK 1986
14	HPV-35	7851	MARICH et al. 1992
15	HPV-39	7833	VOLPERS and STREECK 1991
16	HPV-41	7614	HIRT et al. 1991
17	HPV-42	7917	PHILIPP et al. 1992
18	HPV-47	7726	KIYONO et al. 1990
19	HPV-51	7808	LUNGU et al. 1991
20	HPV-57	7861	HIRSCH-BEHNAM et al. 1990
21	HPV-58	7824	KIRII et al. 1991
22	HPV-63	7348	EGAWA et al. 1993
23	HPV-65	7308	EGAWA et al. 1993

2 Methodology

The DNAs of the different papilloma virus types were available for sequencing at the *Referenzzentrum* in the form of plasmid clones. In the majority of cases, the complete viral genomes were cloned into the vectors as one DNA fragment which had been obtained by linearization of the viral genomes at one restriction site only. Only in a few cases (HPV-7, HPV-10, and HPV-35) were the viral DNAs cloned in two segments. We decided to avoid subcloning these original isolates and to determine the sequence by using exclusively primer-directed sequencing, also called the "primer walking" strategy, which means obtaining the sequences near the cloning site with a primer derived from the vector sequence and all the remaining sequences with primer oligonucleotides derived from the newly determined sequences.

2.1 Sequencing Techniques

Plasmids were grown in *Escherichia coli* bacteria in LB broth. DNA was prepared by alkaline lysis and purified by cesium chloride density centrifugation using standard procedures (SAMBROOK et al. 1989). The sequence determination was done using the dideoxy termination method (SANGER et al. 1977). T7 DNA polymerase (Pharmacia) was used to incorporate ^{35}S-labeled dATP (10 μCi/reaction at a specific activity of 6000 Ci/m*M*). The incorporation of the dideoxynucleotides into the sequencing product was improved by the use of manganese-containing buffer as described by TABOR and RICHARDSON (1989), conditions which lead to more even band intensities. The labeled products were separated after denaturation on 6% polyacrylamide gels (35 × 43 cm, 0.4 mm thickness). A short and a long run were prepared for each sample. Ninety-six tracks could be loaded per gel so that up to 12 reactions could be analyzed per gel. After electrophoresis the gels were transferred to cellulose filter sheets and dried on a vacuum gel drier. Kodak XAR film was exposed, usually for a short exposure of about 3 h and for a long exposure overnight. The sequences were read manually into a PC using an X-Y digitizer (Brühl, Nürnberg) and assembled into a consensus sequence using dedicated software written by one of us (H.D.).

2.2 Clones Used for Sequence Determinations

Table 2 shows a list of the newly determined sequences. In this table the plasmids used for the sequencing of the different types are listed together with the references referring to the description of their origin. Primers of a length of 20 nucleotides were selected from the new sequences without the

Table 2. New sequences

	Type	Total nucleotides	GC (%)	Oligo-nucleotide (*n*)	Nucleotides Oligo-nucleotide	Redundancy	Vector	Cloning site	Subclones (*n*)	Reference
1	HPV-3	7820	45.5	61	256	2.86	pBR322	BamHI	1	Kremsdorf et al. 1983
2	HPV-7	8027	39.5	74	217	3.41	pBR322	HindIII	2	Oltersdorf et al. 1986
3	HPV-9	7434	41.0	57	261	2.84	pBR322	BamHI	1	Kremsdorf et al. 1982
4	HPV-10	7919	45.9	68	233	3.03	PBR322	SalI	2	Kremsdorf et al. 1983
5	HPV-12	7673	42.3	71	216	3.5	pBR322	HindIII	1	Kremsdorf et al. 1983
6	HPV-14D	7439	40.9	68	219	3.35	pBR322	HindIII	1	Kremsdorf et al. 1984[a]
7	HPV-15	7412	39.6	61	253	3.02	pBR322	BamHI	1	Kremsdorf et al. 1984
8	HPV-17	7427	40.9	67	222	3.3	pBR322	BamHI	1	Kremsdorf et al. 1984
9	HPV-19	7685	40.6	59	261	2.93	pEMBL	BamHI	1	Gassenmaier et al. 1984
10	HPV-25	7713	41.8	61	253	3.07	pEMBL	EcoRI	1	Gassenmaier et al. 1984
11	HPV-26	7855	38.6	64			pUC19	BamHI	1	Ostrow et al. 1984[b]
12	HPV-27	7823	48.6	48			pUC8	BamHI	1	Ostrow et al. 1989[b]
13	HPV-30	7852	40.4	60	262	2.85	pBR322	BamHI	1	Kahn et al. 1986
14	HPV-32	7961	41.0	62	257	2.86	pUC19	EcoRI	1	Beaudenon et al. 1987
15	HPV-34	7723	38.2	66	234	2.93	pSP64	BglII/ BamHI	1	Kawashima et al. 1986b
16	HPV-35H	7879	36.9	61	258	2.75	pBR322	BamHI	2	Lorincz et al. 1987
17	HPV-40	7909	43.8	62	255	2.91	pUC19	SacI	1	de Villiers et al. 1989
18	HPV-45	7858	39.6	62	253	2.88	pGEM-1	HindIII	1	Naghashfar et al. 1987
19	HPV-49	7560	41.1	61	248	2.88	pGEM-4	EcoRI	1	Favre et al. 1989b
20	HPV-52	7942	38.6	60	265	2.68	pUC19	EcoRI	1	Shimoda et al. 1988
21	HPV-53	7856	40.1	64	246	3	pUC?	EcoRI	1	Gallahan et al. 1989
22	HPV-56	7844	37.9	64	245	2.94	pT713	BamHI	1	Loerincz et al. 1989a
		170611		1381	Average: 251	Average: 2.9				

[a] This virus was cloned by Kremsdorf et al. (1984) from a patient with a double infection together with HPV-15. The pBR322 clone which has been sequenced contained a HindIII fragment 386 nucleotides in size derived from the HPV-15 E1 ORF at the HindIII cloning site which must have replaced a fragment of similar size in HPV-14 (see matrix analysis, Fig.1).

[b] Part of the sequence had been obtained by using subcloning methods. The entire sequence was confirmed by primer-directed sequencing on the original complete insert.

help of specialized software. We were essentially trying to select primer sequences with stable 3′ ends (GC basepairing) followed by A-containing sequences assuring a sufficient incorporation of label. The position was chosen quite conservatively about 60–80 nucleotides upwards from the 3′ end of the previously determined sequence in order to obtain a good overlap

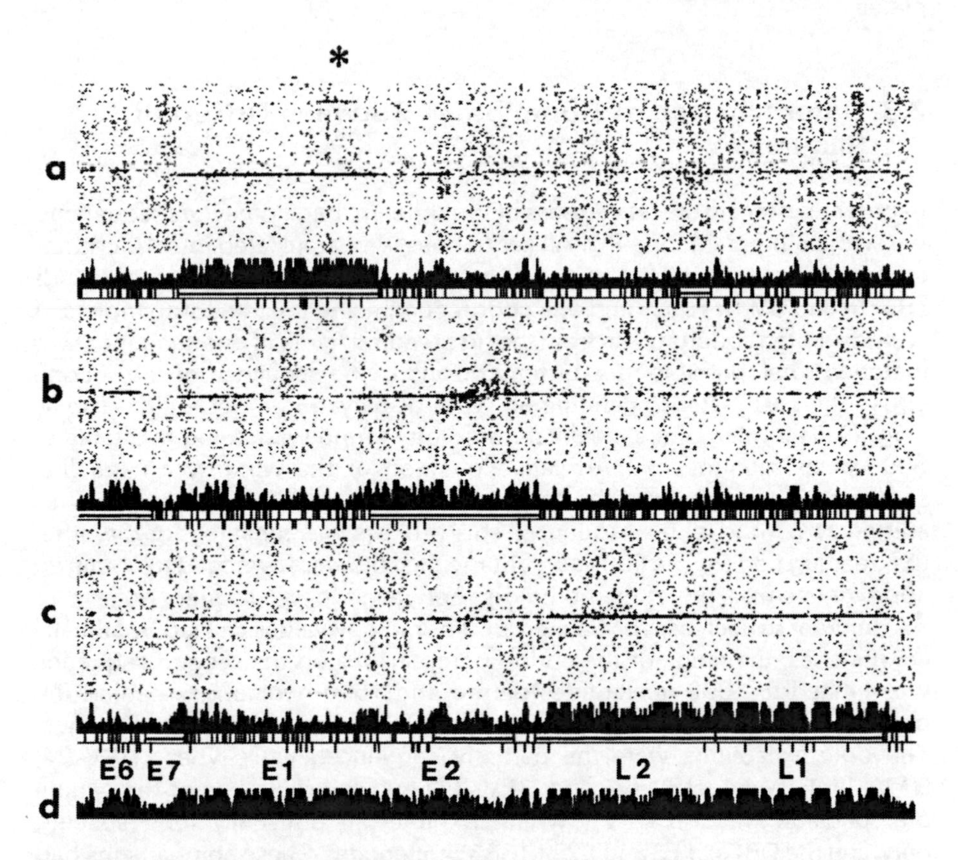

Fig. 1a–d. Matrix plot of homologies between HPV-14 and HPV-25. The three plot fields (**a**– **c**) depict the homologies of 10-codon windows taken from the three different reading frame positions HPV-25 determined on the sequence of HPV-14. The *horizontal axis* represents the length of HPV-25, and the *vertical axis* of the three plot fields the sequence range checked from − 2000 nucleotides to + 2000 nucleotides around the corresponding position on HPV-14 in a compressed form. The *histograms* below the plot fields represent the maximal homologies found. They are plotted together with the maps of ORFs located in the corresponding reading frame (*marks above the line* represent termination codons, *marks below* represent start codons). It can be noted that there is no significant homology observed in part of the E6–E7 region. Instead, there is a duplicate homology observed near the end of the E1 ORF (*star*). This reflects the presence of a fragment derived from the E1ORF of HPV-15 which replaces a fragment of HPV-14 missing from the E6–E7 region in this clone of HPV-14. The sequence of HPV-14 without the HPV-15 segment will be submitted to the database as HPV-14D. The *bottom histogram* (**d**) is based on the maximal homologies determined in any of the three frames. The average of these values was used to compile the tables of homologies between all of the papillomavirus sequences (Tables 4–6). The *lables above the histogram* mark the positions of the major ORFs

between the last sequence stretch and the new one. One in about 15–20 primers failed. About one third of the failures could be traced back to a poor oligonucleotide synthesis, whereas a different primer sequence had to be selected to overcome the remaining failures. All sequences were determined completely on both complementary strands of the double-stranded templates.

2.3 Determination of Homologies Between Different Sequences

In order to estimate the relatedness of the different papillomavirus sequences, we wrote a program for a PC. All sequences were translated into 10 codons comprising groups of triplets coding for functionally related amino acids (SMITH and SMITH 1990) and one codon comprising the three stop triplets. A window of ten codons taken from the first reading frame of one sequence was matched to all possible runs of ten codons in the sequence of the second virus. The best matches determined were displayed in a dot matrix display (Fig. 1). The value of the maximal homology found was stored in a table of 600 values referring to the relative position of the window on the first sequence. The location of the sampling window on the first sequence was shifted across the whole sequence. This process was repeated for the other two reading frames. The 600 values of the best fits were averaged to obtain an approximate measure of the homology between the two sequences.

In order to decrease the background homology values caused by statistical matches, the range of the codon comparisons on the second sequence was limited to 1000 nucleotides before and 1000 nucleotides behind the position corresponding to the location of the window on the first sequence. The only exceptions were the partially sequenced virus types (HPV-24, HPV-38, HPV-54, HPV-64, and HPV-67) and the sequence of the canine oral papillomavirus (COPV), which exhibits an unusually long spacing between the ORFs of E2 and L2 of 1583 nucleotides. These comparisons had to be done across the whole sequences since the deviation from colinearity in these cases exceeded the range of ±1000 nucleotides.

3 Results

The total length and GC content of the newly determined papillomavirus sequences are listed in Table 2. The permutation of the submitted sequences was arbitrarily chosen to position the start of the putative E6 ORFs to nucleotide 102 in the case of the genital viruses and to nucleotide 200 in the case of the epidermodysplasia-related types.

3.1 Editing of Sequences

We have tried to edit the sequences essentially by a comparison of the complementary sequence determinations on both strands of the plasmids. Wherever we found a mismatch between them or wherever we had marked nucleotides as uncertain at the time of the measurement, we tried to resolve the differences by close inspection of the films. In some cases discrepancies were due to band compressions which we tried to clarify by repeating the corresponding sequencing reactions using deaza-dGTP in the reaction mix in order to inhibit hairpin formation during gel electrophoresis. We did not check the integrity of ORFs before we had completely finished this initial stage of editing. Using this procedure we were left with fewer than three errors per viral genome, as evidenced by disrupted ORFs. In two cases, in the sequences of HPV-53 and 56, we were not able to detect possible errors causing the disruption of the E1 ORFs. We assume that in these cases we are dealing either with a cloning artifact (but in these two cases the cloning sites are not directly involved) or with clones which do not necessarily represent an intact viral genome, as is known, from example, from the sequenced prototype of HPV-16 or from the sequence analysis of the original isolate of HPV-41 (HIRT et al. 1991).

3.2 Error Estimates

The error rate according to the criteria described above should be fewer than three errors per 7500 nucleotides or better than one error in about 2500 nucleotides even for the regulatory regions which cannot be checked for the integrity of ORFs.

In the case of HPV-35 we have sequenced a viral DNA which was sequenced by another group at the same time. The sequence has been published by MARICH et al. (1992). We have found about 70 differences to our newly determined sequence. We are therefore submitting the new sequence as a separate entry to the database, with the designation HPV-35H. An alignment of several supposed gene products with related peptides from other HPV types (data not shown) indicates that a better agreement to genes of related viruses is obtained using the new sequence than using the previously published sequence of HPV-35.

Similarly, we observed that a stretch of amino acids in the putative E1 gene product of HPV-25 which had previously been described as being divergent from other HPVs (KRUBKE et al. 1987) does not contain a frame shift in the newly determined complete sequence as had been originally suggested. According to our data, the divergence can be explained by two sequencing errors, an insertion and a deletion causing the local shift in the reading frame.

3.3 Redundancy of Sequence Determinations

The redundancies in the sequence determination given in Table 2 are calculated as the ratio of all bases measured during the sequence determination (including overlaps between long and short sequence runs on the gels and the vector sequences) to the final length of the viral insert sequence. The average sequencing redundancy of only 2.9, which was necessary to obtain complete sequencing runs on both complementary strands, compares favorably with the redundancies of about eight to ten fold which are usually required when complete coverage should be obtained using the "shot-gun" approach, i.e., by sequencing on random subclones of the DNA. The average yield of 251 nucleotides per primer is equivalent to about 125 nucleotides of finished sequence per primer.

3.4 Advantages and Disadvantages of the Primer Walking Strategy

The "primer walking" strategy has the following advantages:

1. The possibility of introducing artifacts during subcloning procedures is avoided.
2. Possible difficulties in obtaining a continuous sequence without gaps are immediately noticed.
3. One batch of template DNA serves for a large number of sequence determinations. A good quality of template can be assured by CsCl density gradient centrifugation.
4. The redundancy of the sequence determinations can be kept low in comparison with a shot-gun approach because the primers can be positioned to obtain small overlaps between contiguous sequence runs.
5. The assembly of sequence runs is very simple. Only a match between the end of the sequence determined last and the new segment has to be found.

Potential disadvantages of this strategy are:

1. The progression through a sequence is rather slow. If a sequence had to be obtained rapidly this could not be easily achieved using this strategy as long as there are no known additional starting points within the insert sequence. In the case of the papillomavirus genomes, we have in several cases successfully used the very well conserved 12-mer (TAAAACGAAAGT) as a primer which has been described by CRAVADOR et al. (1989). On the other hand, when a great number of templates are to be sequenced in parallel, the number of sequencing ends which are available may already be sufficient to allow efficient cycles of primer synthesis, gel electrophoresis, and reading of the new sequences.

2. The cost of oligonucleotide synthesis is relatively high. The chemistry required for oligonucleotide synthesis is rather expensive. In order to save reagents we have adapted the synthesis cycles on our commercial synthesizer to the synthesis of smaller quantities of oligonucleotides. These quantities are still more than 100-fold in excess of what is required for sequencing. In the meantime, we have constructed an oligonucleotide synthesizer specifically designed for the synthesis of small oligonucleotide quantities for their use as sequencing primers.

When considering the total expense of sequencing the cost of synthesizing the primers may be outweighed by the savings obtained by lowering the necessary redundancy in sequencing and by avoiding the preparation of subclones and their DNA.

3.5 Comparison of Homologies Between Sequences of Different Papillomavirus Types

The commonly available computer programs for sequence comparisons are usually dependent on the possibility of obtaining reasonable alignments of nucleotide or amino acid sequence stretches normally corresponding to related ORFs. We did not have suitable software available for a comparison of the relatedness of the complete sequences of the different papillomavirus

Table 3. Sequences

	Type	Approximate nucleotides	Reference
Nonedited			
1	HPV-20	7748	Gassenmaier et al. 1984
2	HPV-21	7778	Kremsdorf et al. 1984
3	HPV-22	7363	Kremsdorf et al. 1984
4	HPV-23	7326	Kremsdorf et al. 1984
6	HPV-28	7955	Favre et al. 1989c
7	HPV-29	7908	Favre et al. 1989a
8	HPV-36	7718	Kawashima et al. 1986a
9	HPV-37	7418	Scheurlen et al. 1986
10	HPV-44	7828	Lörincz et al. 1989b
11	HPV-48	7105	Mueller et al. 1989
12	HPV-50	7183	Favre et al. 1989d
13	HPV-55	7822	Favre et al. 1990
14	HPV-60	7312	Matsukura et al. 1992
15	HPV-61	7986	Matsukura, unpublished
16	HPV-66	7818	Tawheed et al. 1991
Partial			
1	HPV-24	3037	Kremsdorf et al. 1984
2	HPV-38	5356	Scheurlen et al. 1986
3	HPV-54	7141	Favre et al. 1990
4	HPV-64	5981	T. Matsukura (unpublished data)
5	HPV-67	4850	T. Matsukura (unpublished data)

Table 4. Homologies between complete HPV sequences (including nonedited sequences)

HPV-1		HPV-2		HPV-3		HPV-4	
HPV-63	52.51	HPV-27	87.92	HPV-28	77.74	HPV-65	83.07
CRPV	42.31	HPV-57	81.05	HPV-10	76.31	HPV-60	45.79
HPV-17	40.96	HPV-61	51.16	HPV-29	67.55	HPV-48	43.31
HPV-9	40.75	HPV-10	50.03	HPV-2	50.25	HPV-49	42.50
HPV-25	40.57	HPV-3	49.96	HPV-27	50.18	HPV-50	42.25
HPV-14	40.42	HPV-29	49.25	HPV-57	49.98	HPV-14	41.59
HPV-36	40.42	HPV-28	48.72	HPV-61	48.62	HPV-15	41.59
HPV-15	40.40	RhPV-1	46.68	HPV-18	47.57	HPV-9	41.53
HPV-19	40.24	HPV-53	46.05	HPV-51	46.79	HPV-36	41.51
HPV-5B	40.22	HPV-45	45.48	HPV-39	46.68	HPV-37	41.51
HPV-5B		HPV-6B		HPV-7		HPV-8	
HPV-5C	97.27	HPV-11	78.59	HPV-40	80.00	HPV-12	76.29
HPV-36	80.01	HPV-13	67.05	PCPV-1	48.87	HPV-5B	74.81
HPV-47	77.07	HPV-55	63.31	HPV-13	48.87	HPV-5C	74.74
HPV-8	74.20	PCPV-1	63.29	HPV-11	48.83	HPV-47	74.27
HPV-12	73.38	HPV-44	62.48	HPV-6B	47.88	HPV-36	72.98
HPV-25	67.50	HPV-32	48.55	HPV-44	47.44	HPV-25	68.18
HPV-19	67.01	HPV-7	48.31	HPV-55	47.42	HPV-19	67.05
HPV-21	65.81	HPV-40	47.98	HPV-42	46.59	HPV-14	66.81
HPV-14	65.42	HPV-42	47.25	HPV-32	46.40	HPV-21	66.20
HPV-20	64.87	RhPV-1	46.94	HPV-31	45.99	HPV-20	64.81
HPV-9		HPV-10		HPV-11		HPV-12	
HPV-17	67.66	HPV-3	75.50	HPV-6B	78.33	HPV-8	76.37
HPV-37	66.74	HPV-28	75.24	HPV-13	67.20	HPV-5B	74.44
HPV-15	66.66	HPV-29	69.00	PCPV-1	64.22	HPV-5C	74.00
HPV-22	56.57	HPV-27	50.00	HPV-55	64.18	HPV-47	73.29
HPV-23	55.75	HPV-57	49.77	HPV-44	63.24	HPV-36	72.03
HPV-49	53.46	HPV-2	49.61	HPV-7	49.09	HPV-25	66.09
HPV-5C	53.40	HPV-61	48.85	HPV-32	48.79	HPV-19	65.42
HPV-5B	53.24	HPV-18	48.03	HPV-40	48.38	HPV-14	65.05
HPV-47	52.77	HPV-45	47.55	HPV-42	47.48	HPV-21	64.42
HPV-25	52.64	HPV-51	47.18	RhPV-1	46.96	HPV-20	63.38
HPV-13		HPV-14		HPV-15		HPV-16	
PCPV-1	75.72	HPV-21	80.77	HPV-17	78.68	HPV-35H	62.53
HPV-55	71.48	HPV-20	77.50	HPV-37	77.31	HPV-31	62.12
HPV-44	69.66	HPV-25	74.88	HPV-9	66.87	HPV-35	60.85
HPV-11	67.48	HPV-19	73.77	HPV-22	57.62	HPV-33	55.25
HPV-6B	67.16	HPV-5C	66.53	HPV-23	57.44	HPV-52	54.66
HPV-7	49.25	HPV-5B	66.22	HPV-49	54.77	HPV-58	54.27
HPV-32	49.14	HPV-8	66.16	HPV-5C	53.14	RhPV-1	51.50
HPV-40	48.72	HPV-47	65.51	HPV-5B	53.00	HPV-34	50.00
HPV-42	48.62	HPV-36	64.98	HPV-47	52.57	HPV-45	48.72
RhPV-1	46.75	HPV-12	64.44	HPV-8	52.42	HPV-42	48.33
HPV-17		HPV-18		HPV-19		HPV-20	
HPV-37	81.72	HPV-45	74.03	HPV-25	82.22	HPV-14	77.98
HPV-15	78.77	HPV-39	59.11	HPV-14	74.79	HPV-21	77.87
HPV-9	67.79	HPV-35H	48.51	HPV-21	74.77	HPV-25	74.62
HPV-22	58.12	HPV-26	48.40	HPV-20	73.44	HPV-19	73.14
HPV-23	57.01	HPV-31	48.25	HPV-5C	67.92	HPV-5C	65.81
HPV-49	54.31	HPV-53	48.18	HPV-5B	67.74	HPV-5B	65.35
HPV-5B	52.77	HPV-16	47.96	HPV-47	66.88	HPV-8	64.50
HPV-5C	52.75	HPV-10	47.85	HPV-8	66.87	HPV-47	63.92
HPV-47	52.27	HPV-51	47.77	HPV-36	66.25	HPV-36	63.64
HPV-8	52.20	HPV-35	47.48	HPV-12	65.74	HPV-12	63.22

Table 4. (*Continued*)

HPV-21		HPV-22		HPV-23		HPV-25	
HPV-14	81.20	HPV-23	73.61	HPV-22	73.64	HPV-19	81.98
HPV-20	78.16	HPV-17	58.14	HPV-15	57.44	HPV-14	75.75
HPV-25	75.22	HPV-15	57.62	HPV-17	57.14	HPV-21	75.44
HPV-19	74.35	HPV-37	57.22	HPV-37	56.75	HPV-20	75.07
HPV-5C	66.92	HPV-9	56.51	HPV-9	56.31	HPV-5C	68.88
HPV-5B	66.25	HPV-49	52.50	HPV-49	51.87	HPV-8	68.72
HPV-8	66.11	HPV-5C	50.09	HPV-5B	49.64	HPV-5B	68.46
HPV-36	65.00	HPV-5B	50.05	HPV-5C	49.61	HPV-47	67.92
HPV-47	64.70	HPV-19	49.11	HPV-47	49.05	HPV-36	66.44
HPV-12	64.07	HPV-47	49.01	HPV-8	48.72	HPV-12	66.22
HPV-26		HPV-27		HPV-28		HPV-29	
HPV-51	59.25	HPV-2	87.96	HPV-3	76.77	HPV-10	69.09
HPV-30	49.87	HPV-57	80.48	HPV-10	75.50	HPV-3	66.96
HPV-53	49.62	HPV-61	51.50	HPV-29	65.35	HPV-28	65.62
HPV-56	49.40	HPV-10	50.38	HPV-2	48.85	HPV-27	49.66
HPV-45	49.01	HPV-3	50.33	HPV-27	48.68	HPV-2	49.53
HPV-39	48.98	HPV-29	49.55	HPV-57	48.33	HPV-57	48.77
HPV-18	48.70	HPV-28	49.05	HPV-61	47.77	HPV-61	47.70
HPV-35H	48.51	RhPV-1	46.96	HPV-18	46.46	HPV-45	46.57
HPV-16	48.38	HPV-53	45.57	HPV-45	46.16	HPV-18	46.55
HPV-66	48.25	HPV-11	45.53	HPV-53	45.70	HPV-26	46.22
HPV-30		HPV-31		HPV-32		HPV-33	
HPV-53	74.35	HPV-35H	62.87	HPV-42	70.87	HPV-58	76.72
HPV-56	65.03	HPV-16	61.92	PCPV-1	49.18	HPV-52	62.64
HPV-66	63.57	HPV-35	61.66	HPV-13	49.07	HPV-35H	55.75
HPV-26	50.11	HPV-52	54.92	HPV-11	48.85	HPV-16	55.20
HPV-51	49.79	HPV-33	54.42	HPV-6B	48.29	HPV-35	54.44
HPV-45	48.03	HPV-58	53.51	HPV-31	48.24	HPV-31	54.31
HPV-39	47.20	RhPV-1	51.24	HPV-35H	48.12	RhPV-1	51.09
HPV-18	47.12	HPV-18	48.57	HPV-55	48.03	HPV-34	49.74
HPV-10	46.66	HPV-42	48.48	RhPV-1	47.68	HPV-42	48.25
RhPV-1	46.53	HPV-32	48.31	HPV-52	47.53	HPV-32	47.90
HPV-34		HPV-35H		HPV-36		HPV-37	
HPV-35H	50.94	HPV-35	96.03	HPV-5C	80.68	HPV-17	81.31
HPV-16	50.40	HPV-31	62.70	HPV-5B	80.42	HPV-15	77.16
RhPV-1	50.31	HPV-16	62.00	HPV-47	75.33	HPV-9	66.88
HPV-58	50.03	HPV-33	55.51	HPV-8	72.79	HPV-22	57.11
HPV-35	50.01	HPV-52	55.46	HPV-12	71.59	HPV-23	56.31
HPV-33	49.77	HPV-58	55.44	HPV-25	66.40	HPV-49	53.85
HPV-31	48.70	RhPV-1	52.37	HPV-19	65.90	HPV-5C	52.38
HPV-52	48.25	HPV-34	50.29	HPV-14	65.27	HPV-5B	52.29
HPV-26	46.40	HPV-45	49.07	HPV-21	65.27	HPV-47	51.72
HPV-45	46.40	HPV-32	48.88	HPV-20	63.83	HPV-8	51.27
HPV-39		HPV-40		HPV-41		HPV-42	
HPV-45	58.96	HPV-7	80.33	HPV-63	38.24	HPV-32	71.01
HPV-18	58.44	HPV-13	48.27	HPV-1	36.61	HPV-13	48.79
HPV-51	49.18	HPV-11	48.25	CRPV	36.50	PCPV-1	48.38
HPV-26	48.46	PCPV-1	47.85	HPV-8	36.44	HPV-31	48.18
HPV-53	47.64	HPV-6B	47.77	HPV-12	36.40	HPV-16	48.09
HPV-10	47.16	HPV-55	47.57	HPV-10	36.37	HPV-35H	48.03
HPV-30	47.00	HPV-44	47.16	HPV-17	36.35	HPV-33	47.68
HPV-66	46.98	HPV-42	46.72	HPV-49	36.33	RhPV-1	47.61
HPV-35H	46.90	HPV-32	46.37	HPV-47	36.27	HPV-11	47.61
HPV-56	46.88	RhPV1	45.94	HPV-13	36.25	HPV-52	47.51

(Continued)

Table 4. (*Continued*)

HPV-44		HPV-45		HPV-47		HPV-48	
HPV-55	87.59	HPV-18	73.94	HPV-5C	77.90	HPV-50	51.50
HPV-13	70.18	HPV-39	59.20	HPV-5B	77.27	HPV-4	43.98
PCPV-1	66.94	HPV-53	49.07	HPV-36	75.74	HPV-65	43.53
HPV-11	63.42	HPV-26	49.03	HPV-8	74.07	HPV-60	43.12
HPV-6B	63.09	HPV-35H	48.87	HPV-12	72.88	HPV-12	40.35
HPV-7	47.87	HPV-30	48.35	HPV-25	67.59	HPV-14	40.33
HPV-32	47.48	HPV-31	48.33	HPV-19	66.53	HPV-19	40.31
HPV-40	47.37	HPV-16	48.24	HPV-14	66.00	HPV-15	40.27
HPV-42	47.24	HPV-51	48.00	HPV-21	64.48	HPV-20	40.25
HPV-31	46.14	HPV-10	47.94	HPV-20	64.24	HPV-5B	40.24
HPV-49		HPV-50		HPV-51		HPV-52	
HPV-5C	54.42	HPV-48	50.90	HPV-26	59.11	HPV-33	62.42
HPV-15	54.37	HPV-4	42.50	HPV-53	50.42	HPV-58	62.03
HPV-17	54.09	HPV-60	41.96	HPV-39	49.70	HPV-35H	55.18
HPV-21	54.05	HPV-65	41.22	HPV-30	49.24	HPV-31	54.75
HPV-5B	54.03	HPV-17	39.04	HPV-45	48.61	HPV-35	53.94
HPV-36	53.66	HPV-23	39.16	HPV-56	48.53	HPV-16	53.90
HPV-47	53.62	HPV-15	39.11	HPV-18	48.51	RhPV-1	51.62
HPV-12	53.53	HPV-49	39.09	HPV-66	48.48	HPV-42	47.74
HPV-37	53.37	HPV-21	38.87	HPV-16	47.61	HPV-34	47.59
HPV-25	53.35	HPV-37	38.83	HPV-10	47.33	HPV-26	47.50
HPV-53		HPV-55		HPV-56		HPV-57	
HPV-30	74.72	HPV-44	87.66	HPV-66	79.05	HPV-2	81.09
HPV-56	66.90	HPV-13	72.01	HPV-53	66.85	HPV-27	80.27
HPV-66	65.83	PCPV-1	67.59	HPV-30	65.07	HPV-61	51.07
HPV-51	50.22	HPV-11	64.68	HPV-26	49.20	HPV-3	49.88
HPV-26	50.16	HPV-6B	64.16	HPV-51	48.62	HPV-10	49.70
HPV-45	48.70	HPV-32	48.35	HPV-45	48.00	HPV-29	48.92
HPV-18	48.24	HPV-7	48.29	HPV-18	47.51	HPV-28	48.29
HPV-39	48.00	HPV-40	48.11	HPV-39	47.11	RhPV-1	46.92
HPV-16	47.18	HPV-42	47.62	HPV-31	46.55	HPV-18	45.74
HPV-31	47.16	HPV-35H	45.87	HPV-35H	46.38	HPV-53	45.72
HPV-58		HPV-60		HPV-61		HPV-63	
HPV-33	76.57	HPV-4	45.35	HPV-57	51.24	HPV-1	53.85
HPV-52	62.33	HPV-65	45.25	HPV-2	50.96	CRPV	42.87
HPV-35H	55.01	HPV-48	42.79	HPV-27	50.94	HPV-5B	42.01
HPV-16	54.12	HPV-50	41.57	HPV-10	48.81	HPV-5C	42.00
HPV-35	53.62	HPV-47	40.75	HPV-3	48.25	HPV-19	41.62
HPV-31	53.59	HPV-15	40.48	HPV-28	47.81	HPV-49	41.38
RhPV-1	51.20	HPV-8	40.12	HPV-29	47.64	HPV-14	41.31
HPV-34	49.59	HPV-14	40.09	HPV-45	45.35	HPV-9	41.29
HPV-45	47.72	HPV-17	40.09	HPV-18	45.03	HPV-15	41.27
HPV-32	47.07	HPV-12	40.05	HPV-13	44.85	HPV-12	41.20
HPV-65		HPV-66					
HPV-4	83.11	HPV-56	79.01				
HPV-60	45.48	HPV-53	66.09				
HPV-48	43.27	HPV-30	63.83				
HPV-19	42.22	HPV-26	48.44				
HPV-49	41.98	HPV-51	48.29				
HPV-21	41.70	HPV-45	47.87				
HPV-25	41.66	HPV-39	47.62				
HPV-47	41.59	HPV-18	47.01				
HPV-14	41.55	HPV-31	46.42				
HPV-20	41.53	HPV-16	46.07				

CRPV, cottontail rabbit papillomavirus; RhPV, rhesus monkey papillomavirus; PCPV, pygmy chimpanzee papillomavirus.

types. We have therefore written a program which translates the sequences into ten different codons for groups of functionally related amino acids and then determines the best match of a window (10 codons) taken from all three reading frames of one sequence to the second sequence. The best matches are registered and can be displayed in a dot matrix plot as shown in Fig. 1a–c. The maximal value of the three comparisons done at each window position is stored in a histogram which indicates the best possible fit of one sequence segment on the second sequence. These values can be represented in the form of a histogram indicating the locally determined homology. The values of the best fit thus determined in any of the three reading frames were averaged to yield a rough estimate of the relatedness of the two sequences.

Such a comparison will also detect homologies in the different parts of disrupted reading frames and is thus relatively insensitive to occasional errors in the sequence determination. Therefore, a preliminary estimate of relatedness can also be obtained for sequences which have not yet been edited or which have been only partially determined (Table 3).

Calculations have been done for all pairs of complete sequences, including also those which have not yet been completely sequenced on both strands and edited. For each virus type the values for the ten other types showing the highest average homology to it according to the calculation described above are shown in Table 4. In order to make this comparison as general as possible not only the values for the complete HPV sequences have been calculated (Table 4) but also the corresponding values for some partially sequenced virus types and for the animal papillomavirus sequences are shown in Tables 5 and 6, respectively.

Table 5. Homologies of partial HPV sequences (preliminary data)

HPV-24		HPV-38		HPV-54		HPV-64	
HPV-14	53.21	HPV-22	66.59	RhPV-1	51.42	HPV-34	62.64
HPV-8	52.78	HPV-23	65.52	HPV-11	50.38	RhPV-1	44.77
HPV-5C	52.48	HPV-17	61.92	HPV-32	50.35	HPV-58	44.51
HPV-5B	52.19	HPV-15	60.89	HPV-16	50.25	HPV-33	44.38
HPV-19	51.66	HPV-37	60.80	HPV-35H	50.18	HPV-16	44.18
HPV-25	51.63	HPV-9	60.42	HPV-52	49.81	PCPV-1	43.85
HPV-20	51.63	HPV-49	55.87	HPV-13	49.77	HPV-55	43.70
HPV-21	51.63	HPV-5B	55.83	HPV-42	49.70	HPV-35H	43.64
HPV-36	51.46	HPV-5C	55.44	HPV-31	49.68	HPV-45	43.64
HPV-47	51.30	HPV-12	54.99	HPV-35	49.64	HPV-11	43.61
HPV-67							
HPV-33	68.50						
HPV-58	68.36						
HPV-52	66.74						
HPV-35H	59.58						
HPV-35	58.77						
HPV-31	58.52						
HPV-16	57.43						
RhPV-1	56.56						
HPV-34	52.56						
HPV-32	52.49						

Table 6. Homologies of complete sequences of animal papillomaviruses

BPV-1[a]		BPV-2[b]		BPV-4[c]		COPV[d]	
BPV-2	85.05	BPV-1	84.92	HPV-25	40.16	HPV-1	46.14
OPV-1	48.22	OPV-1	47.40	HPV-8	40.14	HPV-63	45.66
EEPV	46.90	EEPV	45.96	HPV-14	40.11	CRPV	44.87
DPV	45.09	DPV	45.22	HPV-5B	40.05	HPV-12	44.64
HPV-8	37.24	HPV-45	37.25	HPV-5C	39.94	HPV-47	44.40
HPV-18	37.24	HPV-10	37.18	HPV-21	39.88	HPV-31	44.38
HPV-12	37.16	HPV-5B	37.11	HPV-20	39.87	HPV-5C	44.18
HPV-15	37.11	HPV-5C	37.07	HPV-9	39.75	HPV-19	44.09
HPV-5B	37.07	HPV-18	37.07	HPV-19	39.66	HPV-49	43.94
HPV-22	37.05	HPV-57	37.07	HPV-49	39.61	HPV-21	43.94
CRPV[e]		**DPV[f]**		**EEPV[g]**		**OPV-1[h]**	
HPV-63	42.11	EEPV	55.29	DPV	56.27	EEPV	50.55
HPV-1	42.01	OPV-1	47.88	OPV-1	50.00	DPV	48.81
HPV-25	39.03	BPV-2	44.70	BPV-1	46.14	BPV-1	48.68
HPV-17	38.98	BPV-1	44.44	BPV-2	45.50	BPV-2	47.74
HPV-14	38.75	CRPV	36.38	CRPV	37.33	HPV-49	37.44
HPV-19	38.64	COPV	36.11	HPV-15	37.16	HPV-10	37.37
HPV-5B	38.53	HPV-17	36.09	HPV-17	37.03	HPV-18	37.33
HPV-13	38.40	HPV-5C	35.94	HPV-1	36.88	HPV-5B	37.31
HPV-3	38.25	HPV-49	35.90	HPV-25	36.70	HPV-26	37.29
HPV-23	38.25	HPV-5B	35.90	HPV-51	36.72	HPV-13	37.29
PCPV-1[i]		**RhPV-1[j]**					
HPV-13	76.22	HPV-35H	51.98				
HPV-55	67.37	HPV-35	51.09				
HPV-44	66.68	HPV-52	50.99				
HPV-11	64.48	HPV-16	50.70				
HPV-6B	63.98	HPV-31	50.48				
HPV-32	49.42	HPV-58	50.42				
HPV-7	49.22	HPV-33	50.35				
HPV-42	48.57	HPV-34	49.53				
HPV-40	48.42	HPV-32	47.09				
RhPV-1	46.03	HPV-11	46.94				

[a] Bovine papillomavirus type 1; CHEN et al. 1982
[b] Bovine papillomavirus type 2; GROFF et al. 1988
[c] Bovine papillomavirus type 4; PATEL et al. 1987
[d] Canine oral papillomavirus; SUNDBERG et al. 1986, H. Delius, MA van Ranst, B. Jenson, H. zur Hausen, and J. Sundberg, in preparation
[e] Cottontail rabbit papillomavirus; GIRI et al. 1985
[f] Deer papillomavirus; GROFF and LANCASTER 1985
[g] European elk papillomavirus; AHOLA et al. 1986
[h] Ovine (sheep) papillomavirus: HAYWARD et al. 1993, H. Delius, H.R.C. Meischke, and P.J. Baird, in preparation
[i] Pygmy chimpanzee papillomavirus; VAN RANST et al. 1992
[j] Rhesus monkey papillomavirus; OSTROW et al. 1991

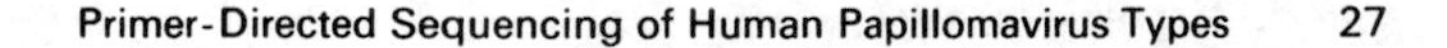

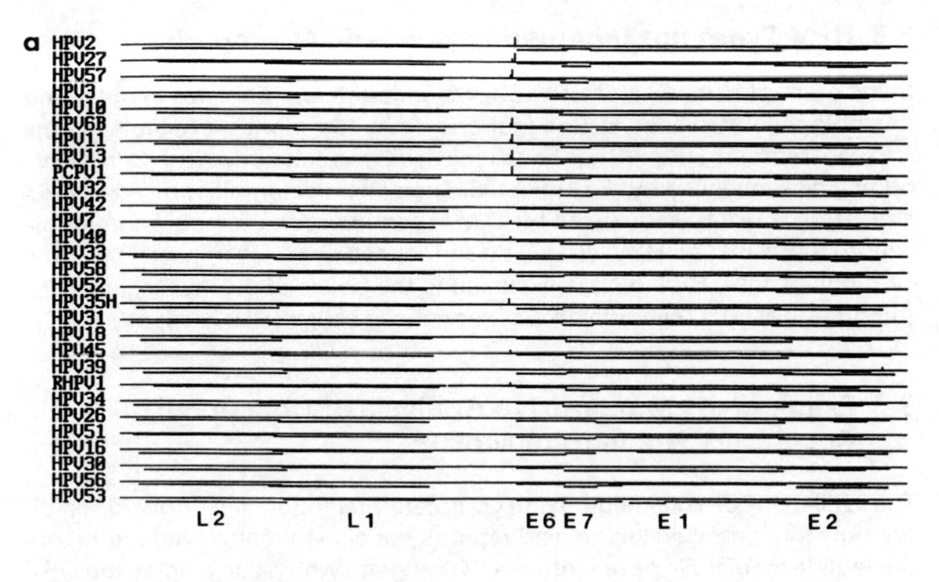

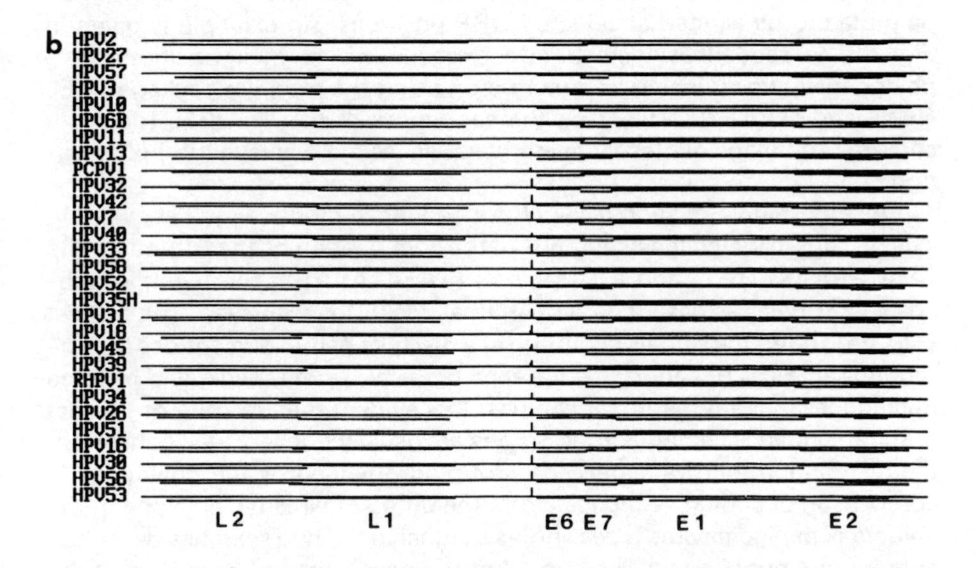

Fig. 2a, b. Maps of the family of mucosal papillomaviruses are shown with the ORFs marked by *lines* in three different reading frames. The arrangement of maps follows the phylogenetic tree constructed using the program package TREE (FENG and DOOLITTLE 1987) based on an alignment of the E1 ORFs. A search was made for nucleotide motifs ACCGAAAACGGT (E2-binding site) and TATATAAA (TATAbox) at a distance of either two (**a**) or three nucleotides (**b**). The *vertical marks* of different lengths designate positions with a match of 19 or all of the 20 nucleotides (*long marks*), 18 (*intermediate*), or 17 out of the 20 nucleotides (*short marks*). The approximate positions of the major ORFs are indicated *below the maps*

3.6 HPV Types not Included in Sequence Comparison

In the cases of HPV-43 and HPV-62, we encountered problems in obtaining unambiguous sequence determinations from the clones available at the *Referenzzentrum*. HPV-46 has been reclassified as subclone HPV-20b. HPV-59 will be sequenced by J. CHOE et al. (personal communication). HPV-67 and HPV-69 will be sequenced by T. MATSUKURA et al. (personal communication). HPV-68 and HPV-70 will be sequenced by G. ORTH et al. (personal communication). HPV-68 is very similar to the isolate of the incomplete HPV integrate in ME 180 cells which was sequenced earlier (REUTER et al. 1991).

3.7 Example of Comparative Analysis of Conserved Regulatory Nucleotide Motifs

The availability of a great number of complete viral sequences allows a search not only for conserved amino acid regions but also for conserved regulatory nucleotide motifs. We have written a PC program which can display the ORF maps of up to 35 different sequences on a monitor screen on which the occurence of selected sequence motifs can be indicated. The maps can be permuted to be aligned at selected ORF positions. An example is given in Fig. 2. The sequences were arranged according to their positions in a phylogenetic tree which was constructed using the program package TREE (FENG and DOOLITTLE 1987) on the sequences of the E1 ORFs (data not shown). The maps of closely related sequences are displayed in juxtaposition.

As an example of such a search for sequence motifs, Fig. 2 shows two sets of maps in which the occurence of the combination of an E2-binding site (ACCGAAAACGGT) and a TATA-box (TATATAAA) in the family of mucosal virus types as described by TAN et al. (1992) is displayed. The first set (Fig. 2a) shows the signals occurring at a distance of two nucleotides and the second set (Fig. 2b) shows the presence of the two motifs separated by three nucleotides in the different virus types. It is evident that, in spite of the fact that the sequences of regulatory regions are usually not very well conserved, the distribution follows largely the grouping obtained by the phylogenetic comparison of coding sequences from the different virus types. None of the epidermal papillomavirus types shows a signal in the two searches described above (data not shown). We hope that the possibility to easily inspect the conserved signals within the otherwise very divergent regulatory regions will help with the identification of significant regulatory nucleotide motifs as well as in a check of general or specific oligonucleotide probes.

Acknowledgments. We gratefully acknowledge the provision of specific HPV clones by Drs. Orth, Matsukura, Ostrow, Lörincz and Shah. We want to thank Dr. J. Gerstner for preparing partial sequences of HPV-26 and HPV-27, B. Drescher for help with the adaptation of the TREE package, Dr. de Villiers for providing the viral clones and for help with the preparation of the manuscript, and Prof. zur Hausen for the support of this work.

References

Ahola H, Bergman P, Stroem AC, Moreno-Lopez J, Pettersson U (1986) Organization and expression of the transforming region from the European elk papillomavirus. Gene 50: 195–205

Beaudenon S, Praetorius F, Kremsdorf D, Lutzner M, Worsaae N, Pehau-Arnaudet G, Orth G (1987) A new type of human papillomavirus associated with focal epithelial hyperplasia. J Invest Dermatol 88: 130–135

Chen EY, Howley PM, Levinson AD, Seeburg PH (1982) The primary structure and genetic organization of the bovine papillomavirus type 1 genome. Nature 299: 529–534

Coggin JR Jr, zur Hausen H (1979) Workshop on papillomaviruses and cancer. Cancer Res. 39: 545–546

Cole ST, Danos O (1987) Nucleotide sequence and comparative analysis of the human papillomavirus type 18 genome. J Mol Biol 193: 599–608

Cole ST, Streeck RE (1986) Genome organization and nucleotide sequence of human papillomavirus type 33, which is associated with cervical cancer. J Virol 58: 991–995

Cravador A, Herzog A, Houard S, d'Ippolito P, Carrol R, Bollen A (1989) Selective detection of human papillomavirus DNAs by specific synthetic DNA probes. Mol Cell Probes 3: 143–158

Danos O, Katinka M, Yaniv M (1980) Molecular cloning, refined physical map and heterogeneity of methylation sites of papilloma virus type 1a DNA. Eur J Biochem 109: 457–461

Dartmann K, Schwarz E, Gissmann L, zur Hausen H (1986) The nucleotide sequence and genome organization of human papilloma virus type 11. Virology 151: 124–130

de Villiers EM, Hirsch-Behnam A, von Knebel-Doeberitz C, Neumann C, zur Hausen H (1989) Two newly identified human papillomavirus types (HPV 40 and 57) isolated from mucosal lesions. Virology 171: 248–253

Egawa K, Delius H, Matsukura T, Kawashima M, de Villiers EM (1993) Two novel types of human papillomavirus, HPV 63 and HPV 65: comparisons of their clinical and histological features and DNA sequences to other HPV types. Virology 194: 789–799

Favre M, Croissant O, Orth G (1989a) Human papillomavirus type 29 (HPV-29), an HPV type cross-hybridizing with HPV-2 and with HPV-3-related types. J Virol 63: 4906

Favre M, Obalek S, Jablonska S, Orth G (1989b) Human papillomavirus type 49, a type isolated from flat warts of renal transplant patients. J Virol 63: 4909

Favre M, Obalek S, Jablonska S, Orth G, (1989c) Human papillomavirus type 28 (HPV-28), an HPV-3-related type associated with skin warts. J Virol 63: 4905

Favre M, Obalek S, Jablonska S, Orth G (1989d) Human papillomavirus (HPV) type 50, a type associated with epidermodysplasia verruciformis (EV) and only weakly related to other EV-specific HPVs. J Virol 63: 4910

Favre M, Kremsdorf D, Jablonska S, Obalek S, Pehau-Arnaudet G, Croissant O, Orth G (1990) Two new human papillomavirus types (HPV54 and 55) characterized from genital tumours illustrate the plurality of genital HPVs. Int J Cancer 45: 40–46

Feng DF, Doolittle RF (1987) Progressive sequence alignment as a prerequisite to correct phylogenetic trees. J Mol Evol 25: 351–360

Fuchs PG, Iftner T, Weninger J, Pfister H (1986) Epidermodysplasia verruciformis-associated human papillomavirus 8: Genomic sequence and comparative analysis. J Virol 58: 626–634

Gallahan D, Mueller M, Schneider A, Delius H, Kahn T, de Villiers E-M, Gissmann L (1989) Human papillomavirus type 53. J Virol 63: 4911–4912

Gassenmaier A, Lammel M, Pfister H (1984) Molecular cloning and characterization of the DNAs of human papillomaviruses 19, 20, and 25 from a patient with epidermodysplasia verruciformis. J Virol 52: 1019–1023

Giri L, Danos O, Yaniv M (1985) Genomic structure of the cottontail rabbit (Shope) papillomavirus. Proc Natl Acad Sci. USA 82: 1580–1584

Goldsborough MD, DiSilvestre D, Temple GF, Lorincz AT (1989) Nucleotide sequence of human papillomavirus type 31: a cervical neoplasia-associated virus. Virology 171: 306–311

Groff DE, Lancaster WD (1985) Molecular cloning and nucleotide sequence of deer papillomavirus. J Virol 56: 85–91

Groff DE et al. (1988) EMBL database

Hayward MLR, Baird PJ, Meischke HRC (1993) Filiform viral squamous papillomas on sheep. Vet Rec 132: 86–88

Hirsch-Behnam A, Delius H, and de Villiers EM (1990) A comparative sequence analysis of two human papillomavirus (HPV) types 2a and 57. Virus Res 18: 81–98

Hirt L, Hirsch-Behnam A, de Villiers EM (1991) Nucleotide sequence of human papillomavirus (HPV) type 41: An unusual HPV type without a typical E2 binding site consensus sequence. Virus Res 18: 179–190

Kahn T, Schwarz E, zur Hausen H (1986) Molecular cloning and characterization of the DNA of a new human papillomavirus (HPV 30) from a laryngeal carcinoma. Int J Cancer 37: 61–65

Kawashima M, Favre M, Jablonska S, Obalek S, Orth G (1986a) Characterization of a new type of human papillomavirus (HPV) related to HPV5 from a case of actinic keratosis. Virology 154: 389–394

Kawashima M, Jablonska S, Favre M, Obalek S, Croissant O, Orth G (1986b) Characterization of a new type of human papillomavirus found in a lesion of Bowen's disease of the skin. J Virol 57: 688–692

Kirii Y, Iwamoto S, Matsukura T (1991) Human papillomavirus type 58 DNA sequence. Virology 185: 424–427

Kiyono T, Adachi A, Ishibashi M (1990) Genome organization and taxonomic position of human papillomavirus type 47 inferred from its DNA sequence. Virology 177: 401–405

Kremsdorf D, Jablonska S, Favre M, Orth G (1982) Biochemical characterization of two types of human papillomaviruses associated with epidermodysplasia verruciformis. J Virol 43: 436–447

Kremsdorf D, Jablonska S, Favre M, Orth G (1983) Human papillomaviruses associated with epidermodysplasia verruciformis. J Virol 48: 340–351

Kremsdorf D, Favre M, Jablonska S, Obalek S, Rueda LA, Lutzner MA, Blanchet-Bardon C, van Voorst Vader PC, Orth G (1984) Molecular cloning and characterization of the genomes of nine newly recognized human papillomavirus types associated with epidermodysplasia verruciformis. J Virol 52: 1013-1018

Krubke J, Kraus J, Delius H, Chow LT, Broker TR, Iftner T, Pfister H (1987) Genetic relationship among human papillomaviruses associated with benign and malignant tumours of patients with epidermodysplasia verruciformis. J Gen Virol 68: 3091-3103

Lörincz AT, Quinn AP, Goldsborough MD, McAllister P, Temple GF (1989a) Human papillomavirus type 56: A new virus detected in cervical cancers. J Gen Virol 70: 3099–3104

Lörincz AT, Quinn AP, Goldsborough MD, Schmidt BJ, Temple GF (1989b) Cloning and partial DNA sequencing of two new human papillomavirus type associated with condylomas and low-grade cervical neoplasia. J Virol 63: 2829–2834.

Lörincz AT, Quinn AP, Lancaster WD, Temple GF (1987) A new type of papillomavirus associated with cancer of the uterine cervix. Virology 159: 187–190

Lungu O, Crum CP, Silverstein S (1991) Biologic properties and nucleotide sequence analysis of human papillomavirus type 51. J Virol 65: 4216–4225

Marich JE, Pontsler AV, Rice SM, McGraw KA, Dubensky TW (1992) The phylogenetic relationship and complete nucleotide sequence of human papillomavirus type 35. Virology 186: 770–776

Matsukura T, Iwasaki T, Kawashima M (1992) Molecular cloning of a novel human papillomavirus (type 60) from a plantar cyst with characteristic pathological changes. Virology 190: 561–564

Mueller M, Kelly G, Fiedler M, Gissmann L (1989) Human papillomavirus type 48. J Virol 63: 4907–4908

Naghashfar ZS, Rosenshein NB, Lörincz AT, Buscema J, Shah KV (1987) Characterization of human papillomavirus type 45, a new type 18-related virus of the genital tract. J Gen Virol 68: 3073–3079

Oltersdorf T, Campo MS, Favre M, Dartmann K, Gissmann L (1986) Molecular cloning and characterization of human papillomavirus type 7 DNA. Virology 149: 247–250

Ostrow RS, Zachow KR, Thompson O, Faras AJ (1984) Molecular cloning and characterization of a unique type of human papillomavirus from an immune deficient patient. J. Invest. Dermatol. 82: 362–366

Ostrow RS, Zachow KR, Shaver MK, Faras AJ (1989) Human papillomavirus type 27: detection of a novel human papillomavirus in common warts of a renal transplant recipient. J Virol 63: 4904

Ostrow RS, LaBresh KV, Faras AJ (1991) Characterization of the complete RhPV 1 genomic sequence and an integration locus from a metastatic tumor. Virology 181: 424–429

Patel KR, Smith KT, Campo S (1987) The nucleotide sequence and genome organization of bovine papillomavirus type 4. J Gen Virol 68: 2117–2128

Philip W, Honore N, Sapp M, Cole ST, Streeck RE (1992) Human papillomavirus type 42: New sequences, conserved genome organization. Virology 186: 331–334

Reuter S, Delius H, Kahn T, Hofmann B, zur Hausen H, Schwarz E (1991) Characterization of a novel human papillomavirus DNA in the cervical carcinoma cell line ME180. J Virol 65: 5564–5568

Sambrook J, Fritsch EF, Maniatis T (1989) Molecular cloning. Cold Spring Harbor Laboratory Press, Cold Spring Harbor

Sanger F, Nicklen S, Coulson AR (1977) DNA sequencing with chain-terminating inhibitors. Proc. Natl. Acad. Sci. USA 74: 5463–5467

Scheurlen W, Gissmann L, Gross G, zur Hausen H (1986) Molecular cloning of two new HPV types (HPV 37 and HPV 38) from a keratoacanthoma and a malignant melanoma. Int J Cancer 37: 505–510

Schwarz E, Dürst M, Demankowski C, Lattermann O, Zech R, Wolfsperger E, Suhai S, zur Hausen H (1983) DNA sequence and genome organization of genital human papillomavirus type 6b. EMBO J 12: 2341–2348

Seedorf K, Krämmer G, Dürst M, Suhai S, Röwekamp WG (1985) Human papillomavirus type 16 DNA sequence. Virology 145: 181–185

Shimoda K, Lörincz AT, Temple GF, Lancaster WD (1988) Human papillomavirus type 52: A new virus associated with cervical neoplasia. J Gen Virol 69: 2925–2928

Smith RF, Smith TF (1990) Automatic generation of primary sequence patterns from sets of related protein sequences. Proc. Natl Acad Sci USA 87: 118–122

Sundberg JP, O'Banion MK, Schmidt-Didier E, Reichmann ME (1986) Cloning and characterization of a canine oral papillomavirus. Am J Vet Res 47: 1142–1144

Tabor S, Richardson CC (1989) Effect of manganese ions on the incorporation of dideoxynucleotides by bacteriophage T7 DNA polymerase and Escherichia coli DNA polymerase I. Proc Natl Acad Sci USA 86: 4076–4080

Tan SH, Gloss B, Bernard HU (1992) During negative regulation of the human papillomavirus-16 E6 promoter, the viral E2 protein can displace Sp1 from a proximal promoter element. Nucleic. Acids Res 20: 251–256

Tawheed AR, Beaudenon S, Favre M, Orth G (1991) Characterization of human papillomavirus type 66 from an invasive carcinoma of the uterine cervix. J Clin Microbiol 29: 2656–2660

van Ranst M, Fuse A, Fiten P, Beuken E, Pfister H, Burk RD, Opdenakker G (1992). Human papillomavirus type 13 and pygmy chimpanzee papillomavirus type 1: Comparison of the genome organizations. Virology 190: 587–596

Volpers C, Streeck RE (1991) Genome organization and nucleotide sequence of human papillomavirus type 39. Virology 181: 419–423

Yabe Y, Sakai A, Hitsumoto T, Kato H, Ogura H (1991) A subtype of human papillomavirus 5 (HPV-5b) and its subgenomic segment amplified in a carcinoma: nucleotide sequences and genomic organizations. Virology 183: 793–798

Zachow KR, Ostrow RS, Faras AJ (1987) Nucleotide sequence and genome organization of human papillomavirus type 5. Virology 158: 251–254

zur Hausen H (1991) Papillomavirus/host cell interactions in the pathogenesis of anogenital cancer In: Brugge J, Curran T, Harlow E, McCormick F (eds) Origins of human cancer: comprehensive review. Cold Spring Harbor laboratory Press, Cold Spring Harbor, pp 685–705

Evolution of Papillomaviruses

H.-U. Bernard[1], S.-Y. Chan[1], and H. Delius[2]

[1] Laboratory for Papillomavirus Biology, Institute of Molecular and Cell Biology, National University of Singapore, Singapore 0511
[2] Angewandte Tumorvirologie, Deutsches Krebsforschungszentrum, Im Neuenheimer Feld 242, 69120 Heidelberg, Germany

Current Topics in Microbiology and Immunology, Vol. 186

1 Introduction

Phylogenetic research in virology addresses the origin and evolution of viruses in time and space, searches for molecular changes that could explain altered pathology or epidemics, and establishes databases for a natural foundation of taxonomy. Some examples are:

1. Influenza virus genome rearrangements and its ability to cross host-species barriers helped to explain influenza epidemics (MURPHY and WEBSTER 1990).
2. Sporadic occurrences of neurovirulence after polio vaccination were caused by rare point mutations of the attenuated virus (MINOR 1992).
3. Evidence that the human immunedeficiency virus-1 is closely related to a retroviral precursor in African primates has been central in the search for the origin of the acquired immunodeficiency syndrome (AIDS) epidemic (SHARP and LI 1988).

Research on papillomavirus evolution is in its infancy. Publications of new papillomavirus sequences or of unusual clinical isolates often contain phylogenetic hypotheses, but systematic studies have started only recently. In spite of this, it is timely to outline the potential of this new discipline because complete or partial sequences of about 60 papillomavirus types and more than 100 variants and subtypes have become available within a few years. This database continues to expand rapidly, and its size could soon make papillomaviruses a model systems for research into molecular evolution and molecular epidemiology.

The following four considerations stand at the focus of recent research:

1. The understanding of the evolution of DNA viruses is lagging far behind that of the evolution of RNA viruses. It is almost certain that studies of DNA viruses will lead to novel findings because they evolve by several orders of magnitude more slowly than RNA viruses: statements about the evolution of RNA viruses are precise only over relatively short periods, e.g., centuries, while statements about DNA virus evolution are apparently possible over millions of years.
2. The large number of genomic variants of papillomavirus types and easy access to their sequences by means of the polymerase chain reaction (PCR) provides novel research opportunities at the interface of phylogeny, genetics, and epidemiology.
3. The phylogenetic relationship of papillomavirus types and variants correlates with pathology. This helps to generalize clinical findings and to make biomedical predictions about novel isolates.
4. The rapidly increasing number of papillomavirus isolates leads to a burgeoning taxonomy. Phylogenetic research generates a rational foundation for organizing all papillomavirus genomes into a natural system.

1.1 Relationship Among Taxonomy, Evolution, and Phylogeny

The scientific undertaking to give names to living things is called taxonomy. The goal is to establish a systematic order whereby individual taxa unite similar organisms. Phylogenetic research analyzes pedigrees and evolutionary histories that led to these taxa. Most often, both lines of research converge and taxa created without phylogenetic knowledge unite organisms that have common ancestors. Traditionally, these disciplines have sampled morphological data and secured supporting evidence from the fossil record. The last 3 decades saw the birth and rapid expansion of molecular research, e.g., the comparison of mitochondrial DNA or ribosomal RNA sequences of different organisms (for references see VIGILANT et al. 1991; STOCK and WHITE 1992). It is now widely accepted that these data not only lead to phylogenetic trees similar to those obtained with traditional methods, but that they even expand our knowledge of evolution beyond the fossil record.

Classical tools are of limited use in virology as viruses have few morphological characteristics. Given this limitation, it is surprising, that molecular data did not require major revisions of the "classical" virus taxonomy (MURPHY and KINGSBURY 1990). It is believed that most of the approximately 20 virus families represent natural relationships.

Different papillomaviruses have such similar genome organizations and extensive sequence similarities that there is no doubt that papillomaviruses are monophyletic. This justifies their present placement into a common genus. At a higher level, however, little genetic relationship is seen to the polyomaviruses, with whom papillomaviruses have been placed into one family. A common origin seems very remote, if indeed there is one. At a lower level, the papillomavirus "type," a concept of nomenclature that was defined on a purely operational basis (COGGIN and ZUR HAUSEN 1979; DE VILLIERS 1989), can today be accepted as a natural taxonomic unit. This is because it unites genomic variants that are as evolutionarily related to one another, as are the members of a biological species (CHAN et al. 1992a, b).

1.2 Tools of Molecular Taxonomy and Molecular Phylogeny

For some virus groups like adenoviruses (SAMBROOK et al. 1980), taxonomy has progressed through to the description of serotypes. In contrast, restriction enzyme digest patterns rather than immunology have become important for the classification of papillomaviruses (LORINCZ 1989; DE VILLIERS 1992; LORINCZ et al. 1992), but this did not give a clear insight into their evolution.

Technical developments outside the papillomavirus field have set the stage for modern molecular texonomy. DNA sequences are now readily obtained after cloning of complete viral genomes or after PCR amplification

(MANOS et al. 1989) of genomic segments and sequencing (SANGER et al. 1977). Equally important are the improvements in computer software for evaluating these DNA sequences. A variety of alignment programs permit us to compare suspected homologies of nucleotide or amino acid sequences. The construction of phylogenetic trees from such alignments has become a strong subdiscipline of biocomputing (for reviews see FELSENSTEIN 1982, 1988), and several widely distributed software packages have sprung from this research, most notably PHYLIP (FELSENSTEIN 1985) and PAUP (SWOFFORD 1991). They contain parsimony, distance matrix and maximum likelihood algorithms that complement one another, and additional validation procedures like the bootstrap (FELSENSTEIN 1985).

1.3 Interpretations

Phylogenetic research analyzes historical processes which are by their very nature not reproducible. Therefore, conservative interpretation is strictly indicated and should only begin after comprehensive sampling of informative data and careful mathematical treatment.

A "selectionist" view of viral evolution says that mutants have changed in frequency because of environmental selection. The alternative "neutralist" view says that the frequency has changed without selection through genetic drift (LI and GRAUR 1991). Our study of genomic diversity within human papillomavirus HPV types (CHAN et al. 1992b; ONG et al., 1993) suggests that variants of HPV types may have evolved predominantly by this latter mechanism. However, it is quite possible that some other findings, such as the abudance of particular HPV types, e.g., HPV-16, may find an explanation in some yet unknown "fitness" of these viruses, e.g., productive shedding. On the other hand, it is not justified to interpret each virus variant and subtype found in association with some particular lesion in terms of environmental selection.

1.4 Direction of Evolution

When two homologous DNA segments differ in a particular nucleotide position, it is difficult to infer which one of the alternative nucleotides is the more ancestral. Similarly, when two sequences differ by the absence or presence of an additional nucleotide segment, one sequence may have changed through a deletion or the other one through an insertion. The decision about these alternatives can only come from circumstantial evidence.

This is the problem in the construction of directed (in time) phylogenetic trees. Mathematical algorithms can identify the relationship of papillomavirus genomes, but determination of their origion, "the root of the tree," must come through alternative assumptions. For example, based on the hypothesis that HPVs evolved from animal papillomaviruses, one could calculate an "unrooted tree" of all HPV types and include one animal papillomavirus. The branch internode of this animal papillomavirus could be interpreted as the root of the HPV tree, and HPV types nearer to this root would be considered as being more closely related to ancient and possibly extinct papillomaviruses.

1.5 Molecular Clock

One facet of phylogenetic research is to identify the time scale of evolutionary changes. The classical approaches require the analysis of fossils. In virology, fossils either do not exist, or, more likely, have not yet been looked for, and evolutionary histories have to be reconstructed from presently existing material.

Nucleotide and amino acid sequences change stepwise with time, such that genes and proteins from remotely related species show more differences than those from close relatives. This observation led to the concept of a molecular clock (ZUCKERKANDL and PAULING 1962). It was hoped that the number of changes would be a linear function of time so that the mismatches between two homologous sequences would reveal the time lapsed since the two genes diverged. This concept is today not accepted in strictest form. There is ample evidence that the molecular clock has different speeds in different organisms and even for different genomic segments in the same organism. In HPV genomes, mutations occur at a faster rate outside genes than within genes, and even more slowly in some highly conserved domains. For each of these different genomic segments, the molecular clock has a different speed.

The speed of the molecular clock can be estimated by independent means, and the objective is for these estimates to converge. For papillomaviruses, we can think of two ways to approximate this goal:

1. Assuming papillomaviruses rarely or never pass host-species barriers (SUNDBERG 1987), the split between two papillomavirus types should be at least as old as the split between their host species. The pigmy chimpanzee papillomavirus (PCPV) is the closest relative of HPV-13 (VAN RANST et al. 1992a). Chimpanzee and human evolution separated about 5 million years ago (SIMONS 1989). All differences between PCPV and HPV-13 must have then accumulated over this period. Twelve nucleotide exchanges occurred in a 132-bp long, highly conserved domain of the L1 gene, and therefore in this segment one nucleotide exchange would correspond to a time span of 400 000 years since the split between the types.

2. A human population that lived in geographical isolation may have become infected by an HPV type only once and most likely at a time before they became isolated. Differences found in the HPVs of such a human population should reflect a time span at least as long as that of the geographical isolation. We did not find sequence differences between most of the HPV-16 and HPV-18 isolates from Japanese and Chinese patients on the one hand and from Finnish, Scottish, German, Greek, and even Indian patients on the other hand, although these populations were isolated or at least lived separately for 1500 to up to 10 000 years. A similar fixation time is suggested by a HPV-18 mutation uniquely found in an Amazonian Indian tribe (CHAN et al. 1992b; C.K. Ong et al., in preparation). We conclude that fixation of one point mutation in the analyzed segments of these viruses takes at least several thousand years and possibly up to 100 000 years (see below).

2 Variability Between Papillomavirus Genomes

Papillomavirus genomes evolve through the accumulation of point mutations, deletions, insertions, and intra-genomic recombinations, but related papillomaviruses share many common alterations. It must be emphasized that phylogenetic calculations are normally based only on point mutational differences, although deletions and insertions also represent an informative evolutionary history.

2.1 Point Mutations

Diversity *within* HPV types occurs mainly through point mutations. Our study of two genomic segments (364 and 321 bp) coding for the transcriptional enhancer of HPV-16 and HPV-18 identified among 400 isolates nearly 80 genomic variants with combinations of 72-point mutations (36% transversions) but only four with deletions, insertions, or rearrangements. Even more important, many of the point mutational variations were ancestors of other variants, while no variant was derived from an insertion/deletion (indel) or rearranged mutant (CHAN et al. 1992b; ONG et al. 1993). But in one particular case we have found that the genomic segments of HPV-18 and HPV-45, coding for the transcriptional enhancer, can be unambiguously aligned after the introduction of two indels. Consequently, these two HPV types probably evolved from a common precursor that was modified by these indels (any many point mutations).

Point mutations also contribute more to the diversity *between* HPV types than indels: most genes of different papillomavirus types can be well aligned if one allows for nucleotide mismatches as well as indels. In such alignments, mismatches are by at least one order of magnitude more frequent than indels.

2.2 Insertions/Deletions

These two genomic alterations are frequently referred to with the term "indel," since only circumstantial evidence can help to decide whether a particular sequence difference between two genomes represents an insertion or a deletion.

Figure. 1 shows an example of informative indels in the multiple alignment of an E7 gene segment of several papillomaviruses. This alignment visualizes that indels are phylogenetically as informative as point mutations, which formed the basis for construction of the tree in Fig. 2 (see below). This statement can be verified by comparing the similarities of indel patterns in Fig. 1 of HPV types that form clusters in Fig. 2. The Rb-binding domain of E7 (PHELPS et al. 1988; DEFEO-JONES et al. 1991) is highlighted to point to an application of this type of alignment in functional studies: it invites hypotheses about conserved segments and permits generalized observations to be made with only one HPV type.

An insertion of unknown function has given rise to the unusually large genome of the canine oral papillomavirus where a 1 kb DNA segment separates the early and the late genes of this virus (H. Delius, M.A. van Ranst, B. Jenson, H. zur Hausen, and J.P. Sundberg, in preparation).

2.3 Genomic Rearrangements

Partial duplications of transcriptional control regions of HPV-6 and HPV-16 have been sporadically observed (KULKE et al. 1989; CHAN et al. 1992b; RUEBBEN et al. 1992). But the rarity of these variants makes it unlikely that rearrangements become established in HPV populations.

It has been proposed that the genes E6 and E7 originated through multimerization of a 33 amino acid segment with zinc finger properties (DANOS and YANIV 1987). If this view is confirmed by future sequence analysis, this potential genomic rearrangement will have brought into being the proteins instrumental for papillomavirus transformation.

Interestingly, the B group bovine papillomaviruses (BPV-3, -4, -6) do not contain the E6 gene but instead a gene, E8, that could give rise to a highly hydrophobic protein similar to the gene E5 of other papillomaviruses (JACKSON et al. 1991). One might speculate whether these two genes are simply analogous or whether they are homologous and originated through genomic rearrangement.

The present taxonomic position of papillomaviruses and polymaviruses expresses the assumption of a monophyletic origin. This hypothesis would

E7: Rb-binding

```
hpv35  MHGEIT TLQ          DYVL DL EPE        AT DLY CYEQLC      DSS   EEEED T   IDGPAGQA      KPDT S   NYNIVTScC   KcEA TLRLCVQSTHIDIRKLEDLLM GTFGIVCPG     cSQRA
hpv16  MHGDTP TLH          EYML DL QPE        TT DLY CYEQLN      DSS   EEED E    IDGPAGQA      EPDR A   HYNIVTFcC   KcDS TLRLCVQSTHVDIRTLEDLLM GTLGIVCPI     cSQKP
hpv31  MRGETP TLQ          DYVL DL QPE        AT DLH CYEQLP      DSS   DEED V    IDSPAGQA      EPDT S   NYNIVTFcC   QcKS TLRLCVQSTQVDIRILQELLM GSFGIVCPN     cSTRL
hpv33  MRGHKP TLK          EYVL DL YPE        PT DLY CYEQLS      DSS   DEDE G    LDRPDGQA      QPAT A   DYYIVTCcH   TcNT TVRLCVNSTASDLRTIQQLLM GTVNIVCPT     cAQQ
hpv58  MRGNNP TLR          EYIL DL HPE        PT DLF CYEQLC      DSS   DEDEI G   LDGPDGQA      QPAT A   NYYIVTCcY   TcGT TVRLCINSTTTDVRTLQQLLM GTCTIVCPS     cAQQ
hpv52  MRGDKA TIK          DYIL DL QPE        TT DLH CYEQLG      DSS   DEEDTDG   VDRPDGQA      EQAT S   NYYIVTYcH   ScDS TLRLCIHSTATDLRTLQQMLL GTLQVVCPG     cARL
hpv6b  MHGRHV TLK          DIVL DLQPPD        PV GLH CYEQLV      DSS   EDEVD E   VDGQDSQP        LK Q   HFQIVTCcC   GcDS NVRLVVQCTETDIREVQQLLL GTLNIVCPI     cAPKT
hpv11  MHGRLV TLK          DIVL DLQPPD        PV GLH CYEQLE      DSS   EDEVD K   VDKQDAQP        LT Q   HYQILTCcC   GcDS NVRLVVECTDGDIRQLQDLLL GTLNIVCPI     cAPKP
hpv13  MHGKYP TLK          DIVL EL TPD        PV GLH CNEQLD      SS    EDEVD E   QATQATQATQHSTLL Q  CYQILTScS   KcCS NVRLVVECTGPDIHDLHDLLL GTLNIVCPL     cAPKS
pcpv1  MHGKYT TLK          DIVL DL SPD        PV GLH CNEQLD      SSE   EDEVD E   QATQATQA       TFT Q   HYQIVTCcG   QcDS NVRLVVDCTGSDIQHLHKLLL GSLNIVCPL     cAPQT
hpv30  MHGKVT TIP          EYIL DLVPQT        EI DLH CYEQLNSS    EEE   DEDEVDN   LQKQPQQA RQEEQH P   CYLINTQcC   RqAS AVQLAVQSPTKELRALQQMLM GALELVCPL     cATRR
hpv53  MHGNVP TLP          QYII ELIPQT        EI DLQ CHEQLNSS    EDE   DEDEVDH   LQEQPQQA RRDEQH P   CYLIETQcC   RcES LVQLAVQSSTKELRILQQMLM GTVELVCPL     cATRR
hpv56  MHGKVP TLQ          DVVL ELTPQT        EI DLQ CNEQLDSS    EDE   DEDEVDH   LQERPQQA RQAKQH T   CYLIHVPcC   EcKF VVQLDIQSTKEDLRVVQQLLM GALTVTCPL     cASSN
hpv26  MHGNII NIE          DVIL DLVPQP        EI DLR CYEQLDYEQFDSS      DEDETDN   MRDQ  QA RQAGQE V   CYRIEAQcC   McNS IVQLAVQSSRQNVRVLEQMLM EDVSLVCHQ     cAAQ
hpv51  MRGNVP QLK          DVVL HLTPQT        EI DLQ CYEQFD      SSE   EEDEVDN   MRDQ LPE RRAGQA T   CYRIEAPcC   RcSS VVQLAVESSGDTLRVVQQMLM GELSLVCPC     cANN
hpv34  MHGKKP SVQ          DIVL DLKPTT        ET DLT CYESLD      NSE   DEDETDS       .HLE RQAEQA W   YRIVTDcS    RcQS TVCLTIESTHADLLVLEDLLM GALKIVCPN     cSRRL
hpv18  MHGPKA TLQ          DIVL HLEPQN     EI PV DLL CHEQLSDSEEEND    EID    G   VNHQH  LPARRAEPQ R  HTMLCMcC    KcEA RIELVVESSADDLRAFQQLFL NTLSFVCPW     cASQQ
hpv45  MHGPRE TLQ          EIVL HLEPQN     ELDPV DLL CYEQLSESEEEND    EAD    G   VSHAQ  LPARRAEPQ R  HKILCVcC    KcDG RIELTVESSAEDLRTLQQLFL STLSFVCPW     cATNQ
hpv39  MRGPKP TLQ          EIVL DLCPYN     EIQPV DLV CHEQLGESEDEID    EPD   HA   VNHQHQLLARRDEPQ R  HTIQCScC    KcNN TLQLVVEASRDTLRQLQQLFM DSLGFVCPW     cATANQ
rhpv1  MIGPKP TLE          DIVL DLQPFP     QPQPV DLM CYEQLSDSSEDED    EVDHHHN   NQQQHHQHARPEVPE DGDCYRIVSDcY  ScGK PLRLVVVSSHEELRVLEDLLM GTLDIVCPS     cASRV
hpv2   MHGNRP SLK          DITL ILDEIP     E  IV DLH CDEQF        DSS   EEENNHQ   LTEPDVQA               YGVVTTcC    KcGR TVRLVVECGQADLRELEQLFL KTLTLVCPH     cA
hpv57  MHGERP SLE          DITL ILEEIP     E  IV DLH CDEQF        DNS   EEDTNYQ   LTEPAVQA               YGVVTTcC    KcHS TVRLVVECGAADIRHLEQLFL NTLTIVCPR     cV
hpv3   MHGPHP TIK          DIEL SLA  P     E  DV PAL CNVQL        D      EDEYIN   AVEPAQQA               YCVYTVcP    KcSS QLRLVVECSHADIRAFEQLLL GTLTVVCPR     cV
hpv10  MHGPHP TVK          DIEL SLA  P     E  DI P V CNVQL        D      EEDYTD   AVEPAQQA               YRVVTEcT    KcSL PLRLVVECSHADIRALEQLLL GTLKLVCPR     cV
hpv7   MHGERP TLG          DIVL DLQPE         PV SLS CNEQLDSSDSEDD      HEQDQLDSSHNRQREQPTQQDLQV NLQSFKIVTHcV  FcHC LVRLVVHCTATDIRQVHQLLM GTLNIVCPN     cAATA
hpv40  MHGERP TLG          DIVL NLHPE         PV CLN CNEQLDSSDSEDD      HEQDQLDSLHSREREQPTQQDLQV NLQSFKVVTRcV  FcQC LVRLAVHCSITDITQFQQLLM GTLHIVCPN     cAATE
hpv32  MRGNAP TLK          DIILYDLPTCDPTTCDTPPV DLY CYEQFD      TSD   EDDEDDDQPIKQDIQR              YRIVCGcT   QcGR SVKLVVSSTGADIQQLHQMLL DTLGIVCPL     cACVE
hpv42  MRGETP TLK          DIVLFDIPTCE       TPI DLY CYEQLD      SSD   EDDQA     KQDIQR              YRILCVcT   QcYK SVKLVVQCTEADIRNLQQMLL GTLDIVCPL     cARVE
shope  MIGRTP KLS          ELVLGETAE          AL SLH CDEALE      NLS   DDDEE     DHQDRQV     FIER P     YAVSVPcK   RcRQ TISFVCVCAPEAIRTLNRLLS ASLSLVCPE     cCN
hpv5b  MIGKEV TVQ          DIIL ELSEVQ     PEVLPV DLF CEE ELP     NE     QETE     EEPDIERI              SYKVIAPcGCRHcEV KLRIFVHATEFGIRAFQQLLT GDLQLLCPD     cRGNCKHDGS
hpv5c  MIGKEV TVQ          DIIL ELSEVQ     PEVLPV DLF CEE ELP     NE     QETE     EEPDNERI              SYKVIAPcGCRNcEV KLRIFVHATEFGIRAFQQLLT GDLQLLCPD     cRGNCKHDGS
hpv8   MIGKEV TVQ          DFVL KLSEIQ     PEVLPV DLL CEE ELP     NE     QETE     EELDIERT              VFKIVAPcGCSCcQV KLRLFVNATDSGIRTFQELLF RDLQLLCPE     cRGNCKHGGS
hpv12  MIGKEV TVQ          DFTL ELSELQ     PEVLPV DLL CEE ELP     NE     QETE     EESDIDRT              VFKIIAPcGCSScEV NLRIFVNATDTGIRTLQDLLI SDLQLLCPE     cRGNCKHGGF
hpv47  MIGKEV TVR          DIVL ELSEVQ     PEVLPV DLF CDE ELP     NE     QQAE     EELDIDRV              VFKVIAPcGCSCcEV KLRIFVNATNRGIRTFQELLT GDLQLLCPE     cRGNCKHGGF
hpv19  MIGKEV ILQ          DIVL ELSELQ     PEVQPV DLF CEE ELP     TE    QQETE     EEPAIERS              AYKVVVLcGC   cKV KLRIFVKATQFGIRTLQDILI EELQLLCPE     cRGNCNHGGV
hpv25  MIGKEV TLQ          DFTL ELSELQ     PEVQPV DLF CEE ELP     AE    HQETE     EEPAIDRT              PYKVVAPcGC   cEV KLRIFVKATDFGIRTLQNLLI EELQLLCPE     cRGNCKHGGS
hpv15  MIGKEA TIP          DIVL ELQEL        VQPT DLH CYE ELS     EE      ETE     EE  PRFI              PYKIVVPc C   FcDS KLRLIVVATPFGIRSQQDLLL EEVKLVCPG     cREKLRHV
hpv17  MIGKEA TIP          DIVL ELQQL        VQPT DLH CYE ELS     EE    ETETE     EE  PRRI              PYKIVAPc C   FcGS KLRLIVLATHAGIRSQEELLL GEVQLVCPN     cREKLRHD
hpv9   MIGKEA TIP          EVVL ELQEL        VQPTADLH CYE ELT     EE      PAE     EE  QCLT              PYKIVAGc G   cGA RLRLYVLATNLGIRAQQELLL GDIQLVCPE     cRGRLRHE
hpv49  MIGKEV TIP          DIIL Q EEF        GQPI DLQ CYE NLTAEAPAEQ      ELEAE     EEL IQGI              PYKVIATcGG  GcGA RLRVFVLATDAAIRSFQELLL EELQFLCPQ     cREEIRNGGR
hpv1   MVGEMP ALK          DLVL QLE        PSVLDL DLY CYE EVP     PD    DIEEE     LVSP  QQ              PYAVVAScA   YcEK LVRLTVLADHSAIRQLEELLL RSLNIVCPL     cTLQRQ
hpv63  MVGEQP NIG          DLVS QEE        PSVLDL N   CYE DIP     AE    EEESE     Y                     PYAIVLPcG   LcDG LLRLTCVSDLSTLTRLEELLL GSLRIVCPL     cAIRHQRH
copv   MIGQCA TLL          DIVL TEQ        PEPIDL Q   CYE QLP     SS    DEEEE     EEEPTEKN              VYRIEAAcG   FcGK GVRFFCLSQKEDLRVLQVTLL  SLSLVCTT     cVQTAKLDHGG
hpv4   MRGAAP TVA          DLNL ELNDL        VLPA NLL SEE VLQ     SS    DDEYE     ITEEESVV              PFR IDTc CYRcEV AVRITLYAAELGLRTLEQLLVEGKLTFCCTA     cARSLNRNGR
hpv65  MRGAAP RVA          DLNL ELNDL        VLPI NLL SEE VLQ     PS    DDESE     APEEE LF              PFR IDTc CYRcEV NVRITLFAVEFGLRALEQLIVDGKLTFCCTT     cARTL RNGR
bpv1   MV QGP NTH          RN LDDSPAGPL LILSPCAGTPTR SPAAPDAPDFRLP CHFGRPTRKRGPTTPPLSSPGKLCATG PRRVYSVTVCcG  NcGK ELTFAVKTSSTSLLGFEHLLN SDLDLLCPR     cESRERHGKR
bpv2   MV QGP TTH          RN LDDSPAGPL LILSPCAGTPTR VPAAPDAPDFRLP CHFGRPTRKRGPSTPPLSSPGKVCATG PRRVYSVTVCcG  HcGK DLTFAVKTGSTTLLGFEHLLN SDLDLLCPR     cESRERHGKR
dpv    MACARP LTG          RT LPADESPCLTLILEPVSGEAAK NST PVVV   DKP   GKPPPKR                         HRRQYNVTVScN  DcDK RLNFSVKTTCSTILTLQQLLT EDLDFLCSF     cEA  KNG
eepv   MVHG P RTK          KH LPPYESPPLTLLLEPVAPVQQT GIQ APQR   KPP   SQKGHKK                       G HKKVYSVTVPcN  GcDK NLEFCARTSSATILTLQNLLL KDLDFLCST     cET  NHG
hpv41  MKGQNV TLQ          DI AI ELEDTI       SPINLH CEE EI            ETEE      VDTPN                 PFAITATcY   AcEC VLRLAVVTSTEEFISTAATAVRQPLSICVQLASKQV FcNRRPERNGP
bpv4   MYGRTVLAVRLIYCLLYCIAV IVRKLLYPVIMRGNSVDLQEIV LVQ QGEVP ENAA VHSGEHSDDEGESEEEREQVQQVPTP RRTLYLVESQcP  FcQA IIRFVCVASNTGIRNLQALLVNSHLDLACHA    cVEQNGVQGLRHRQWQ
```

Fig. 1. Alignment of the E7 proteins of 50 papillomavirus types. The figure exemplifies conserved and variable domains and phylogenetically informative patterns of insertions and deletions. Phylogenetic calculations, as used for the tree in Fig. 2, evaluate amino acid exchanges or point mutations in nucleotide sequences, but disregard insertions and deletions. Most of the nucleotide sequences evaluated in this figure are published by various laboratories and can be obtained from Genbank releases. Some novel sequences were established by our group (H. Delius, in preparation; H. Delius and J.P. Sundberg, in preparation)

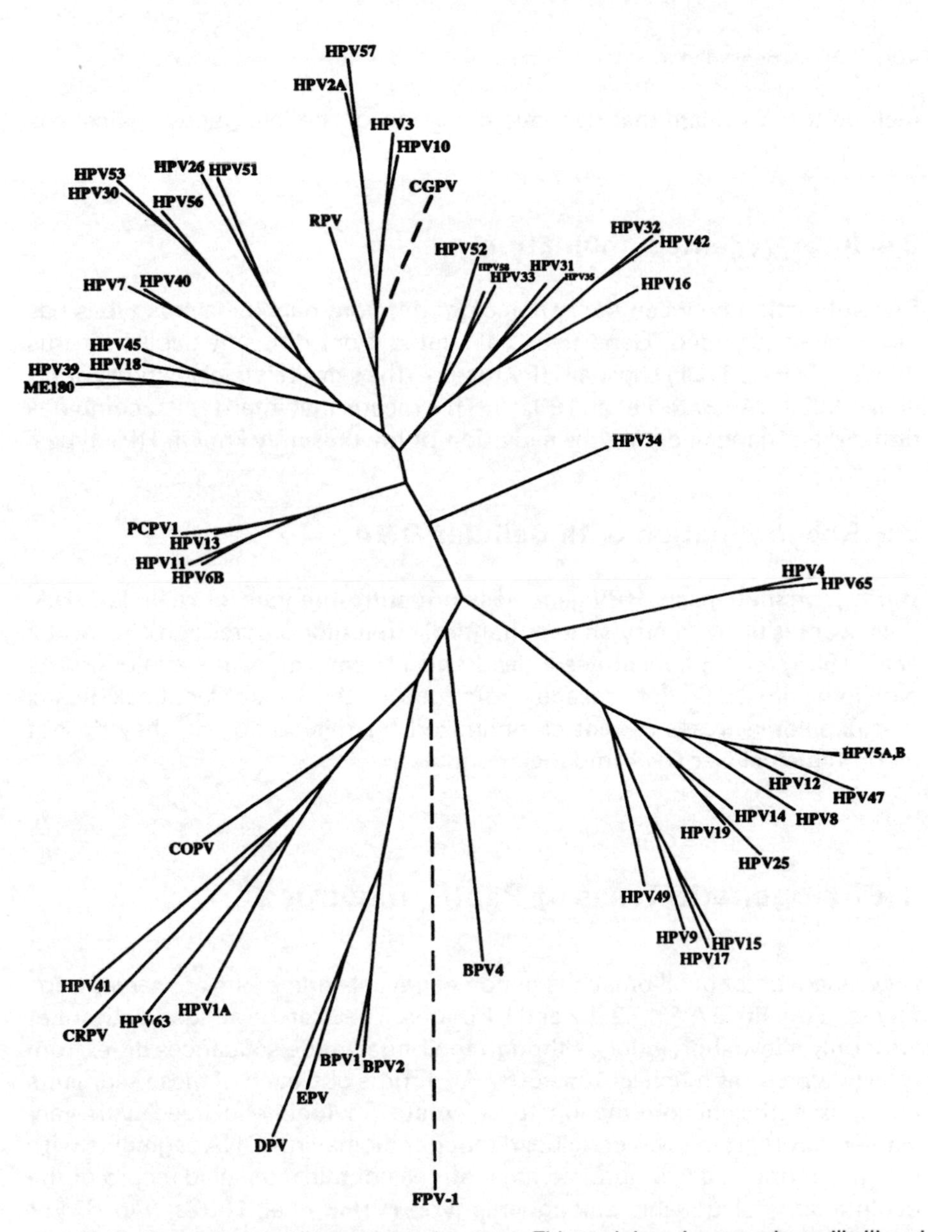

Fig. 2. Phylogenetic tree of 54 papillomavirus types. This tree is based on a maximum likelihood evaluation of two highly conserved segments of the L1 gene with a combined length of 420 bp. These two segments correspond in HPV-16 to the genomic positions 6540–6671 and 6725–7012. A previously published tree (CHAN et al. 1992a) was based only on the 6540/6671 segment. The trees are similar, but the larger fragment confers greater robustness to this tree. The data bases for this tree are published (CHAN et al. 1992a) or were established by our group (H. Delius, in preparation, H. Delius and J.P. Sundberg, in preparation), and the particular sequence alignment for this tree can be obtained from the authors on request. The *dashed lines* identify the relationship of two papillomaviruses for which no L1 sequences were available (CHAN et al. 1992a). BPV, bovine papillomavirus; CGPV, Colobus monkey papillomavirus; COPV, canine oral papillomavirus; CRPV, cottontail rabbit papillomavirus; DPV, deer papillomavirus; EPV, elk papillomavirus; FPV-1, bird (=chaffinch) papillomavirus; PCPV, pigmy chimpanzee papillomavirus; RPV, rhesus monkey papillomavirus

include the postulate that genomic inversion of the late genes against the early genes had occurred in either virus at the split of the two genera.

2.4 Inter-type Recombination

Recombination between the genomes of different papillomavirus types has never been recorded. Gene trees calculated from different papillomavirus genes (E6, E7, E1, L1) show all HPV types in the same relative position (CHAN et al. 1992a; VAN RANST et al. 1992b). This means that inter-type recombination did not happen during the evolution of the presently known HPV types.

2.5 Recombination with Cellular DNA

During carcinogenesis, HPV genomes frequently integrate into cellular DNA. This event is likely to play an important role in tumor progression (SCHWARZ et al. 1985), but it constitutes a dead end for papillomavirus evolution. As papillomaviruses do not "escape" from tumors, this lack of feedback means that papillomaviruses cannot be optimized for this trait. Also, they do not incorporate cellular DNA into their genome.

3 Phylogenetic Trees of Papillomaviruses

A comparison of papillomavirus genomes reveals numerous conserved segments in the E6, E7, E1, E2, L2 and L1 genes. They can be aligned without or with only a few short indels, although their nucleotide sequences differ from type to type by as much as 10%–50%. As it turns out, each of these segments contains sufficient information to calculate phylogenetic trees with very similar topologies. It was concluded that papillomavirus DNA segments with lengths of one to a few hundred nucleotides contain a detailed record of the evolutionary relationship among virus type (CHAN et al. 1992a; VAN RANST et al. 1992b).

3.1 An L1 Tree of Animàl and Human Papillomaviruses

The phylogenetic tree in Fig. 2 is based on sequences of 54 papillomaviruses, all that are presently available to us, and it was calculated on a maximum likelihood evaluation of 420 bp of the L1 gene. This gene is of particular interest because of the substantial number of publications that describe access to L1 sequences with the help of the PCR reaction and consensus primers. Some of these studies found indications of unknown HPV types (SCHIFFMAN et al. 1991; VAN DEN BRULE et al. 1992). Therefore, it seems worthwhile to concentrate taxonomic efforts around this strategy as it could permit us to make phylogenetic and biological predictions about novel viruses, even in the absence of a complete sequence.

3.2 Animal Papillomaviruses

The tree in Fig. 2 contains 11 animal papillomaviruses from a bird, cattle, deer, elk, rabbit, dog, chimpanzee, and two monkeys (see legend to Fig. 2 for letter codes). The relative position of these viruses suggests that papillomaviruses co-evolved with their hosts (CHAN et al. 1992a; ONG et al. 1993). All three non-human primate papillomaviruses (rhesus monkey papillomavirus, RPV; PCPV; Colobus monkey papillomavirus, CGPV) are minor branches of one of the major HPV branches. Four of the five ungulate papillomaviruses (BPV-1; BPV-2; deer papillomavirus, DPV; elk papillomavirus, EPV) from one major branch. The cottontail rabbit papillomavirus (CRPV), the canine oral papillomavirus (COPV), and BPV-4 each forms a separate remote branch. There is some affinity of the human viruses HPV-1a, HPV-41, and HPV-63 to COPV and CRPV, which is likely to represent a remote relationship. The separation of BPV-4, the only keratinocyte-specific ungulate papillomavirus, from the four ungulate (fibro) papillomaviruses, should be understood as biological diversification within a particular host. The only bird papillomavirus (FPV-1, from a European chaffinch), forms the most remote branch (CHAN et al. 1992a), just as the host species would do in a vertebrate tree.

The small number of animal papillomaviruses and the fact that only one virus has been characterized so far in all hosts except cattle almost certainly reflect a sampling bias. Numerous different papillomaviruses are likely to be found in each mammalian species during a careful search.

3.3. Human Papillomaviruses

For obvious reasons, more human than animal papillomaviruses have been sampled and sequenced. The phylogenetic analysis of these HPV types shows that host restrictions are not the only mode of papillomavirus diversification. In the human host they evolved into more than 65 types, varying in tissue tropism, oncogenic potential, and association with anatomically and histologically distinct diseases (DE VILLIERS 1989).

These biological properties must have some genetic basis and therefore ought to reflect evolutionary relationships. Phylogenetic trees such as that in Fig. 2 (CHAN et al. 1992a; VAN RANST et al. 1992b) support these conjectures. They suggest natural relationships that confirm and expand those based on hybridization techniques (PFISTER 1990). When one considers the continuous nature of evolution, it is a puzzling aspect of taxonomy in general and of HPV taxonomy in particular that classifications can be done without much arbitrariness.

The HPV types form four major branches. Three of these confirm previous knowledge, while the fourth, which contains, among others, all "genital" HPVs, has to be analyzed in more detail.

Two major branches contain only a few types each, i.e., HPV-4/HPV-65, and HPV-1a/HPV-63/HPV-41. They had been known to be unrelated to any other HPV type (DE VILLIERS 1989) and had been classified separately (PFISTER 1990). These types are associated with lesions of cutaneous epithelia and often with a distinct disease, e.g., HPV-1a and plantar warts. It must be emphasized that the affinity of HPV-1a, HPV-63, and HPV-41 with one another and with the animal papillomaviruses COPV and CRPV is remote but is likely to represent relationship rather than fortuitous similarities.

All HPV types associated with epidermodysplasia verruciformis (EV) (HPV types 5, 8, 9, 12, 14, 15, 17, 19, 15, 47, 49) form the third branch. Subgroupings (PFISTER 1990), based on analysis of transcription regulatory regions (ENSSER and PFISTER 1990), are also reflected in the minor branchings of this major HPV branch. The sequence of several EV HPV types were not available to us, but it is unlikely that future findings will change the topology of this branch and the classification of this group.

The fourth branch is complex, as one would expect, since it unites HPV types that are associated with very different lesions, such as HPV-2 and -3 (common warts), HPV-6 and -11 (low-risk genital lesions), HPV-16 and -18 (high-risk genital lesions) and HPV-7 (butcher's warts). There can be no doubt about the common origin and close relationship of these different HPV types because trees based either on the genes E6, E1, L1, or on a combination of several genes, have very similar topologies (CHAN et al. 1992a; VAN RANST et al; 1992b). Also, all HPV types of this branch have characteristic epithelial-specific transcription elements (CHONG et al. 1991) and a distinctive alignment of promoter elements (TAN et al. 1992). Both of these regulatory elements are absent or differ in all other papillomaviruses. It must be concluded that these HPV types evolved from a common ancestor that had the potential to evolve into viruses with very different properties.

Minor branches of this major branch, however, unite viruses with similar tropisms and disease patterns, and confirm previous classifications. Examples are HPV-2, -3, -10, and -57, which are mostly associated with cutaneous lesions and were previously grouped together (PFISTER 1990), or HPV-6 and -11, associated with low-risk mucosal lesions. Among the viruses associated with high-risk mucosal lesions (LORINCZ et al. 1992), in particular HPV types 16, 18, 30, 31, 33, 35, 39, 45, 51, 52, 56, and 58, the phylogenetic tree suggests contiguous grades of relationships rather than distinct groups. Certain associations, however, in particular HPV-16/31/35 and HPV-18/45, are pronounced in all trees and reflect close relationship, similar biology/

pathology, and possibly recent common origin. Interesting is the close relationship of HPV-26 and HPV-51, which had been isolated from a common wart and a cervical lesion, respectively (DE VILLIERS 1989).

This fourth major HPV branch also contains two surprising findings. First, all three ape and monkey papillomaviruses CGVP, RPV, and PCPV are closely related to some of the cutaneous HPV types. This seems to require a common origin, most likely at the time of speciation of the hosts, which must have been extinct primate species (ONG et al. 1993). This whole major HPV branch may have an age as old as that of the speciation of primates, i.e., several ten million years.

This HPV branch also includes HPV-7, which is known to be associated with butcher's warts. It may be speculated that this virus is actually an animal papillomavirus that occasionally crosses the host barrier given extensive occupational exposure. The phylogenetic assignment of this virus close to the human virus HPV-40, combined with recent observation of the occurrence of HPV-7 lesions in patients infected with the human immunedeficiency virus (GREENSPAN et al. 1988), rather suggests that HPV-7 is a widespread human virus that leads to disease only under the influence of cofactors, that exist in meat handlers or in AIDS patients.

3.4 Genomic Variability Within the HPV-16 Type

The HPV-16 genome was initially isolated from a German patient (DUERST et al. 1993). We refer to the sequence of this clone (SEEDORF et al. 1985) as the "HPV-16 prototype." This term refers to this isolation priority, and does not indicate any phylogenetic precedence. Subsequent isolates of HPC-16 segments from patients throughout the world had substantial point mutational differences from the prototype. These clones were designated as "variants" (FUJINAGA et al. 1990; HO et al. 1991; CHAN et al. 1992b; ICENOGLE et al. 1991; ESCHLE et al. 1992, HO et al. 1993a). Three observations prove that the mutational patterns of these variants neither originated from frequent random mutations nor do they indicate mutational hot spots:

1. Particular combinations of mutations cluster geographically. For example, a genomic segment between the positions 7478 and 7841 of isolates from Europe has mostly the prototype sequence or a single mutation (G to A transition) at position 7519. The same fragment isolated from Japanese patients often has four changes, namely the same transition at 7519 and three additional mutations at 7728, 7779, and 7840. Isolates from Tanzania are even more complex and show about ten changes against the prototype sequence.
2. Characteristic combinations of mutations of the viral long control region, or of the genes E5, E7, or L1 are linked.

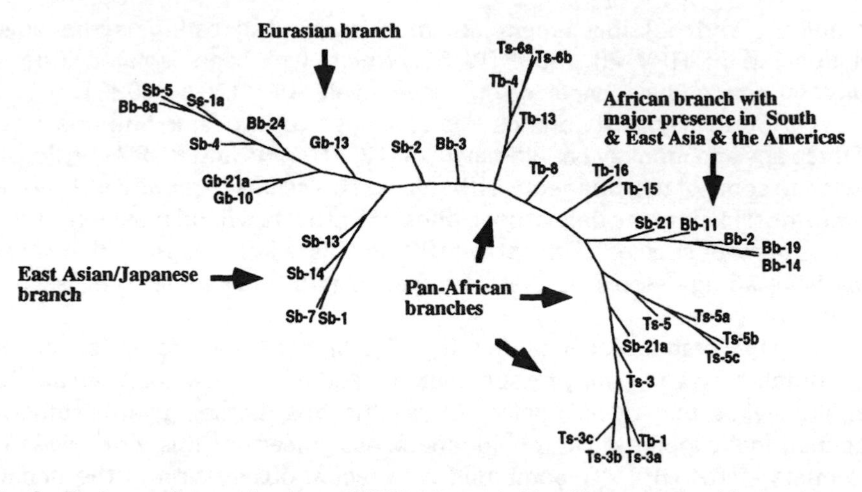

Fig. 3. Phylogenetic tree of variants of HPV-16 sampled in different geographical regions and from different ethnic groups. For mutations and codes, see CHAN et al. 1992b

3. Nearly all mutations form an obvious transformation series and can be incorporated into a phylogenetic tree (Fig. 3)

Our ongoing study of these HPV-16 variants has created a detailed picture of the evolution and prehistoric spread of HPV-16 (CHAN et al. 1992b; HO et al. 1993). The diversity of the 7478/7841 fragment from HPV-16 is limited, and plateaus off at about 50 variants found among 300 isolates from most important geographical regions and ethnic groups of the world. Divergence from the prototype is maximal in Africa (Fig. 3). Here we also found the largest number of different variants and the deepest branches of a phylogenetic tree of the variants, which may mean that HPV-16 originated in this continent. Different estimates for the molecular clock (see above) suggest that HPV-16 diversification started at least 100 000 years or up to 1 million years ago. Genomic diversity of the 7478/7841 segment does not exceed 5%, and genomes that once must have linked HPV-16 with its closest relative, HPV-31, have not yet been found and are most likely extinct. Typical "African" HPV-16 variants "left" Africa very inefficiently (through a very narrow bottleneck) but are present in some ethnic groups in Asia and in native ethnic groups of the New World. In contrast, the "7519 variant" left Africa efficiently, eventually giving rise to the prototype European and particular East Asian variants. Most likely, HPV-16 co-evolved with all populations of *Homo sapiens* since the beginning of our species. In non-immigrant nations throughout the world, HPV-16 variants are still to a large part specific for the infected ethnic group. This contrasts with the recent efficient spread of other sexually transmitted diseases.

3.5 Genomic Variability Within the HPV-18 Type

The phylogenetic tree of HPV-18 variants sampled throughout the world mimics in all details the phylogenetic tree of HPV-16 variants (ONG et al. 1993), and no observations made for the intra-type diversity of HPV-18 require a different interpretation. The prototype was isolated from a biopsy from a patient in northern Brazil (BOSHART et al. 1984). It is very different from European and African isolates but similar to Asian isolates, and could represent a HPV-18 variant from the American Indians.

The phylogenetic trees of the HPV-18 variants could be rooted against HPV-45, the closest related virus, because part of the phylogenetically informative segment of HPV-18 is co-linear to the homologous region of the HPV-45 genome. HPV-45 differs from the HPV-18 prototype by 26 mutations, but seven of these differences found in HPV-45 are identical to those nucleotides found specifically in African HPV-18 variants. In other words, the evolutionary distance between these African variants and HPV-45 is less than that observed relative to the prototype. The distance between the African HPV-18 variants and an African HPV-45 variant is closer still by one mutation. These observations are evidence that African variants of HPV-18 and HPV-45 represent intermediates that form the bridge between these two virus types, and that the "speciation" event that resulted in these two virus types occurred in Africa. We have estimated that this split occurred about 7.6 million years ago and thus predated the evolution of *Homo sapiens.* The intra-type diversity of HPV-18 reflects an age of 1.2 million years, which is approximately the age of the genus *Hom* (ONG et al. 1993).

3.6 Genomic Variability Within the HPV-6 Type

In research that predated the arrival of PCR technology, HPV genomes were detected and typed by Southern blot technologies. The analytical power of this technique was limited to detecting restriction fragment lengths polymorphisms that stem from indels, and loss or creation of restriction sites through point mutations. HPV-6 genomes that were identified in this manner were labeled with a suffix letter, HPV-6a–HPV-6g, to indicate a "subtype" status (GISSMANN et al. 1983; MOUNTS and KASHIMA 1984). Our analysis of some of these isolates (CHAN et al. 1992a) leaves no doubt that their sequences do not differ among themselves more than HPV-16 and HPV-18 variants differ from one another. Also, these mutations gave rise to only minor functional differences, just as in the case of HPV-16 (WU and MOUNTS 1988; CHAN et al. 1992b). From this we conclude that HPV-6 subtypes are taxonomically identical with genomic variants. There are indications that the HPV-6 genome has diversified as much as HPV-16 and HPV-18 (ICENOGLE et al. 1991). It could be very interesting to sample these variants throughout the world and to compare them against a collection of HPV-11 variants. HPV-6 and HPV-11 are as closely related to one another as HPV-18 and HPV-45, and, in analogy to HPV-18 and HPV-45, one may hypothesize that HPV-6 and HPV-11 split off in Africa at the time of the evolution of the hominids.

3.7 Genomic Variability Within the HPV-5 and HPV-8 Types

Independent isolates of HPV-5 and HPV-8 differ from each other by 11% and 3%, respectively, of their E6 gene nucleotide sequences (DEAU et al. 1991), and by up to 6% of the total genomic amino acid sequences (YABE et al. 1991). These numbers exceed by about one order of magnitude the variabilities of HPV-6, HPV-16, and HPV-18 variants, which have roughly 1% intra-genic and only up to 5% extra-genic differences. The studies of HPV-5 and HPV-8 variants did not identify differences between isolates from benign and malignant lesions. This research could lead to a picture of EV HPV evolution that may be quite different from that of the evolution of genital HPV types as these differences mean that HPV-5 and HPV-8 must either be "older" or have evolved faster than genital HPV types.

4 Model for the Evolution and Taxonomy of Papillomaviruses

Comparison of the sequences of about 60 papillomavirus types and of 100 variants of HPV-5, -6, -16, -18 and -45 allows us to propose a model for the molecular evolution and taxonomy of papillomaviruses. We think that the model describes important aspects of the evolution of papillomaviruses quite precisely and serves to generate testable hypotheses.

4.1 HPV Types

The papillomavirus type seems to be a natural taxnomic unit. Evolutionary distances among intra-type isolates (roughly 1% in intra-genic, 5% in extra-genic sequences) are by about one order of magnitude smaller than those found among the comparison of types (more than 10% in intra-genic sequences, extra-genic sequences are normally too divergent for alignment).

This finding has to be reconciled with the fact that HPV types must have evolved from precursors in a process of *continuous* accumulation of mutations. The original definitions of HPV types were operational (COGGINS and ZUR HAUSEN 1979; DE VILLIERS 1992) and could have led to the creation of an artifactual taxonomy, if the genomic continuum between HPV types still existed. However, HPV genomes that link different HPV types have not been found and must either be extinct or exceedingly rare.

The age of HPV types can be roughly estimated with a lower margin of around 100 000 and an upper margin of a few million years (ONG et al. 1993). This, combined with the two assumptions, that animal papillomaviruses do not infect humans and the speed of the molecular clock is similar for all HPV types, leads to some interesting and testable hypotheses. All HPV types must have an African origin, and no HPV type should have originated outside Africa (since the period of human evolution outside Africa may not have sufficed to generate new HPV types). The ancestors of all human ethnic groups were infected by the same HPV types as we carry them today, and all human ethnic groups could potentially still carry all HPV types, unless some were lost during bottlenecks of human evolution. An obvious consequence is that the risk of HPV-associated malignant neoplasia has been our species since its beginnings.

4.2 Minor Branches of Papillomavirus Phylogenies

Minor branches on the phylogenetic tree (e.g., clusters like HPV-6b/11, HPV-16/31/35) unite HPV types that originated from common precursors fairly recently (lower margin a few, upper margin about 10 million years ago). They show extensive nucleotide and polypeptide similarities and consequently have similar biological and pathological properties. These minor branches are identical with groups defined on the basis of hybridization data and they will form a natural basis for future classifications.

The time scale for evolution of minor branches would predict that many novel monkey or ape papillomaviruses should be related to HPV types, while it should not be possible to find papillomaviruses in other animals that are as closely related to some HPV types as PCPV, CGPV, and RPV.

4.3 Major Branches of Papillomavirus Phylogenies

Major branches identify papillomavirus types that are only remotely related to one another. We can presently recognize four major branches among the HPV types and four major animal papillomavirus branches.

If the interpretation of strict separation of papillomavirus evolution by host species barriers is correct, then each new papillomavirus isolate from a different vertebrate order or family may form a new major branch. These major branches may represent tens of millions of years of separate evolution. If the same molecular clock applies to the splits between the four major branches of HPV types, then these splits must have predated primate evolution. The remote relationship of CRPV and COPV with three HPV-types is likely to exemplify this scenario.

4.4 Where Did Papillomaviruses Come from?

According to present taxonomy (MURPHY and KINGSBURY 1990), papillomaviruses and polyomaviruses form two genera in the family papovaviridae. Once taxonomy attempts to suggest natural relationship, it has to be examined whether this placement is qualitatively correct and whether the relationship is sufficiently close to warrant placement in a common taxon.

The present placement is based on the similarity of the viral capsids and genome structure, i.e., a circular double-stranded DNA. However, the different genomic organization and the complete lack of extensive homologies between any polyoma and papillomavirus genes make this placement suspect.

It is conceivable, however, that small domains of viral proteins still hold a record of a common past. These similarities may be inconspicuous during a purely quantitative sequence alignment but may be detected once one understand better the three-dimensional structures of viral proteins and the individual functions of viral domains. We can presently think of two such candidates:

1. The polyomavirus T antigen has similarity to the papillomavirus E1 protein at a potential ATP-binding domain, and both proteins function to initiate DNA replication (CLERTANT and SEIF 1984).
2. Polyoma virus T antigen and the E7 protein of papillomaviruses have similarities in domains that are involved in the binding of the retinoblastoma gene product (PHELPS et al. 1988; DEFEO-JONES et al. 1991).

It may be that phylogenetic trees based on these and future discoveries (CHAN et al. 1992a) will eventually constitute evidence for the monophyletic origin of DNA tumor viruses. On the other hand, the same data may teach us that this split was an event as ancient and as major for virus evolution as the origin of vertebrates was for the evolution of higher organisms.

5 Clinical Aspects of Phylogenetic Information

A large number of variants of each HPV type exists in nations that are composed of immigrants. A cohort of 100 HPV-16-infected patients may carry 30 different HPV-16 variants. This permits conceptually novel epidemiological studies that make use of the fact that each variant can be recognized like an individual, and consequently individual infectious events can be studied. Questions to address include the multiplicity of infections and the monoclonality of HPV lesions, the sexual and/or vertical transmission of HPV, and the differentiation between recurrent disease and reinfection (Ho et al. 1991, 1993b). Also, different variants of all HPV types may have different pathological properties, such as variants of HPV-6 and HPV-16 whose L2 proteins give rise to differential humoral immune responses (Galloway, personal communication).

6 Future Research

Most models established in this review can only be resolved through continuous sampling of HPV genomes. To establish the molecular clock of HPV types, it is necessary to obtain biopsies from aboriginal populations with a precisely known history of isolation, like those of the Americas or Australia. Mucosal lesions in primate and non-primate mammals will yield the answer about the origin of this HPV branch. Papillomavirus isolates from reptiles, amphibians, fish, or even non-vertebrates may help us to approach the root of the papillomavirus phylogeny. For clinical studies, it will be interesting to estimate how many undescribed HPV types may still exist. If this number is large, it may be more practical to attempt an initial survey through partial genomic isolates by PCR. Lastly, it will be of interest to find out whether variants of HPV-16 and HPV-18 or other HPV types are simply of academic interest, or whether they differ in clinically important aspects.

References

Boshart M, Gissmann L, Ikenberg H, Kleinheinz A, Scheurlen W, zur Hausen H (1984) A new type of papillomavirus DNA, its presence in genital cancer biopsies and in cell lines derived from cervical cancer. EMBO J 3: 1151–1157

Chan SY, Bernard HU, Ong CK, Chan SP, Hofmann B, Delius H (1992a) Phylogenetic analysis of 48 papillomavirus types and 28 subtypes and variants: a showcase for the molecular evolution of DNA viruses. J Virol 66: 714–5725

Chan SY, Ho L, Ong CK, Chow V, Drescher B, Duerst M, ter Meulen J, Villa L, Luande J, Mgaya HN, Bernard HU (1992b) Molecular variants of human papillomavirus type 16 from four continents suggests ancient pandemic spread of the virus and its coevolution with humankind. J Virol 66: 2057–2066

Chong T, Apt D, Gloss B, Isa M, Bernard HU (1991) The enhancer of human papillomavirus type 16: binding sites for the ubiquitous transcription factors oct-1, NFA, TEF-1, NF1, and AP-1 participate in epithelial cell specific transcription. J Virol 65: 5933–5943

Clertant P, Seif I (1984) A common function for polyoma virus large-T and papillomavirus E1 proteins. Nature 311: 276–279

Coggin JR, zur Hausen H (1979) Workshop and papillomaviruses and cancer. Cancer Res 39: 545–546

Danos O, Yaniv M (1987) E6 and E7 gene products evolved by amplification of a 33 amino-acid peptide with a potential nucleic-acid-binding structure. In: Steinberg BM, Brandsma JL, Taichman LB (eds) Paillomaviruses. Cold Spring Harbour Laboratory, Cold Spring Harbor, pp 145–149
Deau MC, Favre M, Orth G (1991) Genetic heterogeneity among human papillomaviruses (HPV) associated with epidermodysplasia verruciformis: evidence for multiple allelic forms of HPV-5 and HPV-8 E6 genes. Virology 184: 492–503
Defeo-Jones D, Huang PS, Jones RE, Haskell KM, Vuocolo GA, Hanobik MG, Huber HE, Oliff A (1991). Cloning of cDNAs for cellular proteins that bind to the retinoblastoma gene product. Nature 352: 251–254
de Villiers EM (1989) Heterogeneity of the human papillomavirus group. J Virol 63: 4898–4903
de Villiers EM (1992) Laboratory techniques in the investigation of human papillomavirus infection. Genitourin Med 68: 50–54
Duerst M, Gissmann L, Ikenberg H, zur Hausen H (1983) A papillomavirus DNA from a cervical carcinoma and its prevalence in cancer biopsy sampels from different geographic regions. Proc Natl Acad Sci USA 80: 3812–3815
Ensser A, Pfister H (1990) Epidermodysplasia verruciformis associated human papillomaviruses present a subgenus-specific organization of the regulatory genome region. Nucleic Acids Res 18: 3919–3922
Eschle D, Duerst M, ter Meulen J, Luande J, Eberhardt HC, Pawlita M, Gissmann L (1992) Geographical dependence of sequence variation in the E7 gene of human papillomavirus type 16. J Gen Virol 73: 1829–1832
Felsenstein J (1982) Numerical methods for inferring evolutionary trees. Q Rev Biol 57: 379–404
Felsenstein J (1985) Confidence limits on phylogenies: an approach using the bootstrap. Evolution 39: 783–791
Felsenstein J (1988) Phylogenies from molecular sequences: inference and reliability. Annu Rev Genet 22: 521–565
Fujinaga Y, Okazawa K, Ohashi Y, Yamakawa Y, Fukushima M, Kato I, Fujinaga K (1990) Human papillomavirus type 16 E7 gene sequence in human cervical carcinoma analysed by polymerase chain reaction and direct sequencing. Tumor Res 25: 85–91
Giri I, Yaniv M (1988) Structural and mutational analysis of E2 trans-activating proteins of papillomaviruses reveals three distinct functional domains. EMBO J 7: 2823–2829
Gissmann L, Wolnik L, Ikenberg H, Koldovsky U, Schnurch HG, zur Hausen H (1983) Human papillomavirus types 6 and 11 DNA sequences in genital and laryngeal papillomas and in some cervical cancers. Proc Natl Acad Sci USA 80: 560–563.
Greenspan D, de Villiers EM, Greenspan JS, de Souza YG, zur Hausen H (1988) Unusual HPV types in oral warts in association with HIV infections. J Oral Pathol 17: 482–487
Ho L, Chan SY, Chow V, Chong T, Tay SK, Villa L, Bernard HU (1991) Sequence variants of human papillomavirus type 16 in clinical samples permit verification and extension of epidemiological studies and construction of a phylogenetic tree. J Clin Microbiol 29: 1765–1772
Ho L, Chan SY, Burk RD, Das BC, Fujinaga K, Icenogle JP, Kahn T, Kiviat N, Lancaster W, Mavromara P, Labropoulou V, Mitrani-Rosenbaum S, Norrild B, Pillai MR, Stoerker J, Syrjaenen K, Syrjaenen S, Tay SK, Villa LL, Wheeler CM, Williamson AL, Bernard HU (1993a) The genetic drift of human papillomavirus type 16 is a means of reconstructing prehistoric viral spread and movement of ancient human populations. J Virol 67: 6413–6414
Ho L, Tay SK, Chan SH, Bernard HU (1993b) Sequence variants of human papillomavirus type 16 from couples suggest sexual transmission with low infectivity and polyclonality in genital neoplasia. J Infect Dis 168: 803–809
Icenogle JP, Sathya P, Miller DL, Tucker RA, Rawls WE (1991) Nucleotide and amino acid sequence variation in the L1 and E7 open reading frames of human papillomavirus type 6 and type 16. Virology 184: 101–107
Jackson ME, Pennie, WD, McCaffery RE, Smith KT, Grindlay GJ, Campo MS (1991) The B subgroup bovine papillomaviruses lack an identifiable E6 open reading frame. Mol Carcinog 4: 382–387
Kulke R, Gross GE, Pfister H (1989) Duplication of enhancer sequences in human papillomavirus 6 from condylomas of the mamilla. Virology 173: 284–290
Li WH, Graur D (1991) Fundamentals of molecular evolution. Sinauer, Sunderland, MA, USA.

Lorincz A (1989) Human papillomavirus detection tests. In: Holmes KK et al. (eds) Sexually transmitted diseases. McGraw-Hill, New York, pp 953–959
Lorincz A, Reid R, Jenson AB, Greenberg MD, Lancaster W, Kurman RJ (1992) Human papillomavirus infection of the cervix: relative risk association of 15 common anogenital types. Obstet Gynecol 79: 328–337
Manos MM, Ting Y, Wright DK, Lewis AJ, Broker TR (1989) Use of polymerase chain reaction amplification for the detection of genital papillomaviruses. Cancer Cells 7: 209–214
Minor PD (1992) The molecular biology of poliovaccines. J Gen Virol 73: 3065–3077
Mounts P, Kashima H (1984) Association of human papillomavirus subtype and clinical course in respiratory papillomatosis. Laryngoscope 94: 28–33
Murphy FA, Kingsbury (1990) Virus taxonomy. In: Fields BN, Knipe DM (eds) Virology. Raven, New York, pp 9–36
Murphy BR, Webster RG (1990) Orthomyxoviruses. In: Fields BN, Knipe DM (eds) Virology. Raven, New York, pp 1091–1152
Ong CK, Chan SY, Campo MS, Fujinaga K, Mavromara P, Labropoulou V, Pfister H, Tay SK, ter Meulen J, Villa LL, Bernard HU (1993) Evolution of human papillomavirus type 18: An ancient phylogenetic root in Africa and intratype diversity reflect coevolution with human ethnic groups. J Virol 67: 6424–6431
Pfister H (1990) Molecular biology of genital HPV infections. In: Gross G, Jablonska S, Pfister H, Stegner N (eds) Genital papillomavirus infections. Springer, Berlin Heidelberg New York, pp 37–49
Phelps EC, Yee CL, Muenger K, Howley, PM (1988) The human papillomavirus type 16 E7 gene encodes transactivation and transformation functions similar to those of adenovirus E1A. Cell 53: 539–547
Ruebben, A, Beaudenon S, Favre M, Schmitz W, Spelten B, Grussendarf-Conen EI (1992) Rearrangement of the upstream regulatory region of human papillomavirus type 6 can be found in both Buschke-Lowenstein tumours and in condylomata acuminata. J Gen Virol 73: 3147–3153
Sambrook J, Sleigh M, Engler JA, Broker TR (1980) The evolution of the adenovirus genome. Ann NY Acad Sci 354: 426–452
Sanger F, Nicklen S, Coulson AR (1977) DNA sequencing with chain terminating inhibitors. Proc Natl Acad Sci USA 4: 5463–5467
Schiffman MH, Bauer HM, Lorincz AT, Manos MM, Byrne JC, Glass AG, Cadell DM, Howley PM (1991) Comparison of Southern blot hybridization and polymerase chain reaction methods for the detection of human papillomavirus DNA. J Clin Microbiol 29: 573–577
Schwarz E, Freese UK, Gissmann L, Mayer W, Roggenbuck B, Stremlau A, zur Hausen H (1985) Structure and transcription of human papillomavirus sequences in cervical carcinoma cells. Nature 314: 111–114
Seedorf K, Kraemmer G, Duerst M, Suhai S, Roewekamp WG (1985) Human papillomavirus type 16 DNA sequence. Virology 145: 181–185
Sharp PM, Li WH (1988) Understanding the origins of AIDS viruses. Nature 336: 315
Simons EL (1989) Human origins. Science 245: 1343–1350
Stock DW, White GS (1992) Evidence from 18s ribosomal RNA sequences that lampreys and hagfishes form a natural group. Science 257: 787–789
Sundberg JP (1987) Papillomavirus infections in animals. In: Syrjaenen K, Gissmann L, Koss LG (eds) Papillomaviruses and human disease. Springer, Berlin Heidelberg New York, pp 40–103
Swofford DL (1991) PAUP: Phylogenetic analysis using parsimony, version 3.0, computer program and documentation. Illinois Natural History Survery, Champaign, Illinois, USA
Tan SH, Gloss B, Bernard HU (1992) During negative regulation of the human papillomavirus-16 E6 promoter, the viral E2 protein can displace Sp1 from a proximal promoter element. Nucleic Acids Res 20: 251–256
van den Brule AJC, Snijders PJF, Raaphorst PMC, Schrijnemakers HFJ, Delius H, Gissmann L, Meijer CJLM, Walboomers JMM (1992). General primer polymerase chain reaction in combination with sequence analysis for identification of potentially novel human papillomavirus genotypes in cervical lesions. J Clin Microbiol 30: 1716–1721
van Ranst M, Fuse A, Fiten P, Beuken E, Pfister H, Burk RD (1992a) Human papillomavirus type 13 and pigmy chimpanzee papillomavirus type 1: comparison of the genome organizations. Virology 190: 587–596

van Ranst M, Kaplan JB, Burk RD (1992b) Phylogenetic classification of human papillomaviruses: correlation with clinical manifestations. J Gen Virol 73: 2653–2660
Vigilant L, Stoneking M, Harpending H, Hawkes K, Wilson AC (1991) African populations and the evolution of the human mitochondrial DNA. Science 253: 1503–1507
Wu TC, Mounts P (1988) Transcriptional regulatory elements in the noncoding regions of human papillomavirus type 6. J Virol 62: 4722–4729
Yabe Y, Sakai A, Hitsumoto T, Kato H, Ogura H (1991) A subtype of human papillomavirus 5 (HPV-5b) and its subgenomic segment amplified in a carcinoma: nucleotide sequences and genomic organizations. Virology 183: 793–798
Zuckerkandl E, Pauling L (1962) Molecular disease, evolution, and genetic heredity. In: Kasha M, Pullman B (eds) Horizons in biochemistry. Academic, New York, pp 189–225
zur Hausen H (1991) Viruses in human cancers. Science 254: 1167–1173

Epidemiology of Cervical Human Papillomavirus Infections

M.H. SCHIFFMAN

Executive Plaza North, Room 443, Epidemiology and Biostatistics Program, National Cancer Institute, Bethesda, MD 20892, USA

Current Topics in Microbiology and Immunology, Vol. 186

1 Introduction

This update will summarize recent advances regarding the epidemiology of human papillomavirus (HPV) infections of the cervix. Readers interested in earlier, more comprehensive reviews are referred to two excellent summaries by KOUTSKY et al. (1988) and SCHNEIDER and KOUTSKY (1992). This update will be restricted to cervical HPV infection and will not include recent advances in the epidemiology of cutaneous HPV types or genital types infecting sites other than the cervix (e.g., aerodigestive tract).

Most epidemiologic research on HPV infections of the cervix is still focused on neoplastic outcomes. Epidemiologists are attempting to confirm that genital types of HPV cause most cases of cervical cancer worldwide and to determine etiologic co-factors acting with HPV in cervical cancer pathogenesis. Epidemiologists are just beginning to study cervical HPV from an infectious disease standpoint, in which the genital types of HPV are viewed independently of neoplasia, as a family of common cervical pathogens.

Because the epidemiologic study of HPV as a family of cervical pathogens is new, this update must depend to an unusually large degree on fragmentary data from a few ongoing studies. The speculative and somewhat personal perspective of the update should be recognized from the outset. This update will be outdated as immunologic assays improve and prospective data emerge from the first large natural studies of cervical HPV infection which are now underway.

2 Diagnosis and Definition of Cervical HPV infection

In its infancy, the epidemiologic study of HPV infection has been shaped and limited by available methods of diagnosing and defining infection. Accordingly, this section will briefly summarize the status of DNA diagnostics, serologic assays, and the cytologic diagnosis of cervical HPV infections.

2.1 DNA Detection

The lack of a reliable serologic measure of exposure to HPV has limited the assessment of cumulative lifetime HPV infection rates. With no serologic assays of past infection available, the diagnosis of cervical HPV infection has relied on DNA hybridization methods applied to exfoliated cervical cell specimens in innovative test formats. The detectability of HPV DNA in a single specimen clearly differs from lifetime exposure to HPV because a poorly understood host response mediates the natural history of HPV infection (see the related update by Tindle and Frazer, this volume). Consequently, the reader should be aware that, for precision, the term "HPV infection" must be referred to specific diagnostic criteria.

For the purpose of this update, cervical HPV infection will be primarily defined as the detection of HPV-related DNA sequences in cervical or cervicovaginal specimens, at a level of clinical sensitivity shared by a group of commonly applied HPV DNA assays (e.g., low-stringency Southern blot, L1 consensus primer PCR, commercially available dot blot techniques, and Hybrid Capture). HPV infection could be defined alternatively by reliance on the results of ultra-sensitive type-specific polymerase chain reaction (PCR) assays, in which case viral prevalence estimates would increase to an unknown (and controversial) degree while the correlations with disease outcome would weaken (because disproportionately more controls would be HPV positive). Different definitions of HPV infection are not necessarily contradictory, but care must be taken to compare epidemiologic results only when comparable diagnostic methods and definitions have been used.

The recent history of HPV DNA detection tests can be quickly summarized. Epidemiologic studies (and screening efforts) require large numbers of subjects, reducing the practicality and affordability of using traditional HPV DNA test methods such as high-stringency Southern blot as this is performed in research laboratories. Streamlining of traditional research laboratory approaches took several years, and the early epidemiologic test formats yielded inaccurate results (SCHIFFMAN 1992a, b). DNA hybridization methods permitting reliable large-scale detection of a full spectrum of HPV types, in exfoliated cervical specimens collected at the time of gynecologic examination, have been available for 2–3 years (SCHIFFMAN 1992a, b). The state-of-the-art testing techniques now being used by most epidemiologists tend to yield comparable results when performed expertly on adequate specimens (see Sect. 2.1.2).

2.1.1 Collection of the Cervical Specimen

The method of collecting cervical specimens for HPV testing is still an issue worth discussing. The final result of HPV DNA testing depends on the collection of a sufficient specimen and the performance of the HPV test method (GOLDBERG et al. 1989). Epidemiologists have used cervical swabs, brushings, scrapes, and lavages to obtain cervical cells noninvasively for HPV DNA testing. Full Southern blot analysis or multiple HPV type-specific dot blots require a substantial amount of DNA, and specimens derived from scrapes or swabs may sometimes not permit full and adequate testing for the 15+ HPV types of relevance to the cervix (GUERRERO et al. 1992). As a corollary, scant specimens, such as aliquotted portions of a swab specimen, favor the choice of consensus primer PCR over Southern blot or dot blot because much less specimen is needed for adequate PCR-based detection and typing of the full range of HPV types.

In our experience, however, the collection of DNA-rich specimens (such as lavages) greatly reduces the sensitivity advantage of using consensus primer PCR (SCHIFFMAN et al. 1991). In fact, for reasons that are not understood, the analytic sensitivity of L1 consensus primer PCR may actually be reduced slightly by using a cervicovaginal lavage specimen rather than a direct swab of the cervix (HILDESHEIM et al. 1993). Apart from subtle differences, if ample specimens are obtained, each of the methods mentioned in the section below can be used to obtain comparable epidemiologic results in the populations we have studied. It is possible that, in different study populations, the relative performance of available assays may vary somewhat because of the spectrum and intensities of HPV infections found. Specifically, populations with a high prevalence of very low-level infections, barely detectable at the level of analytic sensitivity of the common tests, may tend to produce method-dependent results (R. Burk, unpublished data).

The list of assays mentioned here is not meant to be exclusionary; other techniques unfamiliar to the author are probably useful as well. As one dogma, however, any technique that fails to detect a wide spectrum of HPV types has correspondingly restricted usefulness for epidemiologic research, as the prevalence of more recently identified, cancer-associated HPV types (e.g., HPV 56, 58) is substantial in some populations (HILDESHEIM et al. 1993; SCHIFFMAN et al. 1993).

2.1.2 Comparability of Currently Available HPV Test Methods

The good comparability of some currently available HPV DNA test methods, when performed expertly, is demonstrated in Table 1. For each of three cytologic diagnoses, overall (all type) HPV DNA positivity was found to be roughly equivalent for four different test methods: L1 consensus primer PCR (MANOS et al. 1989), low-stringency Southern blot (LORINCZ 1992), ViraPap dot blot (LORINCZ 1992), and Hybrid Capture (see below). Moreover, the intermethod agreement for individual specimens was good. The tests were performed, masked to cytologic and other virologic results, using aliquots of

Table 1. HPV DNA positivity by cytologic diagnosis, for four HPV test methods

Cytologic diagnosis	Patients (*n*)	Type spectrum	Consensus PCR (%)	Southern blot (%)	Profile (%)	Hybrid Capture (%)
Normal	100	All types	27	26	4	12
		Cancer associated	13	7	2	7
ASCUS	44	All types	66	55	41	43
		Cancer associated	53	36	32	32
SIL	53	All types	91	87	72	74
		Cancer associated	68	60	62	60

Table 2. Intralaboratory reliability of Hybrid Capture results, on masked retesting

	Result on first test	
	Negative	Positive
Result on repeat test		
Negative	101	2
Positive	1	15

the same 10-ml cervicovaginal lavage, taken at the same examination as the cytologic smears (SCHIFFMAN et al. 1993). The smears were classified using the Bethesda System (NATIONAL CANCER INSTITUTE WORKSHOP 1989).

The major differences in results between the four different methods derived from increased detection of uncharacterized HPV types by the PCR and low-stringency Southern blot. After the test results were unmasked, an attempt was made to resolve typing discrepancies. The results agreed even more closely than indicated in Table 1 when subtle method-related differences in the spectrum of type-specific detection were taken into account, e.g., if probes for rarer cancer-associated types were uniformly included, or if "unknown" types detected only by consensus PCR or nonstringent Southern blotting were excluded (data not shown).

The newest of the techniques, Hybrid Capture, may be unfamiliar to some readers. It is a nonradioactive, rapid, and relatively inexpensive liquid RNA–DNA hybridization technique designed by Digene Diagnostics, Inc. (A. Lorincz, personal communication). The technique permits semiquantitative estimation of type-specific HPV DNA, of one or many HPV types as desired, and can be used with a variety of kinds of cervical specimens. Results from the first masked field studies using the technique have demonstrated excellent intralaboratory reliability, as shown for example in Table 2. The results of interlaboratory reliability studies are pending (M.H. Schiffman et al., unpublished data). Because of its broad spectrum, speed, low cost, and the efforts underway to make the test reliable, the Hybrid Capture technique may be one of the most useful HPV testing techniques for future epidemiologic studies.

2.2 Serologic Assays

The epidemiologist's need for accurate serologic markers of past and current HPV infection was emphasized above. However, HPV serologic assay development has lagged behind advances in DNA testing because of the lack (until recently) of an abundant HPV antigen source and the poorly understood complexities of the host response to HPV (GALLOWAY 1992). Correspondingly, the published serologic literature regarding HPV natural history has been relatively scant and enigmatic. This update on cervical HPV epidemiology will concentrate on infection defined by DNA detection, with the recognition that recent advances in HPV immunology (Tindle and Frazer, this volume) and antigen availability (HAGENSEE et al. 1993; D. Lowy, personal communication) may transform HPV sero-epidemiology in the near future.

A few sero-epidemiologic results are of note, nonetheless. Several epidemiologic studies using serologic assays have observed elevated HPV seropositivity in women with invasive cervical cancer (reviewed in GALLOWAY 1992). The seropositivity observed in cancer patients appeared to be an insensitive biomarker in that over half the patients whose tumors contained DNA of the same viral type being assayed serologically were seronegative (MULLER et al. 1992). Seropositivity was not specific for cancer as up to 20% of control women were seropositive. The few reported sero-epidemiologic studies of CIN have not generally revealed elevated rates of seropositivity compared to control women (GALLOWAY 1992; M.H. Schiffman et al., unpublished data).

The relationship of HPV seropositivity to lifetime infection is unassessable without a reference standard of lifetime infection (which immunoassays of serum or cervical secretions may provide someday). One can assume that the sensitivity of reported assays as biomarkers of lifetime infection rates is not good given that HPV-containing tumors or cervical intraepithelial nesplasia (CIN) lesions do not elicit universal seropositivity using those assays.

Recently, preliminary data derived using newly-synthesized HPV capsids ("virus-like particles) have appeared more promising in defining a useful serologic measure of recent type-specific HPV infection than much of the earlier work using linear epitopes (J. Schiller, unpublished data).

2.3 Cytologic Diagnosis of Cervical HPV Infections

Cervical HPV can be diagnosed in a clinical sense, based on the cytologic diagnosis of characteristic cytopathic effects induced by HPV. Unfortunately, the cytologic diagnosis of HPV is not especially accurate or reliable (SHERMAN et al. 1992). Cytologic evidence of HPV infection has been called by a confusing variety of names, including "koilocytotic atypia," "condylomatous atypia," "flat condyloma," or "HPV effect." There has been an increasing recognition since the late 1970s (MEISELS and FORTIN 1976) that the cytologic effects of HPV infection blend without reproducible distinction into the changes traditionally called "mild dysplasia" or "CIN 1," lesions which are thought by many clinicians and researchers to be (distant) precursors to cervical cancer. In recognition of the cytomorphologic overlap of the two diagnoses, "HPV effect" and "CIN 1" have been combined as "low-grade squamous intraepithelial lesions (LSIL)" in the United States National Cancer Institute-sponsored Bethesda System of cytopathology (NATIONAL CANCER INSTITUTE WORKSHOP 1989). In parallel, the World Health Organization

(WHO) classification of cervical pathology will soon reclassify koilocytotic atypia as CIN 1 (Scully et al., in press).

Though possibly disruptive to current clinical practice (because of the resultant increase in referrals from screening clinics), the concept of LSIL is supported by epidemiologic data demonstrating that the two subsumed diagnoses (koilocytotic atypia and CIN 1) share the same broad HPV type spectrum and have similar demographic characteristics, such as early age peaks (SCHIFFMAN et al. 1990). Moreover, both diagnoses typically have a transient and benign natural history.

The Bethesda System has not gained universal acceptance but, in areas where it is used, the epidemiologic study of HPV infection and cervical neoplasia as defined by cytopathologic diagnosis has been altered. In particular, descriptive studies of HPV (surveillance) have been affected. Typically (in our series at least), the diagnosis of koilocytotic atypia is two to three times more common than CIN 1 (SCHIFFMAN et al. 1990). Therefore, the prevalence of CIN 1 is much increased by inclusion of koilocytotic atypia. The prevalence of the more common diagnosis, koilocytotic atypia, is altered to a proportionally lesser extent by a switch to surveillance of LSIL, but may be greatly altered by concurrent secular trends in the stringency of cytopathologic criteria, for defining HPV effect.

3 Descriptive Epidemiology

3.1 Prevalence of Cervical HPV Infections

With the preceding methodologic section as background, it is easy to see why there is no sensible, simple answer to the frequently asked question: "How common is HPV infection of the cervix?" Each definition of HPV implies a different prevalence. In one study, HPV DNA was detectable in 30%–40+% of young sexually active women when consensus primer PCR capable of detecting a wide spectrum of HPV types was used to test both cervical and introital specimens (BAUER et al. 1991). Dot blot testing of the same specimens yielded a much lower prevalence. Virtually none of the young women had cytologic evidence of HPV infection. The prevalence depends, therefore, on the definition of HPV infection.

Even beyond questions of definition and diagnosis, HPV prevalence varies greatly in different populations, dependent on both demographic and behavioral determinants of infection that will be mentioned below. In particular, young sexually active women appear to experience the highest HPV prevalences (LEY et al. 1991). Older monogamous women are much less likely to be HPV positive.

It is wise, therefore, to avoid simple summary statements about HPV prevalence. Nevertheless, a few points have been established, as shown in the following sections.

3.1.1 Prevalence of HPV in Invasive Cancers

The proportion of invasive cancers containing HPV DNA is high wherever the issue has been intensively studied (reviewed in MUNOZ and BOSCH 1992). A large epidemiologic project is currently studying the proportion and distribution of HPV-negative cervical cancer by assaying 1000 cervical cancers from 50 countries (N. Munoz, IARC). HPV-negative cervical cancer appears to exist as a distinct entity (DE BRITTON et al. 1993), but may be relatively rare.

The relative prevalence of the different HPV types infecting the cervix has been studied most extensively in women with cancer and CIN (FUCHS et al. 1988; LORINCZ et al. 1992; KADISH et al. 1992). Such studies have defined the "cancer-associated" group of HPV types found with appreciable prevalence in invasive cancers (the current list would include at least types 16, 18, 31, 33, 35, 39, 45, 51, 52, 56, 58, and 68). The epidemiologic definition of cancer-associated types (sometimes called "high-risk" types), correlates well with the relative transforming properties of the viral types as defined in vitro, and with "phylogenetic" studies aimed at tracing HPV evolution (CHAN et al. 1992; VAN RANST et al. 1992).

3.1.2 Prevalence of HPV in CIN Lesions

The prevalence of HPV infection (DNA detection) in women with CIN is also very high, whether or not koilocytotic atypia is included in the cytopathologic diagnosis of CIN. In our studies, the prevalence of HPV DNA in definite cases of CIN approaches 100% (SHERMAN et al., submitted manuscript) with cancer-associated HPV types accounting for half or more of infections. In the past, we reported lower prevalence estimates associated with low-grade CIN (NEGRINI et al. 1990; SCHIFFMAN et al. 1990) because, in the earlier studies, we could not test for the full spectrum of HPV types and had not yet minimized cytopathologic misclassification of low-grade CIN. Rigorous cytopathologic review, preferably by a panel of experts, is necessary to exclude benign reactive "lookalikes" from the diagnosis of low-grade CIN. With review, definite cases of low-grade CIN are HPV positive. The nearly universal detection of HPV DNA in verified CIN lesions may be worldwide, or the causes of CIN lesions could theoretically vary by geographic area. Relevant studies incorporating broad-spectrum HPV testing and pathology review have not yet been performed in many areas of the world.

In the united States, the prevalence of HPV infection *as defined by the cytologic diagnosis of koilocytotic atypia* varies from about 10% to less than 1% depending on the type of population studied (M.H. Schiffman, unpublished data). Generally, the cytologic diagnosis of HPV appears to be five to ten times less common than HPV DNA detection in the same populations, i.e., only 10%–20% of HPV DNA-positive women have a cytologic abnormality diagnosed (SCHIFFMAN 1992b).

3.1.3 Prevalence of HPV in Cytologically Normal Women

The prevalence of HPV infection in cytologically normal women, as defined by DNA, varies from 100% to less than 5% (using the types of standard assays discussed above), depending on the demographic and behavioral profile of the study group (M.H. Schiffman, unpublished data). In the general population, HPV DNA at a standard level of detection is not at all ubiquitous, especially among older women with no past medical history of cervical neoplasia (VAN DEN BRULE et al. 1991; SCHNEIDER and KOUTSKY 1992).

The prevalence of HPV infection, as defined by DNA assayed by ultrasensitive PCR assays, is unknown and remains controversial (LORINCZ 1992). The previously reported ubiquity of HPV DNA may be due to testing error or may reflect a biologic truth related to HPV latency or low-level viral persistence. Although controversial views exist, few, if any, laboratory groups are currently working on resolving the true prevalence of HPV infection at the lowest levels of detection (i.e., by directly comparing the results of consensus primer with type-specific PCR).

With regard to specific HPV types, the most important initial task was to define HPV types associated with cervical disease; thus, fewer groups have attempted to define the relative prevalence of the 15–20 known cervical HPV types among cytologically normal women. Table 3 presents previously unpublished data from a survey of a random sample of cytologically normal women screened at Kaiser Permanente health clinics in Portland, Oregon. In this study, as in most others, HPV 16 was observed to be the most common individual type among cytologically normal women. Most HPV types individually accounted for only a small proportion of total infections. Unknown types as a group still accounted for a non-negligible fraction of infections among cytologically normal women. Because still-uncharacterized HPV types are uncommonly observed in cervical cancer, efforts to further characterize the unknown cervical HPV types found in the general population are proceeding slowly in a few centers only.

The prevalence of concurrent, multiple HPV infections is highly dependent on HPV method. In our experience, multiple HPV infections are more commonly observed with consensus primer PCR methods than with Southern blot or dot blot techniques. The prevalence of multiple infections in studies using consensus primer PCR typically approach 20%–30% of all infected women (BAUER et al. 1993; HILDESHEIM et al. 1993; WHEELER et al. 1993). Multiple infections are associated with an increased risk of cytologically evident low-grade CIN (MORRISON et al. 1991; SCHIFFMAN et al. 1993), but the explanations for this are uncertain. The prevalence of multiple infections in high-grade CIN and cancer compared with low-grade CIN is not well

Table 3. Prevalence of individual genital types of HPV infection in cytologically normal women

HPV type	Prevalence (%)
Inadequate	6.1
Adequate, negative[a]	77.5
One HPV type	12.2
⩾Two types	4.3
Individual types	
6/11	1.6
16	2.4
18	0.2
31	0.8
33	0.2
35	0.2
39	0.4
40	0.4
45	0.8
51	0.6
52	1.0
53	1.2
54	1.2
55	0.6
56	1.2
58	0.8
Clinicals	2.0
Unknown	1.8

[a] Indeterminant HPV results grouped as negative; if grouped as positive, these results would have greatly expanded the "type unknown" category.

known; in one small series, the prevalence of multiple infections declined in high-grade CIN compared with low-grade lesions (M.H. Schiffman, unpublished data). Also unclear is whether the distribution of HPV types in multiple infections is random or reveals a tendency of certain types to favor or exclude each other's presence. A formal statistical examination of this topic is underway (Renton et al., personal communication).

3.2 Incidence of Cervical HPV Infections

It is important in natural history studies to distinguish incident (newly acquired) cervical HPV infections from prevalent infections but, lacking serologic markers of lifetime HPV exposure, this has proven to be nearly impossible. To define incident HPV infection requires the knowledge that a woman is uninfected at the start of observation. It may be that we know too little about the possibility of HPV latency and reactivation to be sure that a woman with no HPV DNA detectable is truly uninfected.

It is worth considering how HPV incidence could theoretically be studied. If one accepts that HPV infection is transmitted by sexual activity in

Table 4. Acquisition of new HPV DNA positivity correlated with recent numbers of sexual partners, among women initially HPV negative

Recent sexual partners (*n*)	Women (*n*)	HPV positive at follow-up (%)
0	17	6
1	272	12
2+	41	66

adulthood (see below), one would ideally study HPV incidence in women as they first begin to have sexual intercourse, taking the enrollment HPV measurements prior to first sexual encounters. Such studies are quite difficult, for obvious social and ethical reasons, but projects recruiting college women during their first year at university are attempting to study incident HPV infection (R. Burk, L. Koutsky, and K. Shah, personal communication).

Despite the limitations of the data that exist, we can reasonably estimate that HPV incidence rates are high in sexually active young women. For example, Table 4 shows apparent HPV "incidence" in a group of women found initially to be HPV negative, then retested once 1–2 years later. The acquisition of newly detectable HPV DNA among women with multiple recent sexual partners, if confirmed, would suggest a very high incidence rate related to sexual transmission. However, rates of new detection might give an estimate of HPV incidence that is too high if viral "reactivation" is a common occurrence and if reactivation (whatever that term means for HPV) is related to sexual activity with new partners.

3.3 Age Trends in HPV Prevalence

At the simplest level, it appears that HPV prevalence declines sharply with age, from a peak prevalence at 15–25 years of age (MORRISON et al. 1991; LEY et al. 1991; VAN DEN BRULE et al. 1991). The reasons for this profound age trend, which is seen in most but not all (E. Franco, private communication) data sets, are not known.

As a related issue, little is known regarding the prevalence of HPV infection in women younger than 16 because this is the minimum age of informed consent and inclusion for most epidemiologic studies. Based on scant and controversial data collected on virgins, it is probable that cervical HPV infection is uncommon in virgin girls (LEY et al. 1991; GUTMAN et al. 1992; FAIRLEY et al. 1992; WHEELER et al., in press). If this is true, then the very high prevalence of HPV in sexually active women aged 15–25 appears to describe an "epidemic curve," a rapid rise in prevalence following first (sexual) exposure. The peak in cervical HPV prevalence in 15–25 year olds in the United States, corresponding to the usual age of initiation of sexual intercourse, might be analogous to the peak occurrence of plantar warts in children exposed to HPV 1 at a characteristic age (e.g., the first years of exposure to school gymnasium showers). Under this epidemic curve schema, the profound drop in cervical HPV prevalence in women over 30 might be due to immunologic clearance or suppression of existing infections, combined with less exposure to new HPV types because of fewer new sexual partners. The "sexual acquisition/immunologic clearance" explanation of the HPV age trend is supported by (a) HPV "incidence" data like those in Table 4, suggesting sexual acquisition of HPV; and (b) by prospective data demonstrating that most HPV infections are transiently detectable, suggesting that infections are cleared (see below).

3.4 Time Trends in Cervical HPV Prevalence

As another explanation for the decreasing age trend in HPV prevalence, the trend might represent a "cohort effect," the epidemiologic term for an increase over time in the amount of disease in a population. If the risk of acquiring cervical HPV infection in young women today is higher than it was 10 years ago, and much higher than 20 years ago, etc., then the resultant age pattern in women of all ages observed *today* would be a decrease in HPV prevalence with age. The pattern produced in cross-sectional data by a cohort effect would be indistinguishable from the sexual acquisition/immunologic clearance model just discussed. The possibility of a cohort increase in genital HPV prevalence may be supported by increasing diagnostic surveillance rates for vulvar warts over time (reviewed in KOUTSKY et al. 1988) and by some scant data on increasing prevalence of cervical cytologic diagnoses of koilocytotic atypia (EVANS and DOWLING 1990). However, vulvar wart diagnoses and especially cervical cytologic data are prone to confounding changes in diagnostic criteria. Time trend data regarding HPV infection as measured by DNA are not available because standardized HPV testing is too newly available.

3.5 Geographic Variation in Cervical HPV Prevalence: Correlation Studies

In earlier studies, geographic differences in HPV prevalence did not correlate consistently with geographic differences in risk of cervical cancer (reviewed in KJAER and JENSEN 1992). Geographic correlational data have provided the most puzzling epidemiologic discrepancies in a body of evidence that otherwise strongly supports a central, causal role for HPV. HPV prevalence differences in Greenland versus Denmark still provide the most noteworthy and unexplained puzzle (KJAER and JENSEN 1992). In two well-done surveys using different HPV test methods, Greenlandic women aged 20–39 had a lower prevalence of HPV than Danish women of the same age, despite the much higher Greenlandic population risk of cervical cancer. By analogy, if a country with little tobacco smoking had extremely high rates of lung cancer, how could this be explained? This type of geographic discrepancy warrants continued investigation. Cervical cancer screening and treatment differences could explain the apparent puzzle, but the two regions ostensibly share the same type of health care system. As another theoretical (admittedly a posteriori) explanation, it is possible that the Greenlandic women have been measured "too late" in the natural history of their HPV infections. Perhaps Greenlandic women become infected very early and commonly (consistent with their early ages at first intercourse, multiple sexual partners, and elevated cancer risk), but most suppress DNA detectability by age 20. In other words, Greenlandic women aged 20–39 may be further displaced than Danish women along the descending slope of the epidemic curve of HPV acquisition and suppression by virtue of earlier, repeated exposure and subsequently earlier immunity. This conjectural explanation could be tested by measuring HPV prevalence among very young sexually active Greenlandic women.

In contrast to the still puzzling Greenlandic-Danish comparison, newer geographic correlation studies using sensitive testing methods to detect a wide spectrum of HPV types have observed HPV prevalence to "fit" the population risk of cervical cancer. In a comparison of Colombia and Spain, the Colombian women had higher HPV prevalence when measured by PCR, in accordance with an eight fold higher risk of cancer (MUNOZ et al. 1992). In a pair of studies in the United States using PCR, a higher risk urban clinic population had higher HPV prevalence than a suburban middle class population (BAUER et al. 1993; HILDESHEIM et al. 1993).

More population surveys of cervical HPV prevalence in regions of known high or low cervical cancer risk are needed to establish whether HPV population prevalence is a reliable correlate of subsequent cervical cancer risk among that population. If the geographic correlations turn out not to be strong (as in Greenland), it would suggest that variations in the geographic distribution of (unknown) etiologic co-factors, in addition to the prevalence of HPV infection itself, might be a key population determinant of cervical cancer risk.

3.6 Other Demographic Correlates of Cervical HPV Infection

Based on scant data taking into account self-reported sexual behavior, increasing HPV prevalence in the United States appears to correlate with lower socioeconomic status (BAUER et al. 1993), in accordance with the higher risk of cervical cancer in poorer and less educated populations in the United States. The correlation of HPV prevalence with race is even less clear. To study the correlation of HPV with race demands random, strictly comparable sampling of different racial populations. Clinic-based studies are prone to confounding by correlates of race, such as differences in socioeconomic status and care-seeking behavior. There is, so far, no evidence of racial differences in susceptibility to HPV infection. However, there are preliminary suggestions that specific, racially correlated HLA haplotypes are associated with altered risk of cervical cancer (C.M. Wheeler et al., personal communication), implying that racially assocciated genetic differences in susceptibility to HPV infection may have a biologic basis.

Racial susceptibility studies can trouble those epidemiologists who feel that the lessons learned may not be vital enough to justify perpetuating racial distinctions that are always imprecise and flawed, and risk being offensive. This area of HPV epidemiologic research, not surprisingly, is little studied in the United States, despite well-known "racial" differences in cervical cancer risk that are intertwined with socioeconomic differences in risk (intertwined in turn with a variety of correlated behaviors).

4 Transmission of Cervical HPV Infections

Epidemiologists study the distribution and determinants of disease. The preceding section discussed the distribution of HPV infections, while this section deals with determinants of infections, i.e., risk factors for transmission.

4.1 Evidence for Sexual Transmission of Cervical HPV in Adulthood

Decades ago, the sexual transmission of condyloma acuminatum (venereal warts) was demonstrated (ORIEL 1971). By analogy to exophytic venereal warts, caused by HPV 6 and 11, most researchers have assumed that other genital HPV infections (as defined by HPV DNA testing) are also sexually transmitted. Demonstrating the point with epidemiologic data proved more difficult than expected because the amount of error in early HPV DNA testing methods limited our ability to study determinants of infection (FRANCO 1991). Based on a series of more recent studies using improved HPV tests, it now is clear that cervical HPV infection is sexually transmitted in adulthood, as expected (LEY et al. 1991; BAUER et al. 1993; HILDESHEIM et al. 1993; WHEELER et al. 1993). The prevalence of cervical HPV DNA, along with the prevalence of cytologic effects of HPV, increases with reported numbers of different sexual partners. Type-specific analyses have been limited by small numbers, but the prevalences of both cancer-associated and non-cancer-associated HPV types are associated with sexual history.

It appears that HPV infections are transmitted rather easily to the cervix (and or vagina, since the two sites of infection have been difficult to distinguish), based on a few indirect lines of reasoning. First, the incidence data in Table 4 suggest that having a few new sexual partners leads to high prevalences of cervical HPV within months. Also, in a few epidemiologic studies, the epidemiologic variables "*lifetime* numbers of *different* sexual partners" and "*recent* numbers of *different* sexual partners" predict risk of prevalent cervical HPV infection better than "*lifetime* numbers of *regular* sexual partners," suggesting that long-term association with a partner is not necessary for transmission (BAUER et al. 1993; HILDESHEIM et al. 1993). Moreover, the frequent finding of HPV-related penile lesions in male partners of women with CIN suggests that sexual transmission is common (BARRASSO 1992). Finally, the age curve of cervical HPV prevalence, with a peak already apparent at age 15–25, suggests that the transmission of HPV infection to the cervix occurs soon after the initiation of sexual intercourse (SCHIFFMAN 1992b).

As a possibly related point, the prevalence of HPV infection does not rise linearly in most studies with increasing lifetime numbers of sexual partners; rather, it reaches a "plateau." Typically, the rise in prevalence is most pronounced when comparing lifetime monogamous women to women with more than one partner. Scant data suggest that the plateau of increasing prevalence may be observed at lower numbers of partners in higher-risk populations (HILDESHEIM et al. 1993). The reason for the plateau is not clear, but may involve immunity to infection.

4.2 Difficulty of Measuring HPV Infection in Men

The sexual transmission of cervical HPV infections could be studied more readily if penile HPV infection were more easily measured. HPV-induced penile lesions can be found in the urethral meatus, on the glans and shaft of the penis, and on the scrotum (BARRASSO 1992). Unlike the localized, easily exfoliated mucosal surface of the cervix, the cornified epithelium of the penis and scrotum does not permit a single, standardized collection of relevant cells. Urethral meatal and coronal sulcus swabs have been assayed with some positive results (VAN DOORNUM et al. 1992), but the prevalence estimates of HPV determined by such surveys may not be accurate estimates of true infection rates. In past studies of men, positive results have been more trustworthy than negative results (M.H. Schiffman and L.A. Brinton, unpublished data). This affects the trustworthiness of viral concordance studies. If a woman has HPV type X in her cervicovaginal lavage specimen, while her partner has type Y in his penile specimen, how can we be sure that he does not harbor type X somewhere that was not sampled?

Since the advent of PCR-based HPV test methods, the situation may be more encouraging. A recent study of men, in which specimens were collected by swabbing the entire penile shaft and glans, yielded similar PCR-derived HPV prevalence estimates for men and women attending the same sexually transmitted disease (STD) clinic, suggesting that swabs may collect adequate specimens for HPV DNA detection by PCR (L. Koutskv, personal communication).

Penile–cervical transmission studies of HPV have been rare owing to the technical difficulties of measuring HPV infection in men, combined with practical problems in identifying and recruiting male sexual partners. Here again, an immunologic assay measuring exposure to specific types of HPV would be an asset.

4.3 Non-sexual Transmission of Cervical HPV Infection

Although the sexual transmissibility of cervical HPV infection is clear, important points remain unclear. How important are non-sexual routes of transmission? Fomite transmission has been postulated, based on the detection of HPV DNA in gynecologic settings and on underclothes (FERENCZY et al. 1989). There is an appreciable prevalence of cervical HPV infection in women reporting only one lifetime sexual partner (LEY et al. 1991). How often does HPV infection in reportedly monogamous women reflect their partners' sexual experiences with other partners, versus misreporting of monogamy by the women, versus non-sexual routes of transmission?

Even more importantly, there are only scant data regarding cervical HPV DNA detection in virgins. Some groups, including ours, have found no cervical HPV DNA in virgins (LEY et al. 1991; FAIRLEY et al. 1992), while others have observed some cervical HPV (WHEELER et al. 1993). HPV seropositivity in virgins, even among young children, has raised the possibility of nonsexual (perhaps vertical) transmission of genital HPV types, although not necessarily to the cervix (GALLOWAY 1992). The transmission of laryngeal HPV infections (often types 6 and 11) proves that vertical transmission of genital HPV infections is possible (SHAH et al. 1986). The frequency of vertical transmission of genital types of HPV is still controversial and unknown, despite a number of studies on the topic (SEDLACEK et al. 1989; FIFE et al. 1990; SMITH et al. 1991).

More studies of cervical HPV infection in young virgin girls and virgin women could help clarify the frequency of non-sexual transmission of cervical HPV, but these studies cannot be conducted ethically without compelling clinical rationale. Pediatricians studying children referred for possible sexual abuse might be able, as an ethical part of their clinical responsibilities, to conduct small prevalence surveys of cervical HPV detection in girls who "rule out" in the sexual abuse examination, i.e., who turn out despite initial suspicion not to have been abused (GUTMAN et al. 1992).

4.4 Other Behavioral Risk Factors for Cervical HPV Infection

HPV infection of the cervix appears to be sexually transmitted, but additional factors may increase the risk of infection given exposure. These co-factors promoting infection are less certain and more poorly understood than the sexual and demographic factors described above.

Current pregnancy has been found inconsistently to increase the detection of HPV DNA (reviewed in SCHNEIDER and KOUTSKY 1992; HILDESHEIM et al., in press). Here, the distinction between increased prevalence of DNA detection and increased incidence of infection becomes especially troublesome, as it is difficult to imagine why pregnancy would increase the incidence of infection. In any case, it is doubtful that pregnancy exerts much of an influence on HPV infection, however defined, given the inconsistency of the epidemiologic results. Similarly, oral contraceptive use may weakly increase HPV DNA detection as it has been correlated with HPV prevalence in some but not all studies (LORINCZ et al. 1990; MOSCICKI et al. 1990; LEY et al. 1991; MORRISON et al. 1991; BAUER et al. 1993; HILDESHEIM et al. 1993).

Although the influence of HPV may be weak, the effects of endogenous and exogenous hormones on HPV natural history deserve further study, because parity is a probable independent risk factor for cervical neoplasia (BRINTON 1992). In addition, there is some laboratory support for hormonal influences on HPV.

Although smoking may have a role in later stages of cervical carcinogenesis (BRINTON 1992), smoking is not independently associated with risk of HPV infection once sexual behavior is taken into account (BAUER et al. 1993). Nutritional and immunologic variables influencing susceptibility to HPV infection given exposure to the virus are unknown. The special case of clinically apparent immunosuppression (e.g., due to HIV infection) is an exception (MAIMAN et al. 1991; VERMUND et al. 1991) but one that is again confounded by the distinction between new infection and increased detection.

5 Patterns of Host Response

5.1 HPV DNA Transience Versus Persistence

The most common outcome of cervical HPV infection appears to be clinically inapparent, transient DNA detectability with no clinical sequelae (ROSENFELD et al. 1992; SCHNEIDER and KOUTSKY 1992; Schiffman et al., unpublished data). This presents a research problem to the epidemiologist attempting to define HPV prevalence and natural history. Cross-sectional HPV screening

Table 5. HPV DNA measured at two times in cytologically normal women

HPV DNA result		Women (n)
Enrollment	Follow up	
Negative	Negative	257
Negative	Positive	40
Positive	Negative	40
Positive	Positive	63
Persistent positivity[a]		34
Loss and acquisition[b]		17
Questionable persistence[c]		12

[a] Persistent positivity, detection of least one particular HPV type at both examinations. In case of multiple infections, not all types were necessarily detected twice.
[b] Loss and acquisition, detection of exclusively different HPV types at the two examinations.
[c] Questionable persistence, uncertain concordance between the HPV types detected at the two examinations.

will tend to preferentially detect women whose infections persist long and strong enough to be detected. Persistently detectable infections are likely to be the most serious. By focusing on the most severe "tip of the iceberg" of cervical HPV infection, and missing many benign and unseen outcomes, the tendency might be to overestimate the risk posed by the detection of HPV DNA on a single screen.

In Table 5, evidence is given for the usual transience of HPV infection. Among a group of cytologically normal women measured twice 9–30 months apart, most women with cervical HPV infection at the first measurement were HPV negative at the second. The number of women with "new" infection at the second measurement equalled the number "losing" infection from the first measurement, i.e., the acquisition and loss of infection appeared to be in a dynamic steady state. One other published report supports the hypothesis that most HPV infections are transient. Alternatively, HPV infections could be chronic and persistent but only intermittently detectable; there are some scant repeated testing data to support this possibility as well (SCHNEIDER et al. 1992). Arguing against the chronicity/intermittency interpretation are the data on HPV persistence by time, an extension of the data in Table 5. In that same follow-up study (A. Hildesheim and M.H. Schiffman, manuscript in preparation), the longer the interval between first and second measurement, the smaller the probability of a repeat HPV-positive result with the same HPV type. The time trend data appeared to describe a viral "survival curve," suggesting a rather steady disappearance of viral infection over the 9–30 months of observation. If confirmed, common viral transience would correspond well theoretically to the decrease in HPV prevalence with age and the importance of recent sexual partners in determining prevalence rates.

5.2 Determinants of HPV DNA Persistence in Cytologically Normal Women

If most HPV infections are transient, while few are persistent, then HPV persistence may be an important risk factor for the development of cervical neoplasia. The association of HPV persistence with cervical neoplasia is supported by data from our natural history study (from which the data in Table 5 are drawn). In contrast to the results among cytologically normal women in Table 5, women developing CIN under observation were much more likely to have apparently persistent rather than transient infections (43% of 114 cases, compared to 8% of 400 normal women, submitted manuscript).

The determinants of HPV persistence are largely unknown. Three categories of risk factors seem most plausible: viral factors such as viral type or transcriptional state; host immunologic response, either genetically or environmentally determined; and environmental co-factors such as hormonal factors or smoking. Preliminary data from at least two groups suggest that cancer-associated types of HPV persist longer than non-cancer-associated types (N. Kiviat, personal communication; A. Hildesheim and M.H. Schiffman, manuscript in preparation). The role of copy number (viral load) on persistence is currently under study by several groups.

No published epidemiologic data regarding immune determinants of HPV persistence were found in the preparation of this update. Such studies are feasible now but require difficult immunologic measurements, combined with prospective, highly reliable repeated measurements of HPV. As a possibly relevant point, our preliminary results suggest that older women infected with HPV are more likely to have their infections persist than younger women (A. Hildesheim and M.H. Schiffman, manuscript in preparation). This observation is concordant with the hypothesis that acquisition and loss of HPV typify infection among younger women, while DNA detection in some older women might signal less successful immunologic responses. Many epidemiologic groups are now working with immunologists to explore how to measure relevant immune responses to genital HPV infections.

In a recent exploration of possible behavioral influences on HPV persistence, such as smoking and oral contraceptive use, no statistically significant influences were found (A. Hildesheim and M.H. Schiffman, manuscript in preparation). Environmental factors may influence persistence indirectly via an alteration in HPV immunology; perhaps such factors are subtle and multiple, beyond the reach of epidemiologic questionnaires.

5.3 Early HPV-Induced Cervical Cytologic Changes

5.3.1 Methodologic Concerns

Moving up the natural history "iceberg" of cervical HPV infection toward the more visible peak, it seems that about one tenth to one fifth of infected women on a given day have cervical cytologic changes diagnosible as low-grade squamous intraepithelial lesions (LSIL, including koilocytotic atypia or CIN 1). Attempting to stydy the transition from HPV infection defined by HPV DNA to infection defined cytologically raises important questions of definition and diagnostic accuracy. Epidemiologists studying the relationship of HPV infection (as a risk factor) and LSIL (as a disease outcome) can no longer use a standard epidemiologic approach. The standard two-by-two table that forms one of the bases of epidemiology "breaks down" when the exposure and disease are too intertwined. For example, human T cell leukemia virus 1 (HTLV-1) infection is now tightly linked to the diagnosis of acute T cell leukemia (MANNS and BLATTNER 1991). Analogously, it is difficult to define a trustworthy grouping of HPV-positive women without LSIL (how can we be sure that no koilocytotic cells have been missed by cytopathology?). Similarly, it is difficult to define a grouping of women with HPV DNA-negative LSIL. In series in the United States using reliable, multi-type HPV DNA testing and cytopathologic confirmation of CIN 1, virtually no cases of CIN 1 cases are HPV negative (SHERMAN et al., submitted manuscript). The situation may differ in other areas of the world, but in the United States, the existence of HPV-negative CIN 1 is in doubt because residual errors in measuring all types of HPV and in classifying cytologic smears (i.e., eliminating cytologic look alikes) make it possible that CIN 1 is universally caused by HPV.

To an epidemiologist, once an exposure (HPV) defines a disease (LSIL including koilocytotic atypia and CIN 1), the key question becomes: what causes the transition to cytologically overt LSIL, *given* HPV infection? In epidemiologic jargon, the analysis shifts into the "HPV-exposed stratum," a group that is difficult to define without a serologic measurement of lifetime infection.

To summarize this section, the distinction between HPV infection and cervical neoplasia has recently disappeared, changing the nature of necessary epidemiologic inquiries. In our studies in the United States, it does not make sense to ask what the relative risk of developing CIN 1 is among women with HPV infection compared to unexposed women. The relative risk approaches infinity because CIN 1, as part of LSIL, is a cytopathologic measurement of HPV infection. In place of the old exposure-disease distinction familiar to chronic disease epidemiologists is a series of natural history transitions (conditional probabilities to the statistician), from HPV exposure to HPV infection detectable by DNA diagnostic tests, to the pathologic diagnosis of LSIL, to high-grade SIL, to invasive cancer. The epidemiologic study of HPV infection and cervical neoplasia is beginning to approach the study of multi-step carcinogenesis, making close collaborations with molecular biologists and pathologists essential.

5.3.2 Recent Results

In a recent cross-sectional epidemiologic study examining possible risk factors for cytologically diagnosed CIN 1 among HPV-infected women, we observed that cancer-associated types of HPV were associated with a higher risk of low-grade lesions than were non-cancer-associated types of HPV (SCHIFFMAN et al. 1993). In other words, non-cancer-associated types of HPV were disproportionately common in cytologically normal women. Among HPV-infected women, no apparent associations of risk of overt CIN 1 were seen for smoking, socioeconomic status, lifetime numbers of sexual partners, age at first intercourse, or oral contraceptive use (which appeared weakly protective contrary to expectation). Increasing parity was a risk factor for CIN 1 among both HPV-positive and HPV-negative women.

Prospective studies are preferable to cross-sectional studies in examining the risk factors for cytologic abnormalities among HPV-infected women. In our prospective study, we have observed that the cancer-associated types of HPV impart the highest risk of incident low-grade CIN given infection (unpublished data). At least two additional, large prospective studies of this topic are underway (S. KJAER et al. in Denmark; R. HERRERO et al. in Costa Rica).

5.4 Progression to High-Grade Cervical Neoplasia

The focus of this update is not on neoplastic outcomes, but a few points of current interest will be mentioned. Although prospective studies have demonstrated that women with low-grade CIN are at increased risk of developing high-grade CIN and cervical cancer (KOUTSKY et al. 1992), no one is sure that low-grade *lesions* progress to high-grade lesions. It has been suggested that high-grade lesions might arise from normal-appearing epithelium at the internal margin of low-grade lesions (KIVIAT et al. 1992). Arguing somewhat against this revisionist view is the following: the average age of women with high-grade CIN is 5–10 years older than the average age of women with low-grade CIN (SCHIFFMAN 1992b), most consistent with a progression of low-grade to high-grade lesions (as opposed to a slower rate of development of high-grade CIN than low-grade CIN). On the other hand, supporting a direct progression from HPV infection to high-grade CIN are the results of KOUTSKY et al. (1992), who observed a high absolute risk of rapidly incident CIN 2–3 among women who, at enrollment, were cytologically normal but infected with HPV 16 or 18.

Whether or not high-grade lesions emerge from low-grade lesions, the type of HPV found in the cervix appears to predict the risk of the woman progressing to a diagnosis of high-grade cervical neoplasia. This association has been verified in both cross-sectional and prospective studies (LORINCZ et al. 1992; CAMPION et al. 1986). A crucial current question is "What else besides HPV type predicts the risk of progression to high-grade cervical neoplasia?" The best a priori candidates for co-factors are the established risk factors for cervical cancer that do not appear to be mere proxies for HPV infection. Smoking would seem to be a good candidate for a progression co-factor, and this possibility has some epidemiologic support (HERRERO et al. 1989; SCHIFFMAN et al. 1993). However, smoking was not found to be a co-factor for progression to high-grade CIN in a prospective study (KOUTSKY et al. 1992) or to cancer in two recent large case–control studies (BOSCH et al. 1992; Eluf-Neto et al., manuscript submitted). In one of the case–control studies (BOSCH et al. 1992), oral contraceptive use was the strongest apparent co-factor for invasive cancer among HPV-infected women, and early age at first pregnancy also was associated with increased risk. In the study by Eluf-Neto et al. (manuscript submitted), parity and oral contraceptive use persisted as risk factors for cervical cancer among HPV-positive women. Parity was also observed to be a possible co-factor in the prospective study of KOUTSKY et al. (1992).

In addition to hormonal factors and smoking, the etiologic supporting roles of nutritional deficiencies of vitamin C, carotenoids, and folate deserve more epidemiologic attention (POTISCHMAN et al. 1991; BUTTERWORTH et al. 1992). The role of other concurrently infecting sexually transmitted diseases should also be explored (SCHMAUZ et al. 1989; HILDESHEIM et al. 1991; KOUTSKY et al. 1992). Most importantly, immunologic factors, although still poorly measured, will be the focus of most progression studies now being planned.

5.5 Influence of Setting on the Natural History of Cervical HPV Infection

It is worth considering whether the natural history of HPV infection varies by setting. In this regard, a puzzling contrast is offered by the natural history study of KOUTSKY et al. (1992) compared with our natural history data from the Portland cohort. KOUTSKY et al. observed a high risk of developing apparently new high-grade CIN in 1–2 years among women with HPV 16 or 18 infections who were attending an STD clinic. In contrast, we have observed virtually no new cases of incident high-grade CIN in the Portland cohort, although new cases of low-grade CIN are quite common, and several hundred women are estimated (based on the random sample of the cohort tested so far) to have had HPV 16 or 18 infections at enrollment 2–3 years ago. The diagnostic criteria for high-grade CIN are similar in the two studies, thus diagnostic discrepancies cannot explain the difference in results. One possible explanation for this apparent difference is the influence of setting on risk of progression to high-grade CIN, given infection with a cancer-associated HPV type. The demographic and behavioral profile of the middle-class Portland cohort suggests a low-risk setting for cervical cancer, compared with the high-risk sexually transmitted disease clinic population followed by KOUTSKY et al. Perhaps the risk of developing cervical neoplasia is higher in settings where the co-factors for progression are more prevalent. A group of collaborating epidemiologists is now conducting parallel studies of HPV natural history in regions with widely varying risks of cervical cancer in order to define the distributions and relative influences of viral factors, host factors, and environmental co-factors.

6 Future Directions

Some major objectives of future epidemiologic research on cervical HPV infections will be summarized here.

HPV DNA detection will continue for at least the next few years to be the major definition of HPV infection. It will be important to understand how subtleties of DNA detection, such as the amounts of viral DNA present, reflect and predict natural history. Perhaps high-intensity infection can be used as a cross-sectional predictor of HPV persistence and progression.

Through continued prospective studies, the decrease in cervical HPV prevalence with increasing age should be better understood. The separate contributions to the age trend of cohort effects and immunologic suppression must be distinguished. A strong cohort effect of increasing HPV infection in currently younger women might predict a large increase in invasive cervical cancer in the future. Immunologic suppression as an explanation for the age trend would be a much more reassuring interpretation of the high HPV prevalence in younger women today.

An attempt must be made to study cervical HPV incidence rather than HPV prevalence by enrolling virgin women for repeated measurements in follow-up studies. As a related point, epidemiologists must fully address issues of sexual transmission of cervical HPV infection. These studies will need to be intensive and relatively small until reliable measurements of penile HPV infection are validated.

The importance of cervical HPV persistence must be verified and the determinants of persistence identified. Extremely reliable HPV testing, distinguishing not only HPV types but variants of HPV types, will be required for meaningful repeated measurements (CHAN et al. 1992).

The interactions of multiple HPV types in mixed cervical infections should be clarified as one pathway to understanding HPV immunity.

HPV immunology is likely to occupy epidemiologists studying cervical HPV infection over the next decade. The ultimate goal will be to define the successful immune response to HPV infection, in the hope that immunity can be stimulated by vaccination. In a highly optimistic (but not impossible) scenario, a vaccine to prevent HPV-induced neoplasia will be developed and universally applied, permitting us to witness the beginning and ending of epidemiologic studies of HPV natural history within the span of a career.

References

Barrasso R (1992) HPV-related genital lesions in men. In: Munoz N, Bosch FX, Shah KV, and Meheus A (eds) The epidemiology of human papillomavirus and cervical cancer. Lyon, France, pp 85–92 (IARC Sci Publ 119)

Bauer HM, Ting Y, Greer CE, Chambers JC, Tashiro CJ, Chimera J, Reingold A, Manos MM (1991) Genital Human papillomavirus infection in female university students as determined by a PCR-based method. JAMA 265: 472–477

Bauer HM, Hildesheim A, Schiffman MH, Glass AG, Rush BB, Scott DR, Cadell DM, Kurman RJ, Manos MM (1993) Determinants of genital human papillomavirus infection in low-risk women in Portland, Oregon. Sex Transm Dis 20: 274–278

Bosch FX, Munoz N, de Sanjose S, Izarzugaza I, Gili M, Viladiu P, Tormo MJ, Moreo P, Ascunce N, Gonzalez LC, Tafur L, Kaldor JM, Guerrero E, Aristizabal N, Santamaria M, Alonso de Ruiz P, Shah K (1992) Risk factors for cervical cancer in Colombia and Spain. Int J Cancer 52: 750–758

Brinton LA (1992) Epidemiology of cervical cancer—overview. In: Munoz N, Bosch FX, Shah KV, Meheus A (eds) The epidemiology of human papillomavirus and cervical cancer. Lyon, France, pp 3–23 (IARC Sci Publ 119)

Butterworth CE, Hatch K, Macaluso M, Cole P, Sauberlich H, Soong S-J, Borst M, Baker V (1992) Folate deficiency and cervical dysplasia. JAMA 267: 528–533

Campion MJ, McCance DJ, Cuzick J, Singer A (1986) Progressive potential of mild cervical atypia: prospective cytological, colposcopic, and virological study. Lancet 2: 237–240

Chan S-Y, Ho L, Ong C-K, Chow V, Drescher B, Durst M, ter Meulen J, Villa L, Luande J, Mgaya HN, Bernard H-U (1992) Molecular variants of human papillomavirus type 16 from four continents suggest ancient pandemic spread of the virus and its coevolution with humankind. J Virol 66: 2057–2066

de Britton RC, Hildesheim A, de Lao SL, Brinton LA, Sathya P, Reeves W (1993) Human papillomaviruses and other influences on survival from cervical cancer in Panama. Obstet Gynecol 81: 19–24

Evans S, Dowling K (1990) The changing prevalence of cervical human papilloma virus infection. Aust N Z J Obstet Gynaecol 30: 375–377

Fairley CK, Chen S, Tabrizi SN, Leeton K, Quinn MA, Garland SM (1992) The absence of genital human papillomavirus DNA in virginal women. Int J STD AIDS 3: 414–417

Ferenczy A, Bergeron C, Richart RM (1989) Human papillomavirus DNA in fomites on objects used for the management of patients with genital human papillomavirus infections. Obstet Gynecol 74: 950–954

Fife KH, Bubalo F, Boggs DL, Gaebler JW (1990) Perinatal exposure of newborns to HPV, detection by DNA amplification. UCLA Symp Mol Cell Biol 124: 73–76

Franco EL (1991) The sexually transmitted disease model for cervical cancer: incoherent epidemiologic findings and the role of misclassification of human papillomavirus infection. Epidemiology 2: 98–106

Fuchs PG, Girardi F, Pfister H (1988) Human papillomavirus DNA in normal, metaplastic, prneoplastic and neoplastic epithelia of the cervix uteri. Int J Cancer 41: 41–45

Galloway DA (1992) Serological assays for the detection of HPV antibodies. In: Munoz N, Bosch FX, Shah KV, Meheus A (eds) The epidemiology of human papillomavirus and cervical cancer. Lyon, France, pp 147–161 (IARC Sci Publ 119)

Goldberg GL, Vermund SH, Schiffman MH, Ritter DB, Spitzer C, Burk RD (1989) Comparison of cytobrush and cervicovaginal lavage sampling methods for the detection of genital human papillomavirus. Am J Obstet Gynecol 161: 1669–1672

Guerrero E, Daniel RW, Bosch FX, Castellsague X, Munoz N, Gili M, Viladiu P, Navarro C, Martos C, Ascunce N, Gonzalez LC, Tafur L, Izarzugaza I, Shah KV (1992) A comparison of Virapap, Southern hybridization and polymerase chain reaction methods for human papillomavirus

(HPV) identification in an epidemiological investigation of cervical cancer. J Clin Microbiol 30: 2951–2959
Gutman LT, St Claire K, Herman-Giddens ME, Johnston WW, Phelps WC (1992) Evaluation of sexually abused and nonabused young girls for intravaginal human papillomavirus infection. Am J Dis Child 146: 694–699
Hagensee ME, Yaegashi N, Galloway DA (1993) Self-assembly of human papillomavirus type 1 capsids by expression of the L1 protein alone or by coexpression of the L1 and L2 capsid proteins. J Virol 67: 315–322
Herrero R, Brinton LA, Reeves WC, Brenes MM, Tenorio F, de Britton RC, Gaitan E, Garcia M, Rawls WE (1989) Invasive cervical cancer and smoking in Latin America. J Natl Cancer Inst 81: 205–211
Hildesheim A, Mann V, Brinton LA, Szklo M, Reeves WC, Rawls WE (1991) Herpes simplex virus type 2: a possible interaction with human papillomavirus types 16/18 in the development of invasive cervical cancer. Int J Cancer 49: 335–340
Hildesheim A, Gravitt P, Schiffman MH, Kurman RJ, Barnes W, Jones S, Tchabo JG, Brinton LA, Copeland C, Epp J, Manos MM (1993) Determinants of genital human papillomavirus infection in low-income women in Washington DC. Sex Transm Dis 20: 279–285
Kadish AS, Hagan RJ, Ritter DB, Goldberg GL, Romney SL, Kanetsky PA, Beiss BK, Burk RD (1992) Biologic characteristics of specific human papillomavirus types predicted from morphology of cervical lesions. Hum Pathol 23: 1262–1269
Kiviat NB, Critchlow CW, Kurman RJ (1992) Reassessment of the morphological continuum of cervical intraepithelial lesions: does it reflect different stages in the progression to cervical carcinoma? In: Munoz N, Bosch FX, Shah KV, Meheus A (eds) The epidemiology of human papillomavirus and cervical cancer. Lyon, France, pp 59–66 (IARC Sci Publ 119)
Kjaer SK, Jensen OM (1992) Comparison studies of HPV detection in areas at different risk for cervical cancer. In: Munoz N, Bosch FX, Shah KV, Meheus A (eds) The epidemiology of human papillomavirus and cervical cancer. Lyon, France, pp 243–249 (IARC Sci Publ 119)
Koutsky LA, Galloway DA, Holmes KK (1988) Epidemiology of genital human papillomavirus infection. Epidemiol Rev 10: 122–163
Koutsky LA, Holmes KK, Critchlow CW, Stevens CE, Paavonen J, Beckmann AM, DeRouen TA, Galloway DA, Vernon D, Kiviat NB (1992) Cohort study of risk of cervical intraepithelial neoplasia grade 2 or 3 associated with cervical papillomavirus infection. N Engl J Med 327: 1272–1278
Ley C, Bauer HM, Reingold A, Schiffman MH, Chambers JC, Tashiro CJ, Manos MM (1991) Determinants of genital human papillomavirus infection in young women. J Natl Cancer Inst 83: 997–1003
Lorincz AT, Reid R, Jenson AB, Greenberg MD, Lancaster W, Kurman RJ (1992) Human papillomavirus infection of the cervix: relative risk associations of 15 common anogenital types. Obstet Gynecol 79: 328–337
Lorincz AT (1992) Detection of human papillomavirus DNA without amplification: prospects for clinical utility. In: Munoz N, Bosch FX, Shah KV, Meheus A (eds) The epidemiology of human papillomavirus and cervical cancer. Lyon, France, pp 135–145 (IARC Sci Publ 119)
Lorincz AT, Schiffman MH, Quinn AP, Jaffurs WJ, Marlow J, Temple GF (1990) Temporal assoications of human papillomavirus infection with cervical cytologic abnormalities. Am J Obstet Gynecol 162: 645–651
Maiman M, Tarricone N, Vieira J, Suarez J, Serur E, Boyce JG (1991) Colposcopic evaluation of human immunodeficiency virus-seropositive women. Obstet Gynecol 78: 84–88
Manns A, Blattner WA (1991) The epidemiology of the human T-cell lymphotrophic virus type I and II: etiologic role in human disease. Transfusion 31: 67–75
Manos MM, Wright DK, Lewis AJ, Broker TR, Wolinsky SM (1989) The use of polymerase chain reaction amplification for the detection of genital human papillomaviruses. In: Furth M, Greaves M (eds) Molecular diagnostics of human cancer. Cold Spring Harbor Press, Cold Spring Harbor, pp 209–214 (Cancer Cells 7)
Meisels A, Fortin R (1976) Condylomatous lesions of the cervix and vagina I. Cytologic patterns. Acta Cytol 20: 505–509
Morrison EAB, Ho GYF, Vermund SH, Goldberg Fl, Kadish AS, Kelley KF, Burk RD (1991) Human papillomavirus infection and other risk factors for cervical neoplasia: a case-control study. Int J Cancer 49: 6–13
Moscicki A-B, Palefsky J, Gonzales J, Schoolnick GK (1990) Human papillomavirus infection in sexually active adolescent females: prevalence and risk factors. Pediatr Res 28: 507–513

Muller M, Viscidi RP, Sun Y, Guerrero E, Hill PM, Shah F, Bosch FX, Munoz N, Gissmann L, Shah KV (1992) Antibodies to HPV 16 E6 and E7 proteins as markers for HPV 16 associated invasive cancer. Virology 187: 508–514
Munoz N, Bosch FX (1992) HPV and cervical neoplasia: review of case-control and cohort studies. In: Munoz N, Bosch FX, Shah KV, Meheus A (eds) The epidemiology of human papillomavirus and cervical cancer. Lyon, France, pp 251–261 (IARC Sci Publ 119)
Munoz N, Bosch FX, de Sanjose S, Tafur L, Izarzugaza I, Gili M, Viladiu P, Navarro C, Martos C, Ascunce N, Gonzalez LC, Kaldor JM, Guerrero E, Lorincz A, Santamaria M, Alonso de Ruiz P, Aristizabal N, Shah K (1992) The causal link between human papillomavirus and invasive cervical cancer: a population-based case-control study in Colombia and Spain. Int J Cancer 52: 743–749
National Cancer Institute Workshop (1989) The 1988 Bethesda System for reporting cervical/vaginal cytologic diagnoses. JAMA 262: 931–934
Negrini BP, Schiffman MH, Kurman RJ, Barnes W, Lannom L, Malley K, Brinton LA, Delgado G, Jones S, Tchabo J, Lancaster WD (1990) Oral contraceptive use, human papillomavirus detection, and risk of early cytologic abnormalities of the cervix. Cancer Res 50: 3610–3613
Oriel JD (1971) Natural history of genital warts. Br J Vener Dis 47: 1–13
Potischman N, Herrero R, Brinton LA, Reeves WC, Stacewicz-Sapuntzakis M, Jones CJ, Brenes MM, Tenorio F, de Britton RC, Gaitan E (1991) A case-control study of nutrient status and invasive cervical cancer II. Serological indicators. Am J Epidemiol 134: 1347–1355
Rosenfeld WD, Rose E, Vermund SH, Schreiber K, Burk RD (1992) Follow-up evaluation of cervicovaginal human papillomavirus infection in adolescents. J Pediatr 121: 307–311
Schiffman MH (1992a) Validation of HPV hybridization assays: correlation of FISH, dot blot, and PCR with Southern blot. In: Munoz N, Bosch FX, Shah KV, Meheus A (eds) The epidemiology of human papillomavirus and cervical cancer. Lyon, France, pp 169–179 (IARC Sci Publ 119)
Schiffman MH (1992b) Commentary—Recent progress in defining the epidemiology of human papillomavirus infection and cervical neoplasia. J Natl Cancer Inst 84: 394–398
Schiffman MH, Kurman RJ, Barnes W, Lancaster WD (1990) HPV infection and early cervical cytological abnormalities in 3173 Washington DC women. In: Howley P, Broker TR (eds) Papillomaviruses. Wiley-Liss, New York, pp 81–88
Schiffman MH, Bauer HM, Lorincz AT, Manos MM, Byrne JC, Glass AG, Cadell DM, Howley PM (1991) A comparison of Southern blot hybridization and polymerase chain reaction methods for the detection of human papillomavirus DNA. J Clin Microbiol 29: 573–577
Schiffman MH, Bauer HM, Hoover RN, Glass AG, Cadell DM, Rush BB, Scott DR, Sherman ME, Kurman RJ, Wacholder S, Stanton CK, Manos MM (1993) Epidemiologic evidence showing that human papillomavirus infection causes most cervical intraepithelial neoplasia. J Natl Cancer Inst 85: 958–964
Schmauz R, Okong P, de Villiers E-M, Dennin R, Brade L, Lwanga SK, Owor R (1989) Multiple infections in cases of cervical cancer from a high-incidence area in tropical Africa. Int J Cancer 43: 805–809
Schneider A, Koutsky LA (1992) Natural history and epidemiological features of genital HPV infection. In: Munoz N, Bosch FX, Shah KV, Meheus A (eds) The epidemiology of human papillomavirus and cervical cancer. Lyon, France, pp 25–52 (IARC Sci Publ 119)
Schneider A, Kirchhoff T, Meinhardt G, Gissmann L (1992) Repeated evaluation of human papillomavirus 16 status in cervical swabs of young women with a history of normal Papanicolaou smears. Obstet Gynecol 79: 683–688
Scully RE, Poulson H, Sobin LH (in press) International histological classification and typing of female genital tract tumours. Springer, Berlin Heidelberg New York
Sedlacek TV, Lindheim S, Eder C, Hasty L, Woodland M, Ludomirsky A, Rando RF (1989) Mechanism for human papillomavirus transmission at birth. Am J Obstet Gynecol 161: 55–59
Shah K, Kashima H, Polk BF, Shah F, Abbey H, Abramson A (1986) Rarity of cesarean delivery in cases of juvenile-onset respiratory papillomatosis. Obstet Gynecol 68: 795–799
Sherman ME, Schiffman MH, Erozan YS, Wacholder S, Kurman RJ (1992) The Bethesda system: a proposal for reporting abnormal cervical smears based on the reproducibility of cytopathologic diagnoses. Arch Pathol 116: 1155–1158
Sherman ME, Schiffman MH, Lorincz AT, Manos MM, Scott DR, Kurman RJ, Kiviat NB, Stoler M, Glass AG, Rush BB (in press) Towards objective quality assurance in cervical

cytopathology: correlation of cytopathologic diagnoses with detection of high-risk HPV types. Am J Clin Pathol

Smith EM, Johnson SR, Cripe TP, Pignatari S, Turek L (1991) Perinatal vertical transmission of human papillomavirus and subsequent development of respiratory tract papillomatosis. Ann Otol Rhinol Laryngol 100: 479–483

van den Brule AJ, Walboomers JMM, du Maine M, Kenemans P, Meijer CJLM (1991) Difference in prevalence of human papillomavirus genotypes in cytomorphologically normal cervical smears is associated with a history of cervical intraepithelial neoplasia. Int J Cancer 48: 404–408

van Doornum GJJ, Hooykaas C, Juffermans LHJ, van der Lans SM, van der Linden MM, Coutinho RA, Quint WG (1992) Prevalence of human papillomavirus infections among heterosexual men and women with multiple sexual partners. J Med Virol 37: 13–21

van Ranst M, Kaplan JB, Burk RD (1992) Phylogentic classification of human papillomaviruses: correlation with clinical manifestations. J Gen Virol 73: 2653–2660

Vermund SH, Kelley KF, Klein RS, Feingold AR, Schreiber K, Munk G, Burk RD (1991) High risk of human papillomavirus infection and cervical squamous intraepithelial lesions among women with symptomatic human immunodeficiency virus infection. Am J Obstet Gynecol 165: 392–400

Wheeler CM, Parmenter CA, Hunt WC, Becker TM, Greer CE, Hildesheim A, Manos MM (1993) Determinants of genital human papillomavirus infection among cytologically normal women attending the University of New Mexico student health center. Sex Transm Dis 20: 286–290

Functions of Human Papillomavirus Proteins

M. SCHEFFNER, H. ROMANCZUK, K. MÜNGER, J.M. HUIBREGTSE, J.A. MIETZ, and P.M. HOWLEY

1 Introduction

Cervical cancer is the second leading cause of deaths from cancer among women worldwide with approximately 500 000 deaths annually. Epidemiologic studies have implicated a sexually transmitted agent as a cause of cervical cancer, and molecular virology studies over the past 10 years have established a strong association between specific human papillomavirus (HPV) types and certain anogenital carcinomas, including cervical cancer (reviewed in ZUR HAUSEN and SCHNEIDER 1987). Over 65 different HPV types have now been described, and each is associated with a specific clinical entity (DEVILLIERS 1989). Approximately 20 or 25 HPVs have been associated with anogenital lesions; these HPVs have been further classified as either "low-risk" or "high-risk" types based on the preneoplastic character of the clinical lesions with which they are associated. Low-risk HPVs such as HPV-6 and HPV-11 are generally associated with venereal warts or condyloma acuminata which only rarely progress to malignancy. The high-risk HPVs include HPV-16 and HPV-18 and these are associated with squamous intraepithelial neoplasias which are potentially precancerous. In the cervix,

Laboratory of Tumor Virus Biology, National Cancer Institute, Bethesda, MD 20892, USA

Current Topics in Microbiology and Immunology, Vol. 186

they are associated with cervical intraepithelial neoplasia, or CIN. These CIN lesions are considered preneoplastic in that a small percentage of high-grade CIN lesions will progress to cervical cancer. Approximately 70% of human cervical cancers contain either HPV-16 or HPV-18 DNA (ZUR HAUSEN and SCHNEIDER 1987). Indeed, HPV-16 and HPV-18 DNA were originally isolated from human cervical carcinoma tissues (DÜRST et al. 1983; BOSHART et al. 1984). Other high-risk HPVs, including types 31, 33, 35, 39, 45, 51, and 52, have subsequently been identified and have also been associated with CIN lesions and with invasive cervical carcinomas. All together, approximately 85% of cervical cancers can be shown to contain DNA of one of the high-risk HPV types (RIOU et al. 1990).

The HPV DNA is frequently integrated in cervical cancers and in HPV-positive cervical carcinoma cell lines (ZUR HAUSEN and SCHNEIDER 1987). This contrasts with the extrachromosomal state of viral DNA usually found in benign CIN lesions (DÜRST et al. 1985). Integration of HPV DNA is largely a random phenomenon with respect to the site of integration in the host genome. In some cases, however, integration of the HPV genome occurred in the general vicinity of known oncogenes. In the HeLa cell line which contains HPV-18 DNA, for instance, integration occurred approximately 50 kbp from the c-myc locus on chromosome 8 (DÜRST et al. 1987). Whether integration of the viral genome into the vicinity of an oncogene could provide a selective growth or survival advantage to the cell and, as such, contribute to the progression of a preneoplastic lesion to cancer is uncertain, but it is an attractive hypothesis. There does appear to be some specificity with respect to where the circular viral genome is disrupted as a consequence of the integration. Integration generally occurs within the E1/E2 region of the viral genome, thereby disrupting the integrity of the E1 and E2 open reading frames (Fig. 1) and their expression (BAKER et al. 1987; SCHWARZ et al. 1985). The E1 and the E2 open reading frames each encode DNA binding, regulatory proteins that are involved directly in the regulation of viral transcription and DNA replication (reviewed in LAMBERT 1991; MCBRIDE et al. 1991). Although E2 was initially defined as a transcriptional activator in studies with the bovine papillomaviruses (BPVs), subsequent studies have shown that E2 can also act as a repressor depending on the context of the E2-binding sites near the promoter. For HPV-16 and HPV-18, the E2 proteins act mainly to repress the activity of the promoter from which the E6 and E7 genes are transcribed (ROMANCZUK et al. 1990; THIERRY and YANIV 1987). HPV DNA is transcriptionally active in HPV-positive cervical cancers; however, as a consequence of the integration disrupting the E1 and E2 genes, only the E6 and E7 genes are regularly expressed (BAKER et al. 1987; SCHWARZ et al. 1985; SMOTKIN and WETTSTEIN 1986). Continued expression of the E6 and E7 genes is necessary for the continued proliferation of cervical carcinoma cells. Antisense RNA experiments targeting the E6 and E7 genes suggest a requirement for expression of E6 and E7 for the proliferation of the HPV-18 positive cervical carcinoma cell line C4-1 (VON KNEBEL-DOEBERITZ

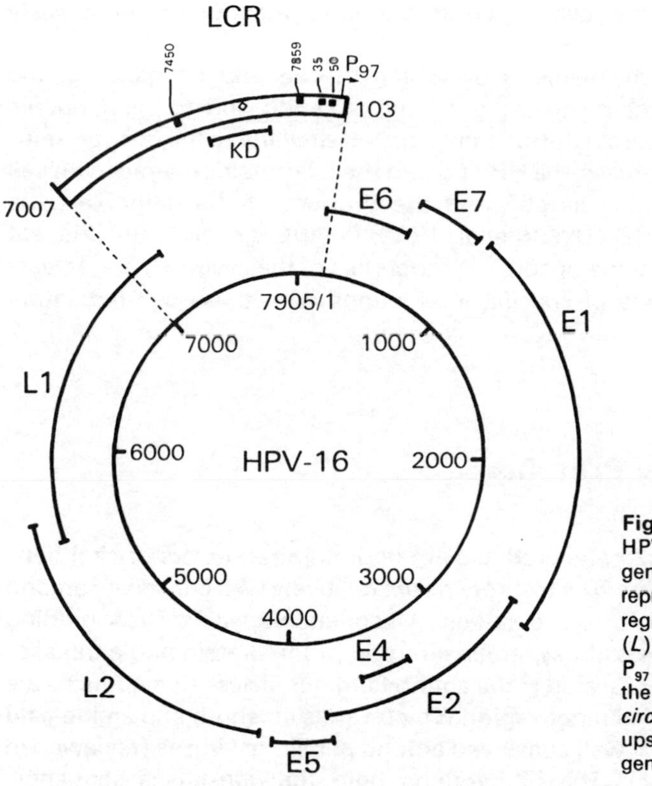

Fig. 1. Genomic map of HPV-16. The HPV-16 genome is schematically represented with the early-region (*E*) and late-region (*L*) open reading frames. The P_{97} promoter (*arrow*) and the E2-binding sites (*closed circles*) within the LCR upstream of the E6 and E7 genes are indicated

et al. 1988). Furthermore, BPV-1 E2 can inhibit proliferation of HeLa cells, and it has been proposed that this occurs through the transcriptional repression of the E6 and E7 genes (THIERRY and YANIV 1987). However, genetic studies examining the effect of mutations of either E1 or E2 on the in vitro immortalization capacity of the HPV-16 genome implicate additional mechanisms in the E2 repression of viral immortalization functions (ROMANCZUK and HOWLEY 1992).

Further support for a role for the E6 and E7 genes in cervical carcinogenesis comes from a variety of in vitro transformation assays that have shown that the E6 and E7 genes encode oncoproteins. The high-risk HPV E7 protein can transform some established rodent cell lines such as NIH 3T3 cells and, in cooperation with an activated ras oncogene, is able to fully transform primary rodent cells (reviewed in MÜNGER et al. 1992). Evidence that E6 was an oncogene came from studies that showed that both the E6 and the E7 protein were necessary and sufficient for efficient immortalization of their natural host cells, primary human squamous epithelial cells (HAWLEY-NELSON et al. 1989; MÜNGER et al. 1989a). The transforming properties of the E6 and E7 genes of the HPVs mirror their clinical associations. In contrast to the immortalizing capacity of the HPV-16 and HPV-18 E6 and E7 proteins, E6 and E7 encoded by the low-risk HPVs are either inactive or only weakly transforming in similar transformation assays.

Insight into the mechanisms by which the E6 and E7 genes of the high-risk HPVs contribute to cellular transformation and to carcinogenic progression has come from the recognition that, similarly to the oncoproteins of other DNA tumor viruses, the HPV E6 and the E7 proteins interact with cell regulatory proteins such as p53 and the product of the retinoblastoma susceptibility gene pRB (DYSON et al. 1989; WERNESS et al. 1990). Recent evidence suggests that the oncogenic properties of these viruses are, at least in part, a consequence of specific interactions with these cell-regulatory proteins.

2 E2 Regulatory Proteins

The papillomavirus E2 gene products are important regulators of viral transcription and replication. The E2 proteins are relatively well conserved among the papillomaviruses in two domains: a sequence-specific DNA-binding domain located in the carboxy terminal region of the protein and a transactivating domain that is located at the amino terminus. These two domains are separated by an internal hinge region which varies in length and amino acid composition and is not well conserved among papillomaviruses (reviewed in MCBRIDE et al. 1991). The E2 proteins bind the consensus sequence, $ACCN_6GGT$, (ANDROPHY et al. 1987; LI et al. 1989) and regulate transcription from promoters containing E2 binding sites (MCBRIDE et al. 1991). E2 binds $ACCN_6GGT$ motifs as a dimer; the dimerization domain has been localized to the carboxy terminus (MCBRIDE et al. 1991).

The E2 proteins have been best studied in the BPV system where three proteins have been identified. The full-length protein (E2TA) can function as a transactivator or a repressor depending on the context of the E2-binding sites within the enhancer/promoter region. It is also necessary for viral DNA replication. Two shorter proteins which contain the DNA binding and dimerization domains of the C terminus can inhibit the transactivation function of E2TA in a dominant manner (reviewed in MCBRIDE et al. 1991). These shorter forms of E2 can inhibit the transcriptional transactivating function of the full-length polypeptide by competing for DNA-binding sites and by forming inactive heterodimers with the full-length transactivator protein (MCBRIDE et al. 1991). The crystal structure of the dimeric DNA-binding domain of BPV E2 has been determined and reveals a previously unobserved structure for DNA-binding protein of a dimeric antiparallel β barrel (HEDGE et al. 1992).

As noted above for the high-risk HPVs, the E2 protein can repress expression of the E6 and E7 genes by binding to E2-binding sites located in close proximity to the TATA box of the E6/E7 promoter and most probably interfering with the assembly of the pre-initiation complex (THIERRY and YANIV 1987; BERNARD et al 1989; ROMANCZUK et al. 1990). For the papillomaviruses, E2 and E1 are the viral encoded factors necessary for efficient viral DNA replication (USTAV and STENLUND 1991; CHIANG et al. 1992; DEL VECCHIO et al. 1992). The role of E2 in DNA replication may be as an auxiliary protein. This has been best studied in the BPV system where E2 has been shown to complex with E1 and strengthen the affinity of E1 for binding to the origin of DNA replication (USTAV et al. 1991; MOHR et al. 1990).

2.1 Transcriptional Regulation of the E6 and E7 Transforming Genes by E2

The E6 and E7 transforming genes of HPV-16 and HPV-18 are transcribed from a major promoter (P_{97} and P_{105}, respectively) contained within the long control region (LCR) of their respective genomes (SMOTKIN and WETTSTEIN 1986; THIERRY and YANIV 1987). Each promoter lies downstream of an enhancer necessary for the cell type-specific expression of the viral genes and a region which can be transactivated or repressed by the viral E2 protein (CRIPE et al. 1987; SWIFT et al. 1987). The interaction of E2 with its sites within the LCR results in the modulation of promoter activity. There are four E2-binding sites within the LCRs of the HPV-16 and HPV-18 genomes. Two E2-binding sites are located immediately adjacent to the TATA box of the P_{97} and P_{105} promoters of HPV-16 and HPV-18, respectively. Analyses of promoter activity in human epithelial cells have shown that the P_{97} promoter of HPV-16 and the P_{105} promoter of HPV-18 each possesses a basal activity which can be repressed by full-length E2 gene products (THIERRY and YANIV 1987; BERNARD et al. 1989; ROMANCZUK et al. 1990; THIERRY and HOWLEY 1991). The basal activity is dependent on the keratinocyte-dependent enhancer contained within the LCR, and E2 repression occurs through binding to its cognate $ACCN_6GGT$ sites in very close proximity to the TATA boxes of each promoter (CRIPE et al. 1987; SWIFT et al. 1987; ROMANCZUK et al. 1990). In the context of the full genome, the modulation of HPV gene expression by E2 is even more pronounced. In immortalization assays of human keratinocytes that require expression of the E6 and E7 genes, E2 can strongly repress the frequency of HPV-16 immortalization. Wild-type HPV-16 DNA immortalizes primary human keratinocytes with low efficiency, whereas a genome with a mutation in the E2 open reading frame is perhaps ten fold more efficient in inducing cellular immortalization (ROMANCZUK and HOWLEY 1992). These results are consistent with the notion that E2 can repress the viral P_{97} promoter. This model, however, cannot fully account for the inhibition of cellular immortalization by E2. Mutation of the E2-binding sites adjacent to the P_{97} promoter only partially alleviate E2-mediated repression, suggesting that an additional mechanism must also be involved. Such a mechanism may also involve the E1 protein, since mutations in the E1 gene also result in an increase in the HPV-16 immortalization efficiency (ROMANCZUK and HOWLEY 1992).

2.2 Role of E2 in Viral DNA replication

Bovine papillomavirus DNA replication requires the origin of DNA replication in cis and the viral E1 and E2 proteins in trans (USTAV and STENLUND 1991; YANG et al. 1991). The similarity in structure of the HPV and BPV-1 E2 proteins suggested that, like BPV-1 E2, HPV E2 may also function in replication, and this has now been shown for the HPVs (CHIANG et al. 1992; DEL VECCHIO et al. 1992). Furthermore, in vitro replication of BPV DNA has been established which is dependent on the presence of both the E1 and E2 proteins for efficient replication, indicating a direct role for both factors (YANG et al. 1991). Because of the genetic similarities of the HPVs to BPV, it is likely that the HPV E2 proteins will function analogously.

The presence of additional promoters within the HPV genomes and analyses of the spliced viral transcripts suggest that shortened and repressor forms of E2 may also be expressed (KARLEN and BEARD 1993; CHOW et al. 1987). If these truncated forms of E2 are functionally similar to their BPV-1 counterparts, they may be able to form inactive heterodimers with full-length E2 proteins, interfering with the full-length E2 replication function as well as its transcriptional regulation functions.

3 E1 Regulatory Protein

The papillomavirus E1 proteins are well conserved among papillomaviruses and in combination with E2 can support the replication of most heterologous papillomavirus DNA replication origins (CHIANG et al. 1992; DEL VECCHIO et al. 1992). The E1 proteins share structural similarities with the large T antigen of simian virus (SV40), notably in the sites involved in ATPase activity, helicase functions, and nucleotide-binding activity (CLERTANT and SEIF 1984; SEO et al. 1993). These similarities suggest a common function for these proteins in the initiation of viral DNA replication. Little is known about the HPV E1 gene products; therefore much of our understanding of the mechanism of E1 regulation of replication is by analogy with BPV-1. The BPV-1 E1 protein is a 68-kDa nuclear phosphoprotein that binds specifically to the origin of replication contained within the viral LCR (WILSON and LUDES-MEYERS 1991). Independently, E1 binds the origin with weak affinity; the additional binding of E2 increases this affinity (USTAV et al. 1991; MOHR et al. 1990). The cellular proteins that interact with E1 and E2 when bound to the origin of replication are not yet known.

3.1 Transcriptional Regulation by E1

Analyses of E1-defective BPV-1 genomes have shown that mutations of the E1 gene result in increased levels of viral transcription and a corresponding increase in viral transformation activity (LAMBERT and HOWLEY 1988; SCHILLER et al. 1989). This phenomenon may be due, at least in part, to the repression of certain BPV-1 promoters by E1, a function that correlates with the ability of E1 to complex with E2 and repress E2 promoter transactivation. (SANDLER et al. 1993). Studies of the HPV-16 E1 functions have shown that mutations in the E1 gene result in the increased immortalization capacity of the viral genome (ROMANCZUK and HOWLEY 1992), a function dependent on efficient E6 and E7 expression (MÜNGER et al. 1989a; HAWLEY-NELSON et al. 1989; HUDSON et al. 1990). These results suggest that HPV-16 E1 may be able to repress the P_{97} promoter, although the mechanism of such a repression is not known. This hypothesis is consistent with the finding that there is often a disruption of the E1 gene in HPV-associated cancers. As discussed above, in HPV cancers there is a selection for the integrity of the E6/E7 coding region with a concomitant loss of E1 and E2 expression.

3.2 E1 Regulation of Viral Replication

Our knowledge of the replication functions of the E1 gene product is derived primarily from studies with the BPV-1 genome. Mutations that disrupt the BPV-1 E1 gene result in a DNA replication-defective mutant. Transient replication assays (USTAV and STENLUND 1991) and in vitro replication systems have substantiated the absolute requirement for E1 in BPV DNA replication (YANG et al. 1991). Similarly, efficient replication of HPV DNA requires E1, as well as E2, and the viral origin of replication (CHIANG et al. 1992; DEL VECCHIO et al. 1992). Given the sequence and functional similarities of E1 and the SV40 large T antigen (CLERTANT and SEIF 1984; SEO et al. 1993), it is believed that E1 may function to coordinate the assembly of an initiation complex at the viral origin of replication. The cellular factors that complete this complex have not yet been identified.

4 E7 Oncoprotein

The HPV-16 E7 protein is a nuclear protein of 98 amino acids, has been shown to bind zinc, and is phosphorylated by casein kinase II (CK II) (reviewed in MÜNGER et al. 1992). Insight into its mechanism of action came from the recognition that E7 had some functional similarities with the adenovirus (Ad) 12S E1A product (PHELPS et al. 1988). It can transactivate the Ad E2 promoter (PHELPS et al. 1988), induce DNA synthesis in quiescent cells (SATO et al. 1989), and cooperate with ras in rodent cell transformation assays (PHELPS et al. 1988; MATLASHEWSKI et al. 1987).

4.1 Interaction of E7 with Host Cellular Proteins

In addition to functional similarities, the HPV-16 E7 protein, the Ad E1A proteins, and the SV40 large tumor antigen (TAg) share regions of significant amino acid sequence similarity (Fig. 2). These conserved regions are involved in the binding to cellular proteins, one of which (pRB) is the product of the retinoblastoma tumor suppressor gene (DECAPRIO et al. 1988; DYSON et al. 1989; WHYTE et al. 1988). Complex formation with pRB involves conserved region 2 of the Ad E1A protein and the corresponding region in the E7 protein and in SV40 large TAg (DECAPRIO et al. 1988; MÜNGER et al. 1992; WHYTE et al. 1989).

The phosphorylation state of pRB is regulated through the cell cycle. It is hypophosphorylated in G0 and G1. Since pRB acts as a negative regulator of cell growth at the G1/S border, it follows that the hypophosphorylated form may represent the active form with respect to cell growth suppression. pRB becomes phosphorylated at multiple serine residues by one or more cyclin-dependent kinases (cdk) at the G1/S boundary and pRB is hyperphosphorylated during S, G2, and early M phase. It becomes hypophosphorylated again in late M phase through the action of a specific phosphatase. Coprecipitation experiments showed that HPV-16 E7, like SV40 TAg, binds

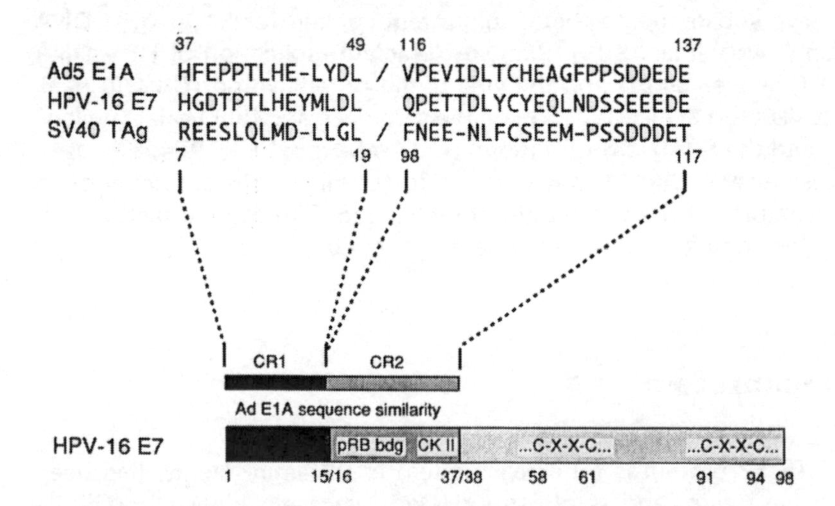

Fig. 2. Amino acid sequence comparison of portions of conserved regions 1 and 2 (*CR1*, *CR2*) of the Ad5 E1A proteins and the regions of SV40 large T antigen (*TAg*) with the amino terminal 38 amino acids of HPV-16 E7. CR2 contains the pRB-binding site and the casein kinase II (*CK II*) phosphorylation site of HPV-16 E7

preferentially to the hypophosphorylated form of pRB. Presumably this interaction results in the functional inactivation of pRB and permits progression of the cell into S phase of the cycle (Fig. 3) (reviewed in COBRINIK et al. 1992). This property of the viral oncoproteins to complex pRB would appear to account, at least in part, for their ability to induce DNA synthesis.

Like Ad E1A protein, the HPV E7 proteins can also bind the cellular proteins p107 and p130 (DYSON et al. 1992). p107 has now been cloned and certain regions of p107 are similar in amino acid composition to regions of pRB. Most of the conserved sequences in p107 correspond to the "pocket" of pRB that is involved in binding the viral oncoproteins (EWEN et al. 1991). Therefore it follows that pR B and p107 might bind a common set of cellular proteins that become displaced by competitive binding by the viral oncoproteins. Indeed, members of the E2F transcription factor family can bind to pRB and p107, and this binding is significantly reduced in the presence of the viral oncoproteins (reviewed in NEVINS 1992). However, the model is not as

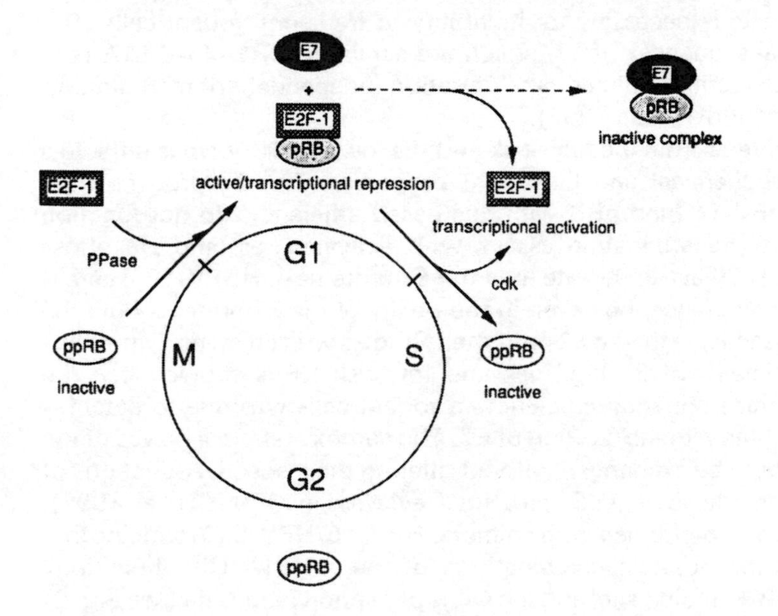

Fig. 3. During the cell cycle, the RB protein is differentially phosphorylated. The under-phosphorylated form (indicated *pRB*) is detected only in the G0/G1 phase of the cell cycle. This form presumably represents the active form of pRB and acts as a negative regulator of cell cycle progression. During the transition to the S phase, the RB protein is phosphorylated (indicated as *ppRB*) at serine residues by cyclin-dependent kinases (*cdk*). This presumably results in the inactivation of RB protein function and may permit entry into S phase. ppRB is a substrate for a serine/threonine phosphatase (*PPase*) during the M phase of the cell cycle and the RB protein becomes dephosphorylated as the cell enters G1. The cellular transcription factor E2F-1 is preferentially bound to the under-phosphorylated form of the RB protein which presumably results in the functional inactivation of E2F-1. Phosphorylation of the RB protein or interaction with a viral oncoprotein such as HPV-16 E7 may result in the release of E2F-1 allowing it to function as a transcriptional activator on target promoters, activating expression of specific genes at the beginning of S phase

simple as it may seem. Whereas Ad E1A disrupts the pRB/E2F complex and the p107/E2F complex with similar efficiencies, the HPV E7 disrupts the pRB/E2F complex more efficiently than the p107/E2F complex (CHELLAPPAN et al. 1992; PAGANO et al. 1992). The physiologic significance of this observation is not presently known but suggests a higher level of complexity. Perhaps different members of the E2F family interact preferentially with the different members of the pRB family of proteins and their activities may be manifest at different times in the cell cycle.

Liked Ad E1A, HPV-16 E7 is able to bind cyclin A. However, this is not likely to be a direct interaction and is probably mediated by p107 (DYSON et al. 1992; TOMMASINO et al. 1993).

4.2 E7: A Modular Multifunctional Protein

Genetic studies of HPV-16 E7 have revealed that an intact and high-affinity pRB binding site is necessary for its ability to transform rodent cells. The amino terminal sequences of E7, which are similar to CR1 of Ad E1A (see Fig. 2), may also effect cellular transformation independent of pRB binding (reviewed in MÜNGER et al. 1992).

The E7 proteins from the high-risk and the low-risk HPV types differ in a number of biochemical and biological properties. The E7 proteins from HPV-6 and HPV-11 bind pRB with decreased efficiency, do not function well in cellular transformation assays with activated ras, and are phosphorylated by CK II at a lower rate than the E7 proteins of HBPV-16. There is no significant difference, however, in the ability of the E7 proteins from the viruses to transactivate the Ad E2 promoter. Studies with chimeric E7 proteins containing domains of E7 high-risk and low-risk HPVs showed that the difference in transformation efficiency in rodent cells was due to determinants in the amino terminal portion of E7. The carboxy terminal halves of the E7 proteins could be exchanged without altering the respective functions of the high-risk and the low-risk E7 proteins (reviewed in MÜNGER et al. 1992). Additional studies performed with chimeric HPV-16/HPV-6 E7 proteins that contained all the possible combinations of the Ad E1A CR1 homology domain, the pRB binding site, and the CK II phosphorylation site (see Fig. 2) revealed that the major determinant for the oncogenic potential of the E7 protein in rodent cells was the pRB-binding site (HECK et al. 1992). Although the CK II sequences of the high-risk E7 proteins constitute a better substrate for CK II phosphorylation, the exchange of the high-risk and low-risk E7 CK II sequences had no significant effects on the transformation capacity of the protein. Sequence comparison revealed a single consistent amino acid sequence difference in pRB-binding sites between the high-risk and the low-risk E7 proteins: an aspartic acid residue (Asp 21 in HPV-16 E7) corresponding to a glycine residue in the low-risk E7 sequence (Gly 22 in HPV-6 E7). Mutation of the appropriate residue in the respective E7 genes revealed that this single amino acid residue was largely responsible for the difference in affinity for complexing pRB and for the transforming capacity of the low-risk and the high-risk E7 proteins (HECK et al. 1992; SANG and BARBOSA 1992).

The function of the carboxy terminus of E7 is not yet clear. Mutations in the carboxy terminal half of E7 that interfere with E7 function have also generally affected the intracellular stability of the protein (EDMONDS and VOUSDEN 1989; PHELPS et al. 1992). Several studies have suggested some specific contributions of the carboxy terminus to E7 functions. Binding of E7 to pRB abrogates the nonspecific DNA-binding properties of pRB, and studies with truncated E7 polypeptides have suggested that carboxy terminal sequences, in addition to the pRB-binding site, may be necessary for this activity (STIRDIVANT et al. 1992). Similarly, the ability of E7 to disrupt the complex between the cellular transcription factor E2F and pRB seems to involve sequences in the carboxy terminus of E7 in addition to the pRB-binding site (HUANG et al. 1993; WU et al. 1993). Recently, a mutant of E7 that was defective in pRB binding was shown to still be competent for immortalization of primary human genital keratinocytes by high-risk HPV DNA, suggesting other properties of E7 may be involved in this function. The only mutation that rendered E7 incompetent for immortalization was located in the carboxy terminus (JEWERS et al. 1992).

5 E6 Oncoprotein

The HPV E6 proteins contain approximately 150 amino acids and have no known intrinsic enzymatic functions. They contain four Cys-X-X-Cys motifs which presumably are involved in zinc binding and have been detected in nuclear and non-nuclear membranes (BARBOSA et al. 1989; GROSSMAN and LAIMINS 1989; GROSSMAN et al. 1989). The E6 proteins from the low-risk and high-risk HPVs can transactivate "minimal" promoters containing a TATA box element only (SEDMAN et al. 1991). Since the E6 proteins of the low-risk HPVs have little or no transformation activity, it seems unlikely that this transcriptional transactivation property of the E6 proteins is linked mechanistically to the transforming function of the E6 proteins of the high-risk HPVs.

The property of some of the oncoproteins of the different small DNA tumor viruses to bind pRB prompted an analysis of whether the high-risk HPV E6 proteins, like the SV40 TAg and the Ad5 E1B 55K protein, might also complex p53. Using in vitro translated human p53 and HPV 56 proteins, it was found that the high-risk HPV E6 proteins could complex with p53, whereas an interaction of p53 with the low-risk HPV E6 proteins was not detected (WERNESS et al. 1990). Structure–function studies on he E6 proteins have been difficult, and it has not yet been possible to define a specific domain of E6 which is involved in complex formation with p53. Unlike the E7/pRB interaction where the pRB binding domain could be mapped to a precise region on E7 (see above), similar studies with E6 have been difficult to interpret (CROOK et al. 1991). Subtle genetic manipulation of E6 often disrupts its functional integrity with respect to p53 binding, suggesting that the tertiary structure of E6 is critical for E6 to complex with p53 (J. Huibregtse, M. Scheffner, and P. Howley unpublished data). Wild-type p53 has the properties of a sequence specific transcriptional transactivator (FUNK et al. 1992; KERN et al. 1992). It is possible that this activity is involved in p53-mediated cell growth and tumor growth suppression. Using CAT reporter constructs containing p53-responsive elements, it has been shown that the high-risk but not low-risk HPV E6 proteins, as well as SV40 TAg and Ad E1B 55K, can efficiently abrogate the transactivation activity of p53 (MIETZ et al. 1992; YEW and BERK 1992). Thus, like SV40 TAg and Ad E1B, HPV E6 can interfere with the negative regulatory function of p53.

Although SV40 TAg, the Ad5 E1B 55-kDa protein, and the high-risk HPV E6 proteins all complex with p53 the consequence of these bindings are different with respect to the stability of the p53 protein. In SV40 and adenovirus-infected cells, levels of p53 are usually quite high (OREN et al. 1981; REICH et al. 1983). In contrast, the levels of p53 in HPV-infected cells are low compared to uninfected primary host cells (SCHEFFNER et al. 1991; HUBBERT et al. 1992). Unlike TAg and the E1B 55-kDa protein, which presumably inactivate p53 by sequestering it in stable complexes, the high-risk HPV E6 proteins appear to inactivate p53 through destabilization. This was first demonstrated in in vitro studies which showed that the high-risk HPV E6 proteins can facilitate the rapid degradation of p53 via the ubiquitin-dependent proteolytic system (SCHEFFNER et al. 1990). The low-risk HPV E6 proteins, which do not bind p53 at detectable levels, have no effect on p53 stability in vitro (SCHEFFNER et al. 1990). Biochemical studies examining the binding of E6 to p53 led to the identification of a cellular protein of 100 kDa that is necessary for complex formation of E6 with p53 (HUIBREGTSE et al. 1991). This cellular protein, designated "E6-AP" for E6-associated protein, binds stably to high-risk HPV E6 proteins in the absence of p53 but not to p53 in the absence of E6. Since both E6 and E6-AP are necessary to detect complex formation with p53, it is not yet possible to determine whether E6, or E6-AP, or both actually contract p53. In correlation with the inability of the low-risk HPV E6 proteins to detectably associate with p53, binding to E6 AP could not be detected. A cDNA encoding E6-AP has been cloned and database searches showed that the C terminal portion of E6-AP contains significant homology to two proteins (NEDD 4 and a rat 100-K protein). However, the function of these proteins is not yet known (HUIBREGTSE et al. 1993). Studies to identify each of the proteins that are necessary in the E6-mediated ubiquitination of p53 have shown that E6-AP is also required for ubiquination as well as for binding of E6 to p53 (SCHEFFNER et al. 1993). The data suggest that E6-AP is indeed a part of the ubiquitin-dependent proteolytic pathway. Therefore an attractive but purely speculative model is that E6-AP may be involved in the regulation of p53 stability and possibly of other cellular proteins, not only in HPV-infected cells but also in uninfected cells.

In addition to the evidence that the high-risk HPV E6 proteins accelerate the in vitro degradation of p53, in vivo E6 also appears to effect the stability of intracellular p53 (BAND et al. 1991; HUBBERT et al. 1992; LECHNER et al. 1992; SCHEFFNER et al. 1991). Levels of p53 in E6 immortalized cells or in HPV-positive cervical carcinoma cells are, on average, two or three fold decreased compared to primary cells whereas in the in vitro system E6 leads to an almost complete removal of p53 (SCHEFFNER et al. 1990, 1991). The half life of p53 is reduced from 3 h to 20 min in human keratinocytes expressing E6 (HUBBERT et al. 1992). Intracellular p53 levels increase significantly in response to DNA damage induced by γ irradiation or mutagenizing agents (KASTAN et al. 1992). The higher levels of p53 are thought to result in growth arrest of the treated cells, which may be a cell defense mechanism that would allow for the DNA damage to be repaired prior to the initiation of a new round of DNA replication. Interestingly, E6-expressing cells do not manifest a p53-mediated cellular response to DNA damage (KESSIS et al. 1993). This might indicate that the ability of E6 to target p53 for degradation is fully active

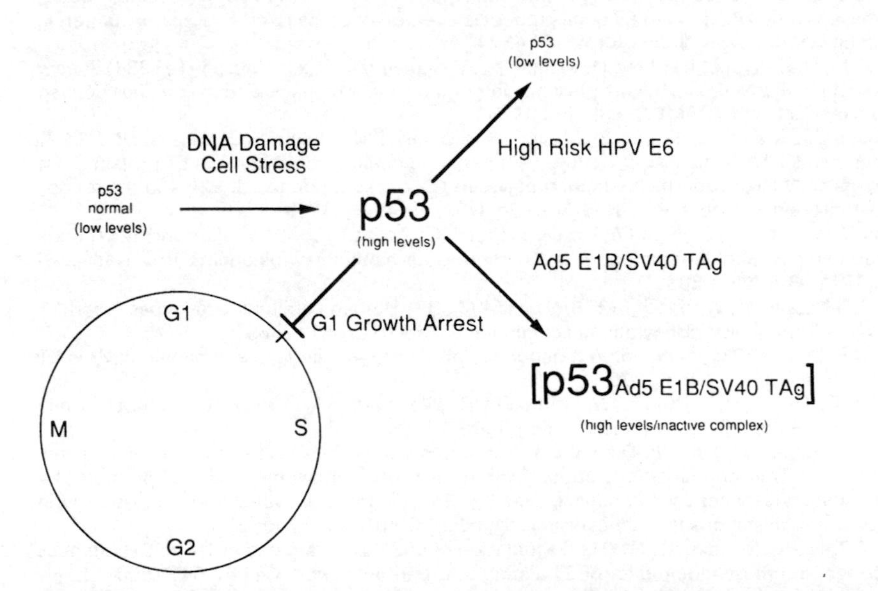

Fig. 4. The level of p53 in primary cells is usually low. Damaging agents increase the level of cellular p53. The elevated levels of p53 can lead to a cell cycle checkpoint arrest in G1 possibly to allow the damaged DNA to be repaired. Viral oncoproteins may interfere with this negative growth regulatory function of p53, either by sequestering p53 into a stable, but non-functional complex (such as with SV40 TAg and the Ad5 E1 B 55 K protein) or by enhanced degradation (as manifest by the high-risk HPV-6 proteins). Both mechanisms result in the functional inactivation of p53

only under certain circumstances, for instance when the amount of p53 rises above a certain threshold level (see Fig. 4). Under DNA-damaging conditions, the E6-stimulated degradation of p53 would presumably abrogate the negative growth regulatory effect of p53 and thereby result in genomic instability. This scenario could, in part, account for the observation that, in general, HPV-immortalized cells or cell lines derived from cervical carcinomas have abnormal karyotypes.

References

Androphy EJ, Lowy DR, Schiller JT (1987) Bovine papillomavirus E2 transactivating gene product binds to specific sites in papillomavirus DNA. Nature 325: 70–73

Baker CC, Phelps WC, Lindgren V, Braun MJ, Gonda MA, Howley PM(1987) Structural and translational analysis of human papillomavirus type 16 sequences in cervical carcinoma cell lines. J Virol 61: 962–971

Band V, DeCaprio JA, Delmolino L, Kulesa V, Sager R (1991) Loss of p53 protein in human papillomavirus type 16 E6-immortalized human mammary epithelial cells. J Virol 65: 6671–6676

Barbosa MS, Lowy DR, Schiller JT (1989) Papillomavirus polypeptides E6 and E7 are zinc-binding proteins. J Virol 63: 1404–1407

Bernard BA, Bailly C, Lenoir M-C, Darmon M, Thierry F, Yaniv M (1989) The human papillomavirus type 18 (HPV18) E2 gene product is a repressor of the HPV18 regulatory region in human keratinocytes. J Virol 63: 4317–4324

Boshart M, Gissmann L, Ikenberg H, Kleinheinz A, Schleurlen W, zur Hausen H (1984) A new type of papillomavirus DNA, its presence in genital cancer biopsies and in cell lines derived from cervical cancer. EMBO J 3: 1151–1157

Chellappan S, Kraus VB, Kroger B, Münger K, Howley PM, Phelps WC, Nevins JR (1992) Adenovirus E1A, simian virus 40 tumor antigen, and human papillomavirus E7 protein share the capacity to disrupt the interaction between the transcription factor E2F and the retinoblastoma gene product. Proc Natl Acad Sci USA 89: 4549–4553

Chiang C-M, Ustav M, Stenlund A, Ho TF, Broker TR, Chow LT (1992) Viral E1 and E2 proteins support replication of homologous and heterologous papillomavirus origins. Proc Natl Acad Sci USA 89: 5799–5803

Chow LT, Nasseri M, Wolinsky SM, Broker TM (1987) Human papillomavirus types 6 and 11 mRNAs from genital condylomata acuminata. J Virol 61: 2581–2588

Clertant P, Seif I (1984) A common function for polyoma virus large-T and papillomavirus E1 proteins. Nature 311: 276–279

Cobrinik D, Dowdy SF, Hinds PW, Mittnacht S, Weinberg RA (1992) The retinoblastoma protein and the regulation of cell cycling. TIBS 17: 312–315

Cripe TP, Haugen TH, Turk JP, Tabatabai F, Schmid PG, Durst M, Gissmann L, Roman A, Turek LP (1987) Transcriptional regulation of the human papillomavirus-16 E6-E7 promoter by a keratinocyte-dependent enhancer, and by viral E2 trans-activator and repressor gene products: implications for cervical carcinogenesis. EMBO J 6: 3745–3753

Crook T, Tidy JA, Vousden KH (1991) Degradation of p53 can be targeted by HPV E6 sequences distinct from those required for p53 building and transactivation. Cell 67: 547–556

DeCaprio JA, Ludlow JW, Figge J, Shew J-Y, Huang C-M, Lee W-H, Marsilío EE, P, Livingston DM (1988) SV40 large tumor antigen forms a specific complex with the product of the retinoblastoma susceptibility gene. Cell 54: 275–283

Del Vecchio AM, Romanczuk H, Howley PM, Baker CC (1992) Transient replication of human papillomavirus DNAs. J Virol 66: 5949–5958

DeVilliers, EM (1989) Heterogeneity of the human papillomavirus group. J Virol 63: 4898-4903

Dürst M, Gissmann L, Idenburg H, Zur Hausen H (1983) A papillomavirus DNA from a cervical carcinoma and its prevalence in cancer biopsy samples from different geographic regions. Proc Natl Acad Sci USA 80: 3812–3815

Dürst M, Kleinheinz A, Hotz M, Gissmann L (1985) The physical state of human papillomavirus type 16 in benign and malignant genital tumors. J Gen Virol 66: 1515–1522

Dürst M, Croce CM, Gissmann L, Schwarz E, Huebner K (1987) Papillomavirus sequences integrate near cellular oncogenes in some cervical carcinomas. Proc Natl Acad Sci USA 80: 3812–3815

Dyson N, Howley PM, Münger K, Harlow E (1989) The human papillomavirus-16 E7 oncoprotein is able to bind to the retinoblastoma gene product. Science 243: 934–937

Dyson N, Guida P, Münger K, Harlow E (1992) Homologous sequences in adenovirus E1A and human papillomavirus E7 proteins mediate interaction with the same set of cellular proteins. J Virol 66: 6893–6902

Edmonds C, Vousden KH (1989) A point mutational analysis of human papillomavirus type 16 E7 protein. J Virol 63: 2650–2656

Ewen M, Xing Y, Lawrence JB, Livingston DM (1991) Molecular cloning, chromosomal mapping, and expression of p107, a retinoblastoma gene product related protein. Cell 66: 1155–1164

Funk WD, Pak DT, Karas RH, Wright WE, Shay, JW (1992) A transcriptionally active DNA-binding site for human p53 protein complexes. Mol Cell Biol 12: 2866-2871

Grossman SR, Laimins LA (1989) E6 protein of human papillomavirus type 18 binds zinc. Oncogene 4: 1089–1093

Grossman SR, Mora R, Laimins LA (1989) Intracellular localization and DNA-binding properties of human papillomavirus type 18 E6 protein expressed with a baculovirus vector. J Virol 63: 366–374

Hawley-Nelson P, Vousden KH, Hubbert NL, Lowy DR, Schiller JT (1989) HPV;6 Et and E7 proteins cooperate to immortalize human foreskin keratinocytes. EMBO J 8: 3905–3910

Heck DV, Yee CL, Howley PM, Münger K (1992) Efficiency of binding to the retinoblastoma protein correlates with the transforming capacity of the E7 oncoproteins of the human papillomaviruses. Proc Natl Acad Sci USA 89: 4442–4446

Hedge RS, Rossman SR, Laimins LA, Sigler PB (1992) Crystal structure at 1.7A of the bovine papillomavirus-1 E2 DNA-binding domain bound to its DNA target. Nature 359: 505–512

Huang PS, Patrick DR, Edwards G, Goodhart PJ, Huber HE, Miles L, Garsky VM, Oliff A, Heimbrook DC (1993) Protein domains governing interactions between E2F, the retinoblastoma gene product, and human papillomavirus type 16 E7 protein. Mol Cell Biol 13: 953–960

Hubbert NL, Sedman SA, Schiller JT (1992) Human papillomavirus type 16 E6 increases the degradation rate of p53 in human keratinocytes. J Virol 66: 6237–6241

Hudson JB, Bedell MA, McCance DJ, Laimins LA (1990) Immortalization and altered differentiation of human keratinocytes in vitro by the E6 and E7 open reading frames of human papillomavirus type 18. J Virol 64: 519–526

Huibregtse JM, Scheffner M, Howley PM (1991) A cellular protein mediates association of p53 with the E6 oncoprotein of human papillomavirus types 16 or 18. EMBO J 10: 4129–4135

Huibregtse JM, Scheffner M, Howley PM (1993) Cloning and expression of the cDNA for E6-AP, a protein that mediates the interaction of the human papillomavirus E6 oncoprotein with p53. Mol Cell Biol 13: 775–784

Jewers RJ, Hildebrandt P, Ludlow JW, Kell B, McCance DJ (1992) Regions of human papillomavirus type 16 E7 oncoprotein required for immortalization of human keratinocytes. J Virol 66: 1329–1335

Karlen S, Beard P (1993) Identification and characterization of novel promoters in the genome of human papillomavirus type 18. J Virol 67: 4296–4306

Kastan MB, Zhan Q, E1 Deiry W-S, Carrier F, Jacks T, Walsh WV, Plunkett BS, Vogelstein B, Fornace A-J Jr (1992) A mammalian cell cycle checkpoint pathway utilizing p53 and GADD 45 is defective in ataxia-telangiectasia. Cell 71: 587–597

Kern SE, Pietenpol JA, Thiagalingam S, Seymour A, Kinzler KW, Vogelstein B (1992) Oncogenic forms of p53 inhibit p53-regulated gene expression. Science 256: 827–830

Kessis TD, Slebos RJ, Nelson WG, Kasta MB, Plunkett BS, Han SM, Lorincz AT, Hedrick L, Cho KR (1993) Human papillomavirus 16 E6 expression disrupts the p53-mediated cellular response to DNA damage. Proc Natl Acad Sci USA 90: 3988–3992

Lambert PF (1991) Papillomavirus DNA replication. J Virol 65: 3417–3420

Lambert PF, Howley PM (1988) Bovine papillomavirus type 1 E1 replication-defective mutants are altered in their transcriptional regulation. J Virol 62: 4009–4015

Lechner MS, Mack DH, Finicle AB, Crook T, Vousden KH, Laimins LA (1992) Human papillomavirus E6 proteins bind p53 in vivo and abrogate p53-mediated repression of transcription. EMBO J 11: 3045–3052

Li R, Knight J, Bream G, Stenlund A, Botchan M (1989) Specific recognition nucleotides and their DNA context determine the affinity of E2 protein for 17 binding sites in the BPV-1 genome. Genes Dev 3: 510–526

Matlashewski G, Schneider J, Banks L, Jones N, Murray A, Crawford L (1987) Human papillomavirus type 16 DNA co-operates with activated ras in transforming primary cells. EMBO J 6: 1141–1146

McBride AA, Romanczuk H, Howley PM (1991) the papillomavirus E2 regulatory proteins. J Biol chem 266: 18411–18414

Mietz JA, Unger T, Huibregtse JM, Howley PM (1992) The transcriptional transactivation function of wild-type p53 is inhibited by SV40 large T-antigen and by HPV-16 E6 oncoprotein. EMBO J 11: 5013–5020

Mohr IJ, Clark R, Sun S, Androphy EJ, ManPherson P, Botchan MR (1990) Targeting the E1 replication protein to the papillomavirus origin of replication by complex formation with the E2 transactivator. Science 250: 1694–1699

Münger K, Phelps WC, Bubb V, Howley PM, Schlegel R (1989a) The E6 and E7 genes of the human papillomavirus type 16 together are necessary and sufficient for transformation of primary human keratinocytes. J Virol 63: 4417–4421

Münger K, Werness BA, Dyson N, Phelps WC, Harlow E, Howley PM (1989b) Complex formation of human papillomavirus E7 proteins with the retinoblastoma tumor suppressor gene product. EMBO J 8: 4099–4105

Münger K, Scheffner M, Huibregtse JM, Howley PM (1992) Interactions of HPV E6 and E7 with tumor suppressor gene products. Cancer Surv 12: 197–217

Nevins JR (1992) A link between the Rb tumor suppressor protein and viral oncoproteins. Science 258: 424–429

Oren M, Maltzman W, Levine AJ (1981) Post-translational regulation of the 54K cellular tumor antigen in normal and transformed cells. Mol Cell Biol 1: 101–110

Pagano M, Dürst M, Joswig S, Draetta G, Jansen-Dürr P (1992) Binding of the human E2F transcription factor to the retinoblastoma protein but not to cyclin A is abolished in HPV-16-immortalized cells. Oncogene 7: 1681–1686

Phelps WC, Yee CL, Münger K, Howley PM (1988) The human papillomavirus type 16 E7 gene encodes transactivation and transformation functions similar to adenovirus Ela. Cell 53: 539–547

Phelps WC, Münger K, Yee CL, Barnes JA, Howley PM (1992) Structure-function analysis of the human papillomavirus E7 oncoprotein. J. Virol 66: 2418–2427

Reich NC, Oren M, Levine AJ (1983) Two distinct mechanisms regulate the levels of a cellular tumor antigen. Mol Cell Biol 3: 2134–2150

Riou G, Favre M, Jeannel D, Bourhis J, Le Doussal V, Orth G (1990) Association between poor prognosis in early-stage invasive cervical carcinomas and non-detection of HPV DNA. Lancet 335: 1171–1174

Romanczuk H, Howley PM (1992) Disruption of either the E1 or E2 regulatory gene of human papillomavirus type 16 increases viral immortalization capacity. Proc Natl Acad Sci USA 89: 3159–3163

Romanczuk H, Thierry F, Howley PM (1990) Mutational analysis of cis elements involved in E2 modulation of human papillomavirus type 16 p97 and type 18 p105 promoters. J Virol 64: 2849–2859

Sandler AB, Vande Pol SB, Spalholz BS (1993) Repression of BPV-1 transcription by the E1 replication protein. J Virol 67: (in press)

Sand B-C, Barbosa MS (1992) Single amino acid substitutions in "low risk" human papillomavirus (HPV) type 6 E7 protein enhance features characteristic of the "high risk" HPV E7 oncoproteins. Proc Natl Acad Sci USA 89: 8063–8067

Sato H, Furuno A, Yoshiike K (1989) Expression of human papillomavirus type 16 E7 gene induces DNA synthesis of rat 3Y1 cells. Virology 168: 195–199

Scheffner M, Werness BA, Huibregtse JM, Levine AJ, Howley PM (1990) The E6 oncoprotein encoded by human papillomavirus types 16 and 18 promotes the degradation of p53. Cell 63: 1129–1136

Scheffner M, Münger K, Byrne JC, Howley PM (1991) the state of the p53 and retinoblastoma genes in human cervical carcinoma cell lines. Proc Natl Acad Sci USA 88: 5523–5527

Scheffner M, Huibregtse JM, Vierstra RD, Howley PM (1993) The HPV-16 E6-AP complex functions as a ubiquitin-protein ligase in the ubiquination of p53. Cell 75 (in press)

Schiller JT, Kleiner E, Androphy EJ, Lowy DR, Pfister H (1989) Identification of bovine papillomavirus E1 mutants with increased transforming and transcriptional activity. J Virol 63: 1775–1782

Schwarz E, Freese UK, Gissmann L, Mayer W, Roggenbuck B, Stremlau A, zur Hausen H (1985) Structure and transcription of human papillomavirus sequences in cervical carcinoma cells. Nature 314: 111–114

Sedman SA, Barbosa MS, Vass WC, Hubbert NL, Hass JA, Lowy DR, Schiller JT (1991) The full-length E6 protein of human papillomavirus type 16 has transforming and trans-activating activities and cooperates with E7 to immortalize keratinocytes in culture. J Virol 65: 4860–4866

Seo Y-S, Müller F, Lusky M, Hurwitz J (1993) Bovine papilloma virus (BPV)-encoded E1 protein contains multiple activities required for BPV DNA replication. Proc Natl Acad Sci USA 90: 702–706

Smotkin D, Wettstein F0 (1986) Transcription of human papillomavirus type 16 early genes in cervical cancer and a cervical cancer derived cell line and identification of the E7 protein. Proc Natl Acad Sci USA 83: 4680–4684

Stirdivant SM, Huber HE, Patrick DR, Defeo-Jones D, McAvoy EM, Garsky VM, Oliff A, Heimbrook DC (1992) Human papillomavirus type 16 E7 protein inhibits DNA binding by the retinoblastoma gene product. Mol Cell Biol 12: 1905–1914

Swift FV, Bhat K, Younghusband HB, Hamada H (1987) Characterization of a cell type-specific enhancer found in the human papilloma virus type 18 genome. EMBO J 6: 1339–1344

Thierry F, Howley PM (1991) Functional analysis of E2-mediated repression of the HPV18 P_{105} promoter. New Biol 3: 90–100

Thierry F, Yaniv M (1987) The BPV-1 E trans-acting protein can be either an activator or repressor of the HPV-18 regulatory region. EMBO J 6: 3391–3397

Tommasino M, Adamczewski JP, Carlotti F, Barth CF, Manetti R, Contorni M, Cavalieri F, Hunt T, Crawford L (1993) HPV16 E7 protein associates with the protein kinase p33CDK2 and cyclin A. Oncogene 8: 195–202

Ustav M, Stenlund A (1991) Transient replication of BPV-1 requires two viral polypeptides encoded by E1 and E2 open reading frames. EMBO J 10: 449–457

Ustav M, Ustav E, Szymanski P, Stenlund, A (1991) Identification of the origin of replication of bovine papillomavirus and characterization of the viral origin recognition factor E1. EMBO J 10: 4321–4329

von Knebel-Doeberitz M, Oltersdorf T, Schwarz E, Gissmann L (1988) Correlation to modify human papillomavirus early gene expression with altered growth properties in C4-I cervical carcinoma cells. Cancer Res 48: 3780–3785

Werness BA, Levine AJ, Howley PM (1990) Association of human papillomavirus types 16 and 18 E6 proteins with p53. Science 248: 76–79

Whyte P, Buchkovich KJ, Horowitz JM, Friend SH, Raybuck M, Weinberg RA, Harlow E (1988) Association between an oncogene and an antioncogene: the adenovirus E1a proteins bind to the retinoblastoma gene product. Nature 334: 124–129

Whyte P, Williamson NM, Harlow E (1989) Cellular targets for transformation by the adenovirus E1A proteins. Cell 56: 67–75

Wilson VG, Ludes-Meyers J (1991) A bovine papillomavirus E1-related protein binds specifically to bovine papillmavirus DNA. J Virol 65: 5314–5322

Wu EW, Clemens KE, Heck DV, Münger K (1993) The human papillomavirus E7 oncoprotein and the cellular transcription factor E2F bind to separate sites on the retinoblastoma tumor suppressor protein. J Virol 67: 2402–2407

Yang L, Li R, Mohr IJ, Clark R, Botchan MR (1991) Activation of BPV-1 replication in vitro by the transcription factor E2. Nature 353: 628–632

Yew PR, Berk A (1992) Inhibition of p53 transactivation required for transformation by adenovirus early 1B protein. Nature 357: 82–85

zur Hausen H (1991) Human papillomaviruses in the pathogenesis of anogenital cancer. Virology 184: 9–13

zur Hausen H, Schneider A (1987) The role of papilloma-viruses in human anogenital cancers. In: Salzman N, Howley PM (eds) The papoviridae, vol 2: the papillomaviruses. Plenum, New York, pp 245–263

Immortalization and Transformation of Human Cells by Human Papillomavirus

J.K. McDougall

1 Introduction

The association of human papillomaviruses (HPVs) with anogenital cancer is now supported by such overwhelming evidence that even the most sceptical critic of a virus/cancer hypothesis must take note. Nevertheless, when the data are examined objectively, it becomes clear that as with so many, perhaps all, cancer etiologies that have been studied, more than one factor is involved in the progression to malignancy. HPV may be the major player in this disease, may be necessary to initiate and maintain the process, but for the eventual manifestation of invasive carcinoma may not be sufficient. In vitro studies of the effects of HPV gene expression on primary human cells in culture have

Fred Hutchinson Cancer Research Center, 1124 Columbia Street, Seattle, WA 98104, USA

Current Topics in Microbiology and Immunology, Vol. 186

provided a basis for evaluation of the phenotypic changes resulting from interaction of the viral gene products with cellular genes.

The detection of HPV DNA in benign genital warts, in dysplasias of the anogenital region, and, most importantly, in invasive carcinomas stimulated rapid progress in the identification of open reading frames (ORF) of HPVs consistently present in the tumor cells and the subsequent use of cloned sequences containing these ORFs to generate immortal and transformed cell lines. These experiments have not only provided much valuable information on the role of specific types of HPV in oncogenesis but have contributed to the recognition of common pathways of immortalization and transformation by the small DNA viruses, i.e., adenoviruses, simian virus 40 and the papillomaviruses.

2 Human Epithelial Cell Lines

Among the many animal papillomaviruses shown to be capable of transformation, bovine papillomavirus (BPV) has been extensively studied (Howley et al. 1986) and provided a basis for investigation of the oncogenic potential of the human viruses. It was demonstrated by Watts et al. (1984) that morphological transformation of rodent cells could be achieved with HPV DNA, but it was not until 1987 that the first reports of immortalization of human cells after transfection with HPV DNA appeared in the literature (Dürst et al. 1987a; Pirisi et al. 1987). These initial successes were with DNA of HPV-16, the virus type most frequently detected in anogenital carcinoma. It was subsequently shown that another type, HPV-18, was equally efficient in immortalization studies (Kaur and McDougall 1988, 1989) and this virus had also been found in a significant number of anogenital tumors and cell lines derived from cervical carcinomas (Yee et al. 1985).

Further studies demonstrated that the results of in vitro experiments reflected differences in the in vivo characteristics of HPV virus infections. Transfection of a number of cloned virus DNA preparations into primary human keratinocytes showed that those viruses, e.g., HPV-16 and -18, commonly detected in invasive carcinoma tissues could immortalize cells, whereas transfection of DNA of viruses usually associated with relatively benign cellular proliferations, e.g., HPV-6 and -11, did not alter the lifespan of the transfected cells which senesced at the same rate as the untransfected controls (Woodworth et al. 1989).

2.1 E6 and E7 ORFs

The results from transcription studies of cervical carcinoma cells (Schwarz et al. 1985) had previously indicated that retention and expression of the E6 and E7 ORFs was a regular feature of HPV-positive tumor cells. Not surprisingly, and reminiscent of earlier studies with subgenomic fragments of adenovirus which persisted in virus-transformed rodent cells (GALLIMORE et al. 1974), it could be shown that the E6/E7 subgenomic fragment was sufficient for transformation of rodent cells (BEDELL et al. 1987). Subsequent experiments established that this same fragment contained the DNA sequences responsible for immortalization of human cells (SCHLEGEL et al. 1988; KAUR et al. 1989). Experiments in primary human keratinocytes by MÜNGER et al. (1989a) indicated that both the E6 and E7 ORFs were required for immortalization and that mutation in either gene resulted in a loss of immortalizing activity. It was in fact shown later by HALBERT et al. (1991) that the E7 gene alone could immortalize cells when the viral sequences were placed under the control of a strong promoter. The E6 gene alone had no effect on cell lifespan under similar conditions, but when transfected together with the E7 gene in a similar retroviral construct increased the efficiency when compared to E7 alone. It was demonstrated by HECK et al. (1992) that the efficiency of binding of the Retinoblastoma (Rb) protein by E7 is reflected in the capacity of different "high-risk" and "low-risk" HPVs to immortalize or transform cells. It was shown that an amino acid difference in the Rb-binding domain was the main determinant of Rb of binding and transformation efficiency.

2.2 E6 in Mammary Epithelial Cells

It is of some interest that, while the above results are consistent in human skin epithelial cells, uterine cervical cells (BLANTON et al. 1992), endothelial cells and kidney epithelium (D.A. Galloway and J.K. McDougall, unpublished), and smooth muscle (PEREZ-REYES et al. 1992), the finding with human mammary tissue cells is the opposite. Epithelial cells derived from reduction mammoplasty were immortalized by transfection with the HPV-16 E6 gene but not by E7 (BAND et al. 1991), highlighting the influence of cell-specific factors. The p53 gene product was sharply reduced in these cells but, unlike the situation in the above-mentioned cell types, the Rb protein appeared to be unaltered. As described in detail in another chapter, it has been shown that the E6 gene of HPV-16 binds and degrades the p53 tumor suppressor gene protein while E7 binds the Rb gene product. It may be that mammary epithelial cells are more susceptible and vulnerable to degradation or mutation of p53 than is the case for other cells and that the Rb gene has less functional significance in mammary epithelial cells. In this context it is worth noting that in Li-Fraumeni syndrome the pathway to breast cancer involves germ-line mutation of p53 (MALKIN et al. 1990) and mutation of p53 has been recorded in about 40% of sporadic breast cancers (COLES et al. 1992).

2.3 E5 and Transformation

The concentration of transformation studies on the E6/E7 genes has in large part been due to the fact that tumors and cell lines have subgenomic HPV sequences integrated into the host cell genome (SCHWARZ et al. 1985) with the result that other ORFs have been deleted (KAUR et al. 1989). Recently, an interest has developed in the HPV E5 gene as a potential mediator of immortalization or transformation of human cells (BANKS and MATLASHEWSKI 1993). The basis for this effort is the established role of the BPV E5 gene in transformation (SCHILLER et al. 1986), interaction of both BPV E5 and HPV E5 gene products with the epidermal growth factor receptor (EGFR) (MARTIN et al. 1989; PIM et al. 1992) and the transformation of rodent fibroblasts by HPV-6 E5 (CHEN and MOUNTS 1990) and HPV-16 E5 (LEPTAK et al. 1991). Upto the date of writing, however, there has been no evidence presented that any HPV E5 gene can immortalize human cells, despite the effects of these genes on signal transduction pathways.

3 Transformation In Vivo

One of the major problems in HPV research has been the inability to produce stocks of virus in the laboratory. In consequence, studies of virus infection, receptors, etc. have been inhibited as has development of serologic assays and vaccine preparation. KREIDER and BARTLETT (1985) developed a method of infection of human epithelial cells by virus in cell-free extracts of condyloma accuminata tissue. The infected epithelial cells were grafted beneath the renal capsule of nude mice. The normal human cells were "transformed," acquiring the histological features of condyloma and were shown to be positive for HPV-11. These studies were a landmark in HPV research, but the system has not been proven to be of general utility for other virus isolates. Recent experiments provide more hope that in vitro virion production is feasible.

The complete virus production program was demonstrated in a modified organotypic culture system by DOLLARD et al. (1992) by explanting fragments of condyloma tissue infected with HPV-11. This experiment used the differentiating conditions of this culture method to allow virus replication to proceed. The virus life cycle of papillomaviruses is closely associated to the differentiation program of the host cell. This feature was successfully exploited by MEYERS et al. (1992) who achieved replication of HPV-31 virions by inducing terminal differentiation of a cervical carcinoma cell line (CIN 612) with 12-*O*-tetradecanoyl phorbol-13-acetate (TPA). It has also been shown that expression of the major and minor (L1, L2) capsid proteins of HPV in a vaccinia virus vector can result in the assembly of viral capsids (ZHOU et al. 1991; HAGENSEE et al. 1993). The next step will be to demonstrate that these capsids can incorporate viral DNA and that virions purified from any of these procedures can infect and immortalize human cells.

4 Characterization of HPV-Immortalized Cells

As with cells transfected with other small DNA virus-transforming genes, cells expressing HPV E6/E7 genes acquire new properties which are generally recognized as the result of altered growth control. Some of the characteristics of the immortalized state are increased mitogenic activity, escape from senescence, growth to higher saturation density, and reduced growth factor requirements. More stringent features, which indicate further progress on the pathway to a fully transformed state, are anchorage-independent growth and the induction of tumors in an animal model. Primary human keratinocytes expressing the high-risk virus types, e.g., HPV-16 or -18, generally satisfy the less stringent criteria but are only rarely anchorage independent or tumorigenic.

Changes in the expression of keratins are definitive markers of alterations in the differentiation program of epithelial cells. For example, in studies on HPV-18-immortalized cells we observed down-regulation of keratins K6a and K14 (KAUR and MCDOUGALL 1989) in comparison to normal non-transfected keratinocytes. Other keratins reduced in expression in HPV-16-immortalized cells in monolayer are K5, K6, K16, and K17, while K7, K8, and K19 are increased (AGARWAL et al. 1991). The immortalized cell lines are also resistant to differentiation stimuli. After prolonged culture in 1.2 m*M* calcium-containing medium, normal keratinocytes do not express Ki67, a nuclear antigen expressed only in proliferating cells (GERDES et al. 1983). In contrast, HPV-immortalized cells grown in 1.2 m*M* calcium continue to express Ki67 and, unlike the normal controls, fail to stratify in most instances. These results are consistent with other reports documenting the resistance of transformed human epithelial cells to differentiation signals such as calcium and TPA. In total, these characteristics are an indication that HPV-immortalized cells are frequently blocked at an early stage of maturation which, combined with a loss of cell cycle control, provides an initial step towards malignant transformation.

5 Differentiation of HPV-Immortalized Cells

The reported high incidence of infection with HPV without the onset of malignant disease (DE VILLIERS et al. 1987) and an apparent long delay between initial infection and the development of malignancy (ZUR HAUSEN

1989) are evidence for the hypothesis that, while these viruses may induce dysplasia and thereby play a role in oncogenesis, other factors are required for fully developed malignancy. HPV-immortalized cells might be expected to mimic the dysplastic phenotype in vitro. The use of an organotypic culture system provides an in vitro model of epithelial cell differentiation (ASSELINEAU et al. 1986) enabling evaluation of the characteristics of the cell lines (REGNIER et al. 1988; STOLER et al. 1988). Studies by MCCANCE et al. (1988) and HUDSON et al. (1990) showed that introduction of HPV-16 or -18 sequences into normal foreskin keratinocytes or an established squamous carcinoma cell line resulted in a loss of differentiation markers in organotypic or "raft" culture.

We analyzed seven HPV-immortalized epithelial cell lines at various passage levels in the epidermal raft system (BLANTON et al. 1991). The cell lines were all derived from calcium phosphate–DNA precipitation on foreskin keratinocytes using HPV-16 or -18 DNA. In comparison with normal keratinocytes, all of the HPV cell lines were abnormal and presented features reminiscent of dysplastic lesions in vivo. All had disorganized tissue architecture, abnormal mitoses, mitotic cells in all stratified layers of viable cells, and enlarged nuclei. In general, these abnormalities were maintained through

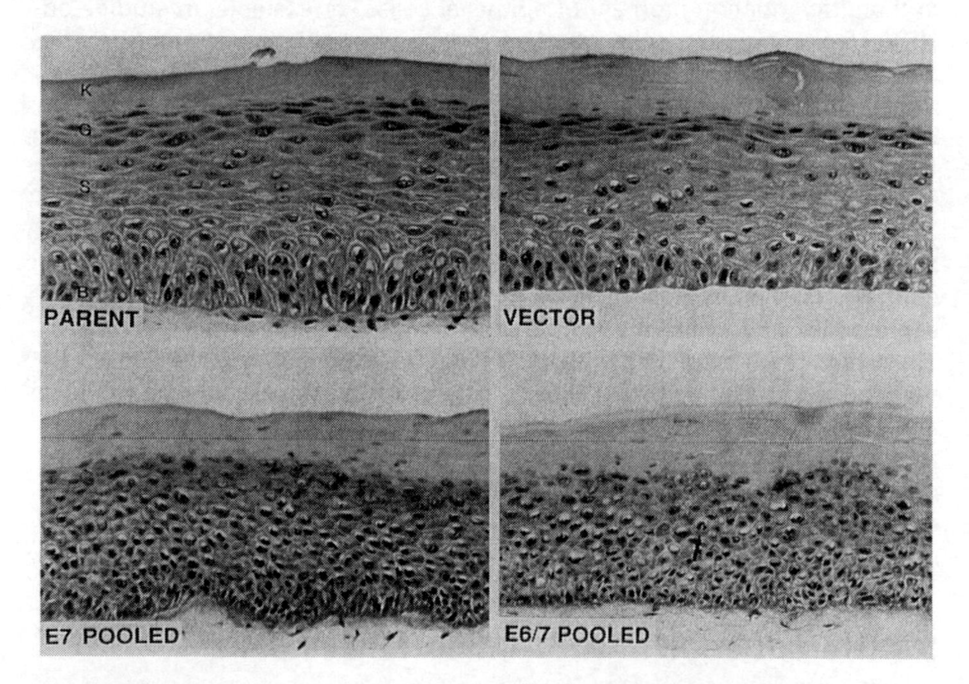

Fig. 1. Sections of organotypic cultures of human epithelial cells stained with hematoxylin and eosin. *PARENT* normal differentiation from basal (*B*) through spinous (*S*), and granular (*G*) layers to keratinization and cornified layer (*K*). Cells from the same primary culture were infected with either the LXSN vector, LXSNE7, or LXSNE6/E7. Expression of the HPV-16 sequences resulted in a dysplastic epithelium but did not inhibit terminal differentiation

multiple passages, but two of the cell lines developed more extreme changes after long-term passage in monolayer culture with the result that none of the characteristics of terminal squamous differentiation was apparent. These experiments demonstrate that HPV-immortalized lines can retain differentiation properties with passage in culture, as has been shown for non-HPV-containing squamous cell carcinoma cell lines (BOUKAMP et al. 1985), and that a subset may progress further on the path to abnormality. These results were obtained using serum-free medium, some variation involving a more rapid loss of differentiation characteristics has been reported in serum-containing medium (MCCANCE et al. 1988). Because the cell lines had been derived through a selection system and therefore in culture for some time, it was difficult to determine which changes were the consequence of HPV gene expression and which might have been acquired over time in culture. In an attempt to answer this question, we used retroviral constructs expressing HPV-16 E6 and E7 genes, separately or together, under the control of a strong promoter (HALBERT et al. 1991). In this way we could define cellular changes occurring in the earliest passages after viral gene expression began. The results from both ectocervical epithelial cells and skin keratinocytes infected with the amphotropic viral constructs were identical (Fig. 1) and clearly demonstrated that expression of the E7 gene or a combination of E6/E7 prevents cells from withdrawing from the cell cycle, consequently mitoses are found in the suprabasal cell layers, but does not inhibit stratification, maturation, or terminal differentiation (BLANTON et al. 1992).

5.1 p53 in E7-Immortalized Cells

Organotypic cultures expressing both E6 and E7 were identical to those with E7 alone, with the exception that p53 was undetectable in both the basal and suprabasal cell layers, as measured by immunohistochemical staining (Fig. 2). Binding and degradation of p53 by HPV-16 E6 has been demonstrated in tissue extracts (SCHEFFNER et al. 1990), the result described by BLANTON et al. (1992) is in agreement with that finding. In contrast, cultures expressing only the HPV-16 E7 gene had nuclear staining for p53 in all layers of the raft, and cultures of normal cells, without HPV genes, were positive in the basal layer only. Therefore stable expression of p53 was detectable in the proliferating layers of both normal and E7-containing cells (Fig. 3). A conclusion from these findings is that proliferation induced by E7 is independent of p53 level, perhaps similar to the situation in E7-transformed NIH 3T3 cells where introduction of a wild-type p53 has no effect on transformation (CROOK et al. 1991a, b). Nevertheless, the lower frequency of immortalization of human cells by E7, compared to E6 and E7, suggests that at least one further genetic change is required to establish immortalization of cells expressing E7 alone. One obvious possibility is that the p53 gene is mutated spontaneously, but it has been shown that in a sample of three E7-immortal-

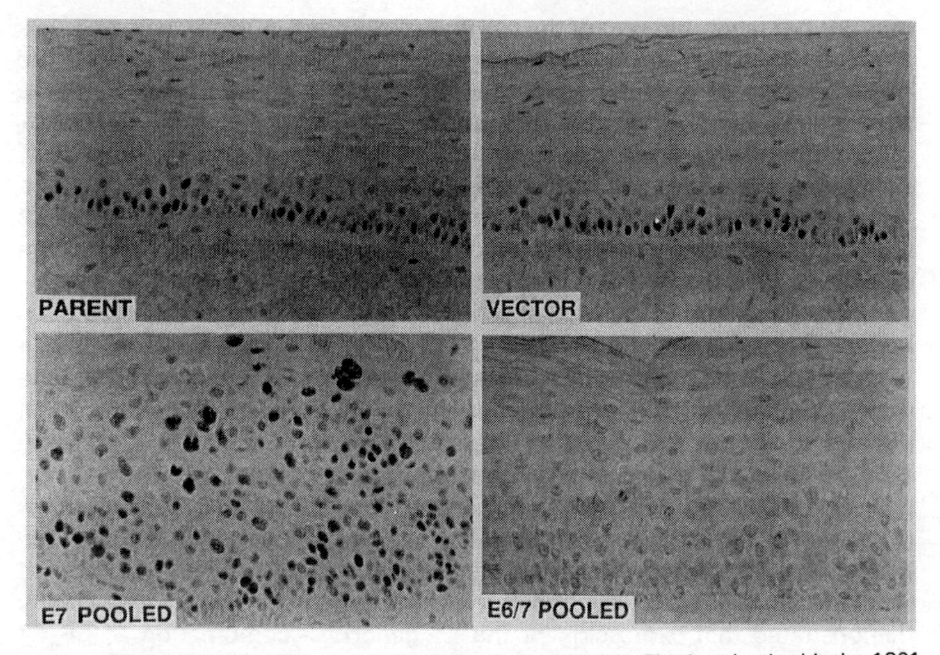

Fig. 2. Serial sections from organotypic cultures as shown in Fig. 1 stained with the 1801 antibody to p53

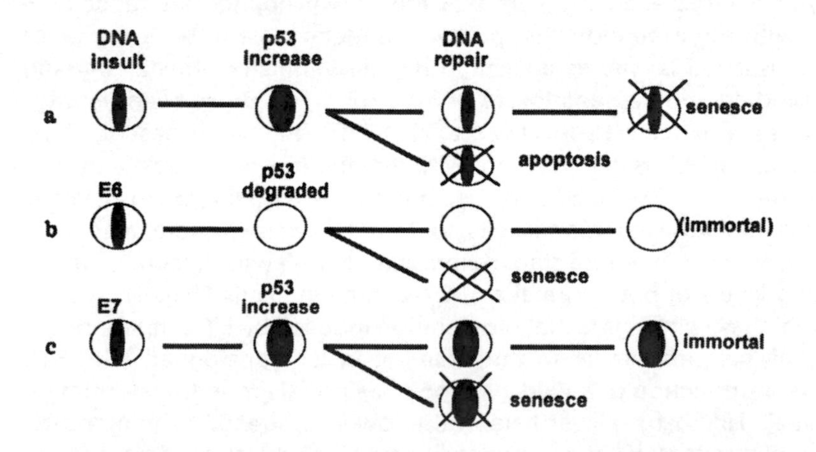

Fig. 3 a–c. The effect on p53 expression and cell survival of (**a**) a DNA-damaging insult; (**b**) expression of HPV-16 E6; and (**c**) expression of HPV-16 E7. The only human cell type shown to be immortalized by E6 alone is of mammary origin (BAND et al. 1991). The p53 increase observed with E7-expressing cells may represent increased stability of the protein

ized keratinocyte lines only one had mutant p53 detected (W. Demers and D.A. Galloway, personal communication). This result was of interest since it has been reported that HPV-negative cervical carcinoma cell lines have mutant p53 (WREDE et al. 1991; CROOK et al. 1991), therefore one might have expected that E6-negative cell lines would be similar. Examination of seven primary tumors negative for HPV by FUJITA et al. (1992) did not, however, provide support for an hypothesis that HPV-negative tumors would be expected to have mutant p53 genes.

5.2 Retinoic Acid

Retinoic acid and HPV influence similar processes in epithelial cells, including inhibition of normal growth and differentiation (ROBERTS and SPORN 1984). Retinoic acid has also been used to inhibit or treat squamous cell carcinoma (JETTEN et al. 1990; LIPPMAN et al. 1992). The effects of retinoic acid on HPV-immortalized cells have been studied in both monolayer (AGARWAL et al. 1991; PIRISI et al. 1992) and organotypic culture (FUCHS and GREEN 1981; KOPAN et al. 1987; MERRICK et al. 1993). Ten- to thirty-fold higher concentrations of retinoic acid were required to block terminal differentiation in organotypic cultures of HPV-immortalized human keratinocytes compared to control cells (MERRICK et al. 1993). The expression of differentiation specific markers, e.g., keratin K1, was shown by Northern analysis to be maintained at the 30-fold higher concentration of retinoic acid in HPV-16 and HPV-18 cell lines. The same result was found in cells early after infection with an amphotropic virus expressing HPV-16 E6/E7 genes, indicating that the retinoic acid-resistant phenotype is induced by expression of HPV genes. No significant differences in expression of the retinoic acid receptors or binding proteins were recorded in these experiments. Although the retinoic acid-resistant phenotype was a reproducible characteristic of all the HPV cell lines studied, it contrasts with some results reported previously showing sensitivity to retinoic acid (AGARWAL et al. 1991; PIRISI et al. 1992). One explanation could be that the growth-inhibitory properties of retinoic acid are mediated via a different pathway than that operating in the differentiation response.

6 Cooperating Factors in Transformation

Cervical carcinoma tissues have been reported to have increased expression of c-myc and ras oncogenes (RIOU et al. 1985, 1987; OCADIZ et al. 1987), with gene rearrangements and amplification described in a significant proportion of tumors. Increased expression of c-myc was observed in cell lines from cervical carcinomas, e.g., HeLa and C4-1 cells, by DÜRST et al. (1987b) who proposed that HPV integration in the region of c-myc might result in *cis*-activation of the oncogene with progression to malignancy. Despite the attractiveness of this hypothesis, a study by IKENBERG et al. (1987) found no amplification of c-myc in a study of 28 HPV-positive invasive genital tumors.

Transformation to malignancy has been achieved by cooperation between activated ras and HPV-16 in primary rodent cells (MATLASHEWSKI et al. 1987) and in human cells (DÜRST et al. 1989). These results demonstrate the similarity in function between adenovirus E1A and the HPV-16 E7 gene, a similarity predicted at the amino acid sequence level (PHELPS et al. 1988) of the conserved domains 1 and 2 of adenovirus E1A polypeptides, the papovavirus large T antigen, myc, and yeast CDC 25. The sequences shared between E7 and adenovirus E1A are those shown to be involved in the induction of host cell DNA synthesis, cooperation with ras for full transformation, and the repression of transcription.

Based upon previous studies proposing a role for herpes simplex virus type 2 (HSV-2) in the etiology of cervical carcinoma, DIPAOLO et al. (1990) examined the possibility that a subgenomic region of HSV-2 DNA might act as a cofactor and convert HPV-immortalized cells to malignancy. Transfection of the Bgl II N region into HPV-16-immortalized cells converted them to tumorigenic squamous cell carcinoma cell lines, but, as described previously for HSV-2 transformation experiments, the Bgl II N sequences were eventually lost. The possibility that other sexually transmitted disease agents, e.g., chlamydia, cytomegalovirus, human immunodeficiency virus, in addition to SV-2, may interact with or facilitate progression of HPV-induced lesions is worth further consideration.

The effects of hormones on the transformation process have been examined in a number of studies. It was shown by DÜRST et al. (1989) that glucocorticoid enhanced the transformation of HPV-immortalized human epithelial cells by an activated ras oncogene. In a study of primary baby rat kidney cells, PATER and PATER (1991) showed that hormone-dependent transformation by HPV-16 and ras could be inhibited by a hormone antagonist. Glucocorticoid has also been shown to affect the transcription rate of HPV E6/E7 genes (GLOSS et al. 1987), and results from VON KNEBEL DOEBERITZ et al. (1991) suggest that the efficiency of glucocorticoid regulation of transcription may depend upon the chromosomal integration site of the viral sequences. Of four cervical carcinoma cell lines examined, two had increased transcription rates of integrated E6/E7 genes, one was unaltered, and the other decreased after treatment with dexamethasone.

Numerous studies have indicated that smoking may be an additional risk factor in the development of anogenital cancer (HOLLY et al. 1989; WINKELSTEIN 1990). The presence of carcinogens in the cervical secretions of women with cervical dysplasia has been noted (MCCANN and IRWIN 1992), and particular attention has been given to tobacco-specific nitrosamines (FISCHER et al. 1990). A study by GARRETT et al. (1993) used a model in vitro system to examine the potential of nitrosomethylurea (NMU) to convert HPV-immortalized human cells to malignancy. Animal studies have provided evidence for the co-carcinogenic effects of papillomaviruses and chemical carcinogens in malignant progression (ROUS and FRIEDWALD 1944; JARRETT 1980). A human keratinocyte cell line, previously immortalized with HPV-18 DNA by transfection, was treated with NMU at a very low dose (250 ng/1 million cells) and TPA before subcutaneous inoculation into nude mice. Cells with HPV DNA, NMU, and TPA treatment induced squamous cell carcinoma, omission of any one of these exposures resulted in no tumor induction. This in vitro model thus provides a method to analyze the mechanisms of interaction between a virus and chemical carcinogens that may be of importance in human disease.

7 Progression to Malignancy

As the evidence for an association between HPV and some human tumors becomes irrefutable, so does the hypothesis that other changes, not necessarily directly virus-mediated, are required for the expression of malignancy. In general, HPV-immortalized human cell lines are not tumorigenic when inoculated into the nude mouse, the exceptions have occurred after extensive in vitro passaging (PECORARO et al. 1991; HURLIN et al. 1991). Some cell lines which do not progress to form invasive tumors have produced cysts after inoculation into nude mice (DÜRST et al. 1991). The encapsulating membrane of the cyst is formed of human epithelial cells differentiating towards the center. Many cysts formed in this manner eventually regress.

Inoculation of cells from successive passages of an HPV-18-immortalized foreskin keratinocyte cell line resulted in the formation of invasive tumors only after the cells had been passaged more than 60 times (HURLIN et al. 1991). Prior to 30 in vitro passages there was no evidence of tumor or cyst development but at levels between approximately 30–60 passages there was in vivo growth of the human cells, forming nodules which eventually regressed. No change in copy number of the integrated HPV-18 genome or in the level of expression of the viral gene products was detected between tumorigenic and non-tumorigenic cells. Analysis of cells at early, middle, and late passage revealed cytogenetic changes that segregated with cell passage and tumor induction (see below). Organotypic cultures performed in parallel with monolayer passaging showed that the cell line became progressively more dysplastic and disorganized until in later cultures there was no stratification or cornification. The tumors formed were poorly differentiated squamous cell carcinomas which were transplantable and invasive into muscle and adipose tissues. Similar findings were reported by PECORARO et al. (1991) for HPV-16- and -18-immortalized cervical epithelial cells, and these cell lines were used in an attempt to establish a genetic basis for progression (CHEN et al. 1993). Somatic cell hybrids were made between an HPV-18-immortalized human cell line and normal human epithelial or fibroblast cells and only one of 19 clones retained an immortal phenotype. This was not related to a loss of viral gene expression. Further conclusions, based upon fusion of HPV-immortalized cells with each other, were that an increase in expression of p53 and Rb was not responsible for the loss of immortality and that inactivation of these tumor suppressor genes by HPV was not sufficient for immortalization. These results are consistent with the hypothesis that the phenotypes examined as characteristic of HPV immortalization or transformation are recessive and are not a direct consequence of HPV gene expression. Activation of cellular oncogenes or loss of heterozygosity of tumor suppressor genes can be important steps in the progression to malignancy; further study may indicate the contribution of these factors.

8 Chromosome Aberrations in Immortalized and Tumor Cells

Human papillomavirus immortalized cells and cell lines derived from anogenital carcinoma tissues generally have aneuploid karyotypes (DÜRST et al. 1987a; SMITH et al. 1989; TEYSSIER 1989), with both structural and numerical abnormalities. Cytogenetic abnormalities have also been observed in direct preparations of cervical carcinoma tissue. The aberrations recorded from numerous studies suggest a random effect with limited evidence for non-random changes on chromosomes 1 and 3 (TEYSSIER 1989; ATKIN and BAKER 1982) and chromosome 5 (POPESCU et al. 1987a, b; JAMES et al. 1989; HERZ et al. 1977). HASHIDA and YASUMOTO (1991) investigated the relationship between HPV-16 E6 and E7 expression and changes in ploidy induced in mouse and human keratinocytes. The conclusion from this study was that development of aneuploidy was associated with the E7 gene and that only minor chromosomal change resulted from E6 expression. One difficulty with this approach is that the number of population doublings after DNA transfection and isolation of clones does not allow for the chromosomes to be examined very early after onset of viral gene expression. To avoid this problem my colleagues (A.J. Klingelhutz, P.P. Smith) and I used the gene transfer vectors described above to enable the examination of mass cultures from the second population doubling onwards. The results of these experiments indicated that until approximately 40 population doublings human keratinocytes expressing HPV-16 E6/E7, E6 alone, or E7 alone maintained an essentially diploid karyotype. After an apparent "crisis" period the cells were aneuploid and clonal cell lines isolated at this point varied in chromosomal constitution.

8.1 Integration

The question of specificity of integration of HPV sequences into host cell chromosomes of cells immortalized or transformed in vitro has been the topic of a number of studies. Integration of the viral sequences is consistently found in cell lines although this is not always the case in anogenital carcinoma. Many different integration sites have been recognized and some studies have indicated that these may coincide with proto-oncogene loci (COUTURIER et al. 1991) or with chromosomal fragile sites (POPESCU and DIPAOLO 1989; SMITH et al. 1992). If integration into host chromosomes is random, that is not the situation for the viral genome. The circular virus DNA molecule is opened between the E1 and E2 ORFs in cervical carcinoma cell lines (SCHWARZ et al. 1985; YEE et al. 1985) with frequent loss or mutation of ORF sequences other than those coding for the E6/E7 genes (KAUR et al. 1989).

8.2 Telomeres

An important factor in genetic instability associated with immortalization of cells may be the stability of telomere sequences. These sequences, in the form of TTAGGG repeats, are located at the end of chromosomes (ZAKIAN 1989) and may play an important role in cell senescence (GREIDER 1991). There is shortening of telomeres as a function of age (ALLSOPP et al. 1992) and as cells are passaged in vitro (HARLEY et al. 1990). An extension of the study of aneuploidy described above measured telomere length during passages prior to and after HPV-16 immortalization of foreskin and cervical epithelial cells (KLINGELHUTZ et al. 1993a). The telomeres of normal and pre-crisis HPV-expressing cells shortened until they senesced or reached crisis. Cells surviving crisis then showed a gradual recovery in telomere length which was maintained in the immortalized cell lines derived from this experiment. A key factor in the establishment of immortalized cells may be the reactivation of a telomerase-like enzyme. Passages early after crisis were mostly tetraploid with translocations and with some metaphases showing telomere fusions, later passages had only minimal further cytogenetic changes. Increased stability of the karyotype therefore appears to be associated with increased telomere length.

9 Tumor Suppressor Genes

The underlying theme of this chapter is that more changes are required for full transformation of human cells than those consequent on the binding of Rb by E7 or the degradation of p53 by E6. One mechanism which could contribute to the progression of cells to malignancy is the loss or mutation of other tumor suppressor genes. Cytogenetic analysis of the HPV-18-immortalized cells exposed to NMU, as described above, detected a new chromosome aberration in the tumor cells. The original cell line, identified as 1811 in SMITH et al. (1989), is essentially diploid with a number of abnormal chromosomes. Of two separate NMU transformants examined, one had a deletion of part of the long arm of chromosome 18 (18q-) and the other deletion of an entire chromosome 18 (18-). Neither of these aberrations was present in any other passage of clone of 1811. Since the DCC tumor suppressor gene maps to 18q (FEARON et al. 1990), RFLP analysis was used to demonstrate that both NMU-treated cell lines had lost one allele of the DCC gene (Fig. 4) (KLINGELHUTZ et al. 1993). Expression of DCC from the remaining allele was greatly reduced, consistent with the hypothesis that both alleles were affected in these cell lines. The DCC gene product is a cell surface protein which has homology to neural cell adhesion molecules (N-CAM) and could be important in cell adhesion, differentiation, and invasion. In experiments in rodent cells antisense RNA to DCC has been reported to inhibit adhesion and differentiation (NARAYANAN et al. 1992; LAWLOR and NARAYANAN 1992).

Studies with cell hybrids have indicated that the tumorigenic phenotype can be suppressed by a gene (or genes) located on chromosome 11 (SRIVATSAN et al. 1986; SAXON et al. 1986). Human embryonic fibroblasts with a deletion on the short arm of chromosome 11 have been reported to have increased transcription of HPV-16 early genes and to be transformed by HPV-16, whereas normal diploid fibroblasts are resistant (SMITS et al. 1988). Direct suppression of E6/E7 gene expression has been shown in hybrids between HeLa and normal fibroblasts (BOSCH et al. 1990). A role for chromosome 11 in intracellular surveillance of tumorigenicity by suppression of viral transcription has been proposed by ZUR HAUSEN (1991).

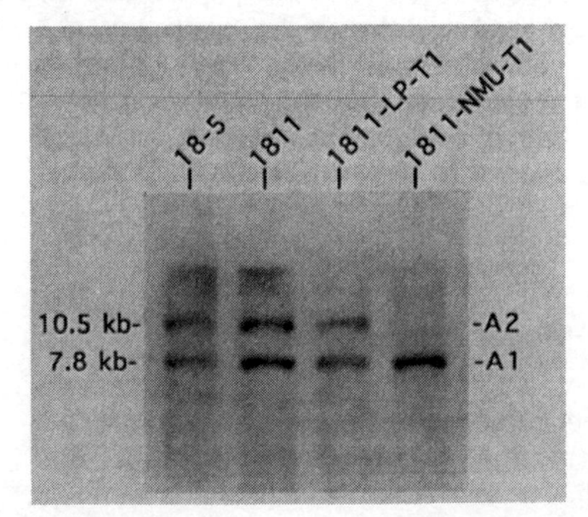

Fig. 4. Allelic deletion of DCC gene in NMU-transformed 1811 cell line. DNA digested with Msp I and hybridized with p15-65 DCC probe. Cell line 18-5 was derived from same primary human keratinocyte cell batch as 1811. Described in KLINGELHUTZ et al. (1993)

10 Conclusions

The property of immortalization of primary human cells parallels the association of HPVs with anogenital neoplasia. Expression of two early genes is sufficient for immortalization in vitro, and these same genes are consistently expressed in tumors positive for HPV. The major, perhaps the only significant, effect of E6/E7 is to abrogate cell cycle controls, thereby initiating a process through which further changes accumulate leading to a malignant phenotype. A role for some HPV types in the causation of anogenital cancer is supported by the results of in vitro immortalization studies.

Acknowledgments. Studies from this laboratory were supported by NIH grants CA42792 and AI29363. I thank Marci Wright for manuscript preparation.

References

Agarwal C, Rorke EA, Irwin JC, Eckert RL (1991) Immortalization by human papillomavirus type 16 alters retinoid regulation of human ectocervical epithelial cell differentiation. Cancer Res 51:3982–3989

Allsopp RC, Vaziri H, Patterson C, Goldstein S, Younglai V, Futcher AB, Greider CW, Harley CB (1992) Telomere length predicts replicative capacity of human fibroblasts. Proc Natl Acad Sci USA 89:10114–10118

Asselineau D, Bernard BA, Bailly C, Darmon M, Prunieras M (1986) Human epidermis restructured by culture: Is it normal? J Invest Dermatol 86:181–186

Atkin NB, Baker MC (1982) Non-random chromosome changes in carcinoma of the cervix uteri. Cancer Genet Cytogenet 7:209–222

Band V, De Caprio JA, Delmolino L, Kulesa V, Sager R (1991) Loss of p53 protein in human papillomavirus type 16 E6-immortalized human mammary epithelial cells. J Virol 65:6671–6676

Banks L, Matleshewski G (1993) Cell transformation and the HPV E5 gene. Papillomavirus Rep 4:1–3

Bedell MA, Jones KH, Laimins LA (1987) The E6–E7 region of human papillomavirus type 18 is sufficient for transformation of NIH 3T3 and rat-1 cells. J Virol 61:3635–3640

Blanton RA, Perez-Reyes N, Merrick DT, McDougall JK (1991) Epithelial cells immortalized by human papillomaviruses have premalignant characteristics in organotypic culture. Am J Pathol 138:673–685

Blanton RA, Coltrera MD, Gown AM, Halbert CL, McDougall JK (1992) Expression of the HPV 16 E7 gene generates proliferation in stratified squamous cell cultures which is independent of endogenous p53 levels. Cell Growth Differ 3:791–802

Bosch FX, Schwarz E, Boukamp P, Fusenig NE, Bartsch D, zur Hausen H (1990) Suppression in vivo of human papillomavirus type 18 E6–E7 gene expression in nontumorigenic HeLa X fibroblast hybrid cells. J Virol 64:4743–4754

Boukamp P, Rupniak HT, Fusenig NE (1985) Environmental modulation of the expression of differentiation and malignancy in six human squamous cell carcinoma cell lines. Cancer Res 45:5582–5592

Chen SL, Mounts P (1990) Transforming activity of E5a protein of human papillomavirus type 6 in NIH 3T3 and C127 cells. J Virol 64:3226–3233

Chen T-M, Pecoraro G, Defendi V (1993) Genetic analysis of in vitro progression of human papillomavirus-transfected human cervical cells. Cancer Res 53: 1167–1171

Coles C, Condie A, Chetty U, Steel CM, Evans HJ, Prosser J (1992) p53 mutations in breast cancer. Cancer Res 52:5291–5298

Couturier J, Sastre-Garau X, Schneider-Maunoury S, Labib A, Orth G (1991) Integration of papillomavirus DNA near myc genes in genital carcinomas and its consequences for proto-oncogene expression. J Virol 65: 4534–4538
Crook T, Fisher C, Vousden KH (1991a) Modulation of immortalizing properties of human papillomavirus type 16 E7 by p53 expression. J Virol 65: 505–510
Crook T, Wrede D, Vousden KH (1991b) p53 point mutation in HPV negative human cervical carcinoma cell lines. Oncogene 6: 873–875
de Villiers EM, Wagner D, Schneider A, Wesch H, Miklaw H, Wahrendorf J, Papendick U, zur Hausen H (1987) Human papillomavirus infections in women with and without abnormal cervical cytology. Lancet 2: 703–706
DiPaolo JA, Woodworth CD, Popescu NC, Koval DL, Lopez JV, Doniger J (1990) HSV-2-induced tumorigenicity in HPV 16-immortalized human genital keratinocytes. Virology 177: 777–779
Dollard SC, Wilson JL, Demeter LM, Bonnez W, Reichman RC, Broker TR, Chow LT (1992) Production of human papillomavirus and modulation of the infectious progression in epithelial raft cultures. Genes Dev 6: 1131–1142
Dürst M, Dzarlieva-Petrusevska RT, Boukamp P, Fusenig NE, Gissmann L (1987a) Molecular and cytogenetic analysis of immortalized human primary keratinocytes obtained after transfection with human papillomavirus type 16 DNA. Oncogene 1: 251–256
Dürst M, Croce, CM, Gissmann L, Schwarz E, Huebner K (1987b) Papillomavirus sequences integrate near cellular oncogenes in some cervical carcinomas. Proc Natl Acad Sci USA 84: 1070–1074
Dürst M, Gallahan D, Jay G, Rhim JS (1989) Glucocorticoid-enhanced neoplastic transformation of human keratinocytes by human papillomavirus type 16 and an activated ras oncogene. Virology 173: 767–771
Dürst M, Bosch FX, Glitz D, Schneider A, zur Hausen H (1991) Inverse relationship between human papillomavirus (HPV) type 16 early gene expression and cell differentiation in nude mouse epithelial cysts and tumors induced by HPV-positive human cell lines. J Virol 65: 796–804
Fearon ER, Cho KR, Nigro JM, Kern SE, Simons JW, Ruppert JM, Hamilton SR, Preisinger AC, Thomas G, Kinzler KW, Vogelstein B (1990) Identification of a chromosome 18q gene that is altered in colorectal cancers. Science 247: 49–56
Fischer S, Spiegelhalder B, Eisenbarth J, Preussmann R (1990) Investigations on the origin of tobacco-specific nitrosamines in mainstream smoke of cigarettes. Carcinogenesis 11: 723–730
Fuchs E, Green H (1981) Regulation of terminal differentiation of cultured human keratinocytes by vitamin A. Cell 25: 617–625
Fujita M, Inoue M, Tanizawa O, Iwamoto S, Enomoto T (1992) Alterations of the p53 gene in human primary cervical carcinoma with and without human papillomavirus infection. Cancer Res 52: 5323–5328
Gallimore PH, Sharp PA, Sambrook J (1974) Viral DNA in transformed cells. II. A study of the sequences of adenovirus 2 DNA in nine lines of transformed rat cells using specific fragments of the viral genome. J Mol Biol 89: 49–72
Garrett LR, Perez-Reyes N, Smith PP, McDougall JK (1993) Interaction of HPV-18 and nitrosomethylurea in the induction of squamous cell carcinoma. Carcinogenesis 14: 329–332
Gerdes J, Schwab U, Lemke H, Stein H (1983) Production of a mouse monoclonal antibody reactive with a human nuclear antigen associated with cell proliferation. Int J Cancer 31: 13–20
Gloss B, Bernard HU, Seedorf K, Klock G (1987) The upstream regulatory region of the human papilloma virus-16 contains an E2 protein-independent enhancer which is specific for cervical carcinoma cells and regulated by glucocorticoid hormones. EMBO J 6: 3735–3743
Greider CW (1991) Telomeres. Curr Opin Cell Biol 3: 444–451
Hagensee M, Yaegashi N, Galloway DA (1993) Self-assembly of HPV-1 capsids by expression of the L1 protein alone or by co-expression of the L1 and L2 capsid proteins. J Virol 67: 315–322
Halbert CL, Demers GW, Galloway DA (1991) The E7 gene of human papillomavirus type 16 is sufficient for immortalization of human epithelial cells. J Virol 65: 473–478
Harley CB, Futcher AB, Greider CW (1990) Telomeres shorten during aging of human fibroblasts. Nature 345: 458–460

Hashida T, Yasumoto S (1991) Induction of chromosome abnormalities in mouse and human epidermal keratinocytes by the human papillomavirus type 16 E7 oncogene. J Gen Virol 72: 1569–1577
Heck DV, Yee CL, Howley PM, Münger K (1992) Efficiency of binding the retinoblastoma protein correlates with the transforming capacity of the E7 oncoproteins of the human papillomaviruses. Proc Natl Acad Sci USA 89: 4442–4446
Herz F, Miller OJ, Miller DA, Auersperg N, Koss LG (1977) Chromosome analysis and alkaline phosphatase of C41, a cell line of human cervical origin distinct from HeLa. J Cancer Res 37: 3209–3213
Holly EA, Whittemore AS, Aston DA, Ahn DK, Nickoloff BJ, Kristiansen JJ (1989) Anal cancer incidence: genital warts, anal fissure or fistula, hemorrhoids, and smoking. J Natl Cancer Inst 81: 1726–1731
Howley PM, Yang YC, Spalholz BA, Rabson MS (1986) Papillomavirus transforming functions. Ciba Found Symp 120: 39–52
Hudson JB, Bedell MA, McCance DJ, Laiminis LA (1990) Immortalization and altered differentiation of human keratinocytes in vitro by the E6 and E7 open reading frames of human papillomavirus type 18. J Virol 64: 519–526
Hurlin PJ, Kaur P, Smith PP, Perez-Reyes N, Blanton RA, McDougall JK (1991) Progression of human papillomavirus type 18-immortalized human keratinocytes to a malignant phenotype. Proc Natl Acad Sci USA 88: 570–574
Ikenberg H, Schwörer D, Pfleiderer A, Polack A (1987) Lack of c-myc gene amplification in genital tumours with different HPV status. Lancet 2: 577
James GK, Kalousek DK, Auersperg N (1989) Karyotypic analysis of two related cervical carcinoma cell lines that contain human papillomavirus type 18 DNA and express divergent differentiation. Cancer Genet Cytogenet 38: 53–60
Jarrett WF (1980) Bracken fern and papilloma virus in bovine alimentary cancer. Br Med Bull 36: 79–81
Jetten AM, Kim JS, Sacks PG, Rearick JI, Lotan D, Hong WK, Lotan R (1990) Inhibition of growth and squamous-cell differentiation markers in cultured human head and neck squamous carcinoma cells by β-all-trans retinoic acid. Int J Cancer 45: 195–202
Kaur P, McDougall JK (1988) Characterization of primary human keratinocytes transformed by human papillomavirus type 18. J Virol 62: 1917–1924
Kaur P, McDougall JK (1989) HPV-18 immortalization of human keratinocytes. Virology 173: 302–310
Kaur P, McDougall JK, Cone R (1989) Immortalization of primary human epithelial cells by cloned cervical carcinoma DNA containing human papillomavirus type 16 E6/E7 open reading frames. J Gen Virol 70: 1261–1266
Klingelhutz AJ, Barber SA, Smith PP, Dyer K, McDougall JK (1993a) Restoration of telomeres in HPV-immortalized human anogenital epithelial cells. Mol Cell Biol (in press)
Klingelhutz AJ, Smith PP, Garrett LR, McDougall JK (1993) Alteration of the DCC tumor-suppressor gene in tumorigenic HPV-18 immortalized human keratinocytes transformed by nitrosomethylurea. Oncogene 8: 95–99
Kopan R, Traska G, Fuchs E (1987) Retinoids as important regulators of terminal differentiation: examining keratin expression in individual epidermal cells at various stages of keratinization. J Cell Biol 105: 427–440
Kreider JW, Barlett GL (1985) Shope rabbit papilloma-carcinoma complex. A model system of HPV infections. Clin Dermatol 3: 20–26
Lawlor KG, Narayanan R (1992) Cell Growth Differ 3: 609–616
Leptak C, Ramon Kulke R, Horwitz BH, Riese DJ, Dotto GP, DiMaio D (1991) Tumorigenic transformation of murine keratinocytes by the E5 genes of bovine papillomavirus type 1 and human papillomavirus type 16. J Virol 65: 7078–7083
Lippman SM, Kavanagh JJ, Paredes-Espinoza M, Delgadillo-Madrueno F, Paredes-Casillas P, Hong WK, Holdener E, Krakoff IH (1992) 13-cis-retinoic acid plus interferon α-2a: highly active systemic therapy for squamous cell carcinoma of the cervix. J Natl Cancer Inst 84: 241–245
Malkin D, Li FP, Strong LC, Fraumeni JF Jr., Nelson CE, Kim DH, Kassel J, Gryka MA, Bischoff FZ, Tainsky MA (1990) Germ line p53 mutations in a familial syndrome of breast cancer, sarcomas, and other neoplasms. Science 250: 1233–1238
Martin P, Vass WC, Schiller JT, Lowy DR, Velu TJ (1989) The bovine papillomavirus E5

transforming protein can stimulate the transforming activity of EGF and CSF-1 receptors. Cell 59: 21–32
Matlashewski G, Schneider J, Banks L, Jones N, Murray A, Crawford L (1987) Human papillomavirus type 16 DNA cooperates with activated ras in transforming primary cells. EMBO J 6: 1741–1746
McCance DJ, Kopan R, Fuchs E, Laimins LA (1988) Human papillomavirus type 16 alters human epithelial cell differentiation in vitro. Proc Natl Acad Sci USA 85: 7169–7173
McCann MF, Irwin DE (1992) Nicotine and cotinine in the cervical mucus of smokers, passive smokers, and nonsmokers. Cancer Epidol Biomark Prev 1: 125–129
Merrick DT, Gown AM, Halbert CL, Blanton RA, McDougall JK (1993) HPV-immortalized keratinocytes are resistant to the effects of retinoic acid on terminal differentiation. Cell Growth Differ 4:831–840
Meyers C, Frattini MG, Hudson JB, Laimins LA (1992) Biosynthesis of human papillomavirus from a continuous cell line upon epithelial differentiation. Science 257: 971–973
Münger K, Phelps WC, Bubb V, Howley PM, Schlegel R (1989a) The E6 and E7 genes of the human papillomavirus type 16 together are necessary and sufficient for transformation of primary human keratinocytes. J Virol 63: 4417–4421
Münger K, Werness BA, Dyson N, Phelps WC, Harlow E, Howley PM (1989b) Complex formation of human papillomavirus E7 proteins with the retinoblastoma tumor suppressor gene product. EMBO J 8: 4099–4105
Narayanan R, Lawlor KG, Schaapveld RQJ, Cho KR, Vogelstein B, Bui-Van Tran P, Osborne MP, Telang NT (1992) Antisense RNA to the putative tumor-suppressor gene DCC transforms Rat-1 fibroblasts. Oncogene 7: 553–561
Ocadiz R, Sauceda R, Cruz M, Graef AM, Gariglio P (1987) High correlation between molecular alterations of the c-myc oncogene and carcinoma of the uterine cervix. Cancer Res 47: 4173–4177
Pater MM, Pater A (1991) RU486 inhibits glucocorticoid hormone-dependent oncogenesis by human papillomavirus type 16 DNA. Virology 183: 799–802
Pecoraro G, Lee M, Morgan D, Defendi V (1991) Evolution of in vitro transformation and tumorigenesis of HPV16 and HPV18 immortalized primary cervical epithelial cells. Am J Pathol 138: 1–8
Perez-Reyes N, Halbert CL, Smith PP, Benditt EP, McDougall JK (1992) Immortalization of primary human smooth muscle cells. Proc Natl Acad Sci USA 89: 1224–1228
Phelps WC, Yee CL, Münger K, Howley PM (1988) The human papillomavirus type 16 E7 gene encodes transactivation and transformation functions similar to those of adenovirus E1A. Cell 53: 539–547
Pim D, Collins M, Banks L (1992) Human papillomavirus type 16 E5 gene stimulates the transforming activity of the epidermal growth factor receptor. Oncogene 7: 27–32
Pirisi L, Yasumoto S, Feller M, Doniger J, DiPaolo JA (1987) Transformation of human fibroblasts and keratinocytes with human papillomavirus type 16 DNA. J Virol 61: 1061–1066
Pirisi L, Batova A, Jenkins GR, Hodam JR, Creek KE (1992) Increased sensitivity of human keratinocytes immortalized by human papillomavirus type 16 DNA to growth control by retinoids. Cancer Res 52: 187–193
Popescu NC, DiPaolo JA (1989) Preferential sites for viral integration on mammalian genome. Cancer Genet Cytogenet 42: 157–171
Popescu NC, DiPaolo JA, Amsbaugh SC (1987a) Integration sites of human papillomavirus 18 DNA sequences on HeLa cell chromosomes. Cytogenet Cell Genet 44: 58–62
Popescu NC, Amsbaugh SC, DiPaolo JA (1987b) Human papillomavirus type 18 DNA is integrated at a single chromosome site in cervical carcinoma cell line SW756. J Virol 61: 1682–1685
Regnier M, Desbas C, Bailly C, Darmon M (1988) Differentiation of normal and tumoral human keratinocytes cultured on dermis: reconstruction of either normal or tumoral architecture in vitro cell. Dev Biol 24: 625–632
Riou G, Barrois M, Dutronquay V, Orth M (1985) Presence of papillomavirus DNA sequences, amplification of c-myc and c-Ha-ras oncogenes and enhanced expression of c-myc in carcinomas of the uterine cervix. In: Howley P, Broker T (eds) Papillomaviruses. Liss, New York, pp 47–56
Riou G, Barrois M, Le MG, George M, LeDoussal V, Haie C (1987) C-myc proto-oncogene expression and prognosis in early carcinoma of the uterine cervix. Lancet 1: 761–763

Roberts AB, Sporn MB (1984) Cellular biology and biochemistry of retinoids. In: Sporn MB, Roberts AB, Goodman DS (eds) The retinoids vol 2. Academic, New York, pp 209–286
Rous P, Friedwald WF (1944) The effect of chemical carcinogens on virus-induced rabbit papillomas. J Exp Med 79: 511–522
Saxon PJ, Srivatsan ES, Stanbridge EJ (1986) Introduction of human chromosome 11 via microcell transfer controls tumorigenic expression of HeLa cells. EMBO J 5: 3461–3466
Scheffner M, Werness BA, Huibregtse JM, Levine AJ, Howley PM (1990) The E6 oncoprotein encoded by human papillomavirus types 16 and 18 promotes the degradation of p53. Cell 63: 1129–1136
Schiller JT, Vass WC, Vousden KH, Lowy DR (1986) E5 open reading frame of bovine papillomavirus type 1 encodes a transforming gene. J Virol 57: 1–6
Schlegel R, Phelps WC, Zhang YL, Barbosa M (1988) Quantitative keratinocyte assay detects two biological activities of human papillomavirus DNA and identifies viral types associated with cervical carcinoma. EMBO J 7: 3181–3187
Schwarz E, Freese UK, Gissmann L, Mayer W, Roggenbuck B, Stremlau A, zur Hausen H (1985) Structure and transcription of human papillomavirus sequences in cervical carcinoma cells. Nature 314: 111–114
Smith PP, Bryant EM, Kaur P, McDougall JK (1989) Cytogenetic analysis of eight human papillomavirus immortalized human keratinocyte cell lines. Int J Cancer 44: 1124–1131
Smith PP, Friedman CL, Bryant EM, McDougall JK (1992) Viral integration and fragile sites in HPV immortalized human keratinocyte cell lines. Genes Chrom Cancer 5: 150–172
Smits HL, Raadsheer E, Rood I, Mehendale S, Slater RM, van der Noordaa J, ter Schegget J (1988) Induction of anchorage-independent growth of human embryonic fibroblasts with a deletion in the short arm of chromosome 11 by human papillomavirus type 16 DNA. J Virol 62: 4538–4543
Srivatsan ES, Benedict WF, Stanbridge EJ (1986) Implication of chromosome 11 in the suppression of neoplastic expression in human cell hybrids. Cancer Res 46: 6174–6179
Stoler A, Kopan R, Duvic M, Fuchs E (1988) Use of monospecific antisera and cRNA probes to localize the major changes in keratin expression during normal and abnormal epidermal differentiation. J Cell Biol 107: 427–446
Teyssier JR (1989) The chromosomal analysis of human solid tumors. A triple challenge. Cancer Genet Cytogenet 37: 103–125
von Knebel Doeberitz M, Bauknecht T, Bartsch D, zur Hausen H (1991) Influence of chromosomal integration on glucocorticoid-regulated transcription of growth-stimulating papillomavirus genes E6 and E7 in cervical carcinoma cells. Proc Natl Acad Sci USA 88: 1411–1415
Watts SL, Phelps WC, Ostrow RS, Zachow KR, Faras AJ (1984) Cellular transformation by human papillomavirus DNA in vitro. Science 225: 634–636
Winkelstein W Jr (1990) Smoking and cervical cancer—current status: a review. Am J Epidemiol 131: 945–957
Woodworth CD, Doniger J, DiPaolo JA (1989) Immortalization of human foreskin keratinocytes by various human papillomavirus DNAs corresponds to their association with cervical carcinoma. J Virol 63: 159–164
Wrede D, Tidy JA, Crook T, Lane D, Vousden KH (1991) Expression of RB and p53 proteins in HPV-positive and HPV-negative cervical carcinoma cell lines. Mol Carcinog 4: 171–175
Yee C, Krishnan-Hewlett I, Baker CC, Schlegel R, Howley PM (1985) Presence and expression of human papillomavirus sequences in human cervical carcinoma cell lines. Am J Pathol 119: 361–366
Zakian VA (1989) Structure and function of telomeres. Annu Rev Genet 23: 579–604
Zhou J, Sun XY, Stenzel DJ, Frazer IH (1991) Expression of vaccinia recombinant HPV 16 L1 and L2 ORF proteins in epithelial cells is sufficient for assembly of HPV virion-like particles. Virology 185: 251–257
zur Hausen H (1989) Papillomaviruses in anogenital cancer as a model to understand the role of viruses in human cancers. Cancer Res 49: 4677–4681
zur Hausen H (1991) Human papillomaviruses in the pathogenesis of anogenital cancer. Virology 184: 9–13

Protein Phosphatase 2A and the Regulation of Human Papillomavirus Gene Activity

J. TER SCHEGGET and J. VAN DER NOORDAA

1 Introduction

Oncogenic human papillomavirus (HPV) types are considered to be causally involved in the pathogenesis of cervical cancer (ZUR HAUSEN 1991a). DNA of the oncogenic HPV types has been detected in more than 90% of cervical carcinomas (ZUR HAUSEN 1989). HPV-16 and HPV-18 are the most common HPV types present in cervical carcinomas and their E6 and E7 genes are able to immortalize but not to oncogenically transform primary human keratinocytes (PIRISI et al. 1988; WOODWORTH et al. 1989). Proliferation of cervical carcinoma cell lines appears to be dependent on E6/E7 expression, indicating an important role of the E6 and E7 genes in the maintenance of the oncogenic phenotype of these cancer cells (VON KNEBEL DOEBERITZ et al. 1991). Oncogenic conversion of HPV-16- and HPV-18-immortalized keratinocytes is probably dependent on specific host cell DNA modifications occuring only after a long time in in vitro cultivation (HURLIN et al. 1991; PECORARO et al. 1991). A minority of cervical intra-epithelial neoplasias,

University of Amsterdam, Faculty of Medicine, Department of Virology, Academic Medical Center, Meibergdreef 15, 1105 AZ Amsterdam, The Netherlands

causally associated with the same oncogenic HPV types as cervical cancer, develops into cancer after a long latency period of generally more than 10 years. This strongly suggests that the expression of the HPV E6/E7 genes is necessary but not sufficient for the maintenance of the malignant phenotype of cervical cancer cells (ZUR HAUSEN 1989; VOUSDEN 1989).

ZUR HAUSEN (1986) postulated a model in which the development of human cancer is proposed as a failure of host cell control of viral gene expression. The expression of the E6/E7 genes of the oncogenic HPV types probably results in chromosomal instability, possibly due to the inactivation of the products of two tumor suppressor genes, the retinoblastoma gene and the p53 gene. The cellular mutations as a result of this instability could be important in the pathogenesis of premalignant lesions and tumor development (ZUR HAUSEN 1991b).

2 Chromosome 11 Influences the Oncogenicity of Cervical Carcinoma-Derived Cell Lines

The importance of chromosome 11 in the development of cervical cancer has been indicated by cytogenetic studies (ATKIN and BAKER 1984, 1988; SREEKANTAIAH et al. 1991) and more convincingly by the suppression of the oncogenic phenotype of cervical carcinoma-derived cell lines (SiHa and HeLa) upon the reintroduction of a normal chromosome 11 into these cell lines (KOI et al. 1989; SAXON et al. 1986).

A suppressive effect of host cell genes on the oncogenic phenotype was earlier shown in somatic cell hybrids that were established by fusion of tumorigenic HeLa cells with normal human fibroblasts (see review by STANBRIDGE 1990). After addition of the demethylating agent 5-azacytidine to these cell hybrids, the HPV-18 transcription is down-regulated. Addition of 5-azacytidine to tumorigenic segregants, which had lost one chromosome 11, did not, however, result in the suppression of the viral transcription, suggesting that a locus on chromosome 11 is responsible for the transrepression of the viral transcription (RÖSL et al. 1988).

A subset of Wilms' tumors (WT), a malignancy of the kidney, is associated with mutations on the short arm of chromosome 11, suggesting a tumor suppressor on 11p13 (RICCARDI et al. 1980). In our laboratory we have generated evidence for the involvement of a suppressor of viral transcription located in the region of the short arm of chromosome 11 containing the WT locus (SMITS et al. 1988).

3 Chromosome 11 and the Regulation of HPV Transcription

Experiments performed in our laboratory indicated that human fibroblasts with a deletion in the short arm of chromosome 11 (11p11.11p15.1) (del-11 cells) could be transformed by HPV-16 DNA, in contrast to normal diploid human fibroblasts. Transformation could be established by the HPV-16 early region expressed from the homologous enhancer–promoter (long control region; LCR). The E2 gene contained a translation termination linker abolishing its transcription repressor function (SMITS et al. 1988).

The high susceptibility of del-11 cells, compared with diploid cells, to HPV-16-induced transformation was shown to be correlated with the strength of the HPV-16 enhancer–promoter (LCR). The HPV-16 LCR was shown to be active in del-11 fibroblasts and inactive in diploid human fibroblasts. Since the HPV-16 enhancer cloned upstream of the simian virus 40 (SV40) promoter is active in both cell types, the target for chromosome 11-regulated HPV transcription is most likely located in the HPV promoter region (SMITS et al. 1990). In bandshift experiments we have shown that the TATA-box is probably involved in the chromosome 11-mediated regulation of HPV-16 transcription (SMITS et al. 1993). Our hypothesis is that a tumor suppressor gene located between 11p11 and 11p15.1 (e.g., the WT1 tumor suppressor gene) suppresses the HPV-16 promoter in diploid cells so that these cells are resistent to HPV-16-induced transformation.

4 Activation of the HPV-16 Promoter by Deletion of Loci on the Short Arm of Human Chromosome 11 Can Be Mimicked by SV40 Small t Antigen

Simian virus 40 (SV40) codes for two transforming proteins, large T antigen (LT) and small t antigen (ST), which are derived from two alternatively spliced transcripts of the SV40 early region. LT and ST contain identical amino terminal regions and unique carboxy termini (GRIFFIN 1980). In most cell systems LT has been previously shown to be sufficient to initiate and maintain SV40-induced transformation. SV40 ST will exert an auxillary transformation function under certain conditions, e.g., in cells arrested in Go (BIKEL et al. 1987). ST has been shown to enhance transformation when limiting concentrations of LT are present (BIKEL et al. 1987).

Transformation experiments in our laboratory have shown that SV40 ST is essential for SV40-induced transformation of human diploid fibroblasts (DE RONDE et al. 1989). More recently we have shown that SV40 ST is dispensable for SV40-induced transformation of del-11 cells (SMITS et al.

1992a). We concluded from these experiments that del-11 cells provide a cellular factor that functionally mimics SV40 ST by complementing SV40 LT in cell transformation.

The question we asked was: is this "SV40 ST-like factor" present in del-11 cells also modulating the HPV-16 promoter? Since LOEKEN et al. (1988) had shown that in rodent cells SV40 ST is able to transactivate selected RNA polymerase II and III promoters, we investigated whether SV40 ST could transactivate the HPV-16 promoter. For this reason we performed transient expression experiments using chloramphenicol acetyl transferase (CAT) as reporter gene cloned downstream of the HPV-16 enhancer–promoter. SV40 ST clearly transactivated the HPV-16 enhancer–promoter in these experiments (SMITS et al. 1992b).

We tentatively concluded that, as a result of the deletion of loci on the short arm of chromosome 11 in del-11 cells, a cellular analogue of SV40 ST is activated or induced. This ST-like factor present in del-11 cells is as well able to transactivate the HPV-16 enhancer–promoter as to complement SV40 LT in cell transformation. The activation of the HPV-16 enhancer–promoter results in susceptibility of del-11 cells to transformation by HPV-16 DNA containing the HPV-16 enhancer–promoter since the HPV-16 E6/E7 genes are strongly expressed. The next question was: what is the nature of this cellular SV40 ST-like factor?

5 SV40 ST Binds to Protein Phosphatase 2A

Polyomavirus medium T and the ST antigens of SV40 and polyomavirus form complexes with the serine/threonine-specific protein phosphatase 2A (PP2A). SV40 LT does not associate with PP2A (PALLAS et al. 1990; WALTER et al. 1990).

The enhancing effect of ST on transformation is possibly mediated by the effect on PP2A. PP2A consists at least of three subunits, the 36-kDa catalytic subunit (C) and two additional subunits of 65 and 55 kDa, also termed subunits A and B, respectively (MAYER et al. 1991; COHEN 1989). The 65- and 55-kDa subunits do not have phosphatase activity but probably regulate the activity of the 36-kDa catalytic subunit. SV40 ST associates in vivo with the 36- and 65-dDa subunits (PALLAS et al. 1990). The purified 36/65-kDa heterodimer forms a complex in the absence of ST with the 55-kDa regulatory subunit of PP2A, PR55, or with a 72-kDa subunit (for review see COHEN 1989). PR55 is the best-studied regulatory subunit. It has been shown to reduce the phosphatase activity of the dimeric complex of C and A (YANG et al. 1991). By in vitro studies it has been shown that SV40 ST binds to purified PP2A through interaction with the A subunit and that the interaction inhibits enzyme activity, suggesting that ST and PR55 have similar effects on the PP2A enzyme activity (YANG et al. 1991). Recently the cDNAs of two isoforms of PR55 were sequenced, the α and β isoforms, coded for by two different genes. PR55α expression had the characteristics of a housekeeping gene, PR55β was found to be highly expressed in a cell line of neuronal origin, the neuroblastoma-derived cell line LA-N-1. The analysis of PR55β mRNA in porcine tissues revealed only expression in brain (MAYER et al. 1991). The expression in human tissues has not been studied.

Since SV40 ST and PR55β are functionally related, we hypothesized that PR55β could be the ST-like factor in del-11 cells. Therefore we investigated by Northern blot hybridization the mRNA levels of PR55α and PR55β in diploid and del-11 cells by using two gene-specific probes. The steady-state level of PR55β mRNA was much higher in del-11 cells than in diploid cells thus supporting our hypothesis. In contrast, PR55α mRNA was highly expressed in both cell types, in accordance with its nature as a housekeeping gene (SMITS et al. 1992b).

A plasmid encoding PR55β, like SV40 ST, was able to transactivate the HPV-16 enhancer–promoter, also in accordance with our assumption that PR55β represents the ST-like factor. Small t antigen inhibits dephosphorylation of p53, Rb, and LT (for review see FUJIKI 1992). Since the hyperphosphorylated form of the Rb tumor suppressor protein is less active as growth-suppressing protein, it is possible that the inhibition of the dephosphorylation of this protein is relevant to the accessory function of ST in transformation (SCHEIDTMANN et al. 1991). We are currently investigating the phosphorylation state of these tumor suppressor proteins in del-11 cells and diploid fibroblasts.

6 Okadaic Acid Inhibits PP2A

Okadaic acid (OA) is a polyether compound of a 38-carbon fatty acid, isolated from a black sponge, *Halichondria okadai* and a potent mouse skin tumor promoter (FUJIKI 1992). OA is a potent inhibitor of protein phosphatase 2A activity (HESCHELER et al. 1988) and probably exerts its effect by binding to the 36-kDa catalytic subunit. At high concentration it also inhibits protein phosphatase 1.

If the previously described transcriptional activity of the HPV-16 enhancer–promoter in del-11 cells is mediated by inhibition of the enzyme activity of PP2A by PR55β, then inhibition of PP2A in diploid cells by OA should also result in transactivation of the HPV-16 enhancer–promoter. OA (100 n*M*) resulted in transactivation of the HPV-16 enhancer–promoter, further strengthening our hypothesis (SMITS et al. 1992b).

Okadaic acid causes sustained and increased phosphorylation of proteins induced by various serine/threonine protein kinases. At present, expression of various widely different genes has been reported to be stimulated by OA: e.g., c-fos, c-jun, human collagenase, the human immunodeficiency virus long terminal repeat (FUJIKI 1992). AP-1, a family of transcription factors consisting of heterodimers of the fos and jun family and the AP-1 consensus sequence is required for the induction of the collagenase gene (KIM et al. 1990). Furthermore, it has been shown that activation of the promoter of HPV-16 by phorbol esters, which lead to the induction of protein kinase C, is mediated by AP-1 (CHAN et al. 1990). Therefore it would be interesting to investigate whether the transactivation of the HPV-16 enhancer–promoter in del-11 cells and in diploid cells by OA is also mediated by AP-1. We have shown that the HPV-16 enhancer, which contains AP-1 sites, was as active in diploid as in del-11 cells when cloned upstream of the SV40 promoter (SMITS et al. 1990). This indicates that the activation of the HPV-16 LCR is not only mediated by AP-1 sites in the enhancer. The previously mentioned bandshift experiments indicated that the regulation of HPV-16 transcription in del-11 cells is also mediated by the TATAAA motif of the HPV-16 promoter (SMITS et al. 1993).

7 Effects of Mutations in a DKGG Motif Common to SV40 ST and the PR55β Subunit of Protein Phosphatase 2A

As both SV40 ST and the PR55β subunit of PP2A transactivate the HPV-16 LCR and bind to the dimeric 36/63-kDa complex of PP2A in a mutually exclusive manner, we supposed that structural homology could exist between these proteins. Comparison of the primary amino acid sequence did not show extensive homology, as also observed by others (PALLAS et al. 1990). However, a stretch of four amino acids (DKGG), appeared to be present at 42–43 amino acids downstream from the first methionine in both proteins. The DKGG motif is also present in polyoma middle T and ST, which also has been shown to bind the dimeric form of PP2A (WALTER et al. 1990; PALLAS et al. 1990). We mutated this motif in SV40 ST to KKGG, DKGR, and EKGG, and observed that each of these mutant STs had lost the ability to transactivate the HPV-16 LCR. This lack of transacting ability appeared not to be due to the instability of these mutant proteins (SMITS et al. 1992b). As the motif is also present in SV40 LT, which does not bind to PP2A, the DKGG motif is not sufficient for binding. We are currently investigating whether this motif is involved in the binding to PP2A.

8 Conclusion

The results described in this review strongly suggest that deletion of loci on the short arm of human chromosome 11 leads to activation of the HPV-16 LCR. This activation is mediated by inactivation of PP2A as a result of high expression of PR55β. It favors the concept that phosphorylation of cellular proteins, e.g., transcription factors, is involved in the regulation of the HPV-16 promoter. Recent evidence also indicates that several transcription factors are induced by OA, such as the cellular immediate-early genes c-fos and EGR-1 (CAO et al. 1992), which rapidly respond to mitogenic and differentiative stimuli. Interestingly, the WT locus zinc finger protein might act in an antagonistic manner to EGR-1 by binding to the same DNA recognition sequence (RAUSCHER et al. 1990). It will be interesting to compare the effects of the products of the WT-1 gene, which is located at chromosome 11p13, and EGR-1 on HPV transcription.

Acknowledgment. The authors like to thank Dr. Henk L. Smits for critical reading of the manuscript.

References

Atkin NB, Baker MC (1984) Non-random chromosome changes in carcinoma of the cervix uteri I Nine near-diploid tumours. Cancer Genet Cytogenet 47: 106–107

Atkin NB, Baker MC (1988) Deficiency of all or part of chromosome 11 in several types of cancer: significance of the reduction in the number of normal chromosomes 11. Cytogenet Cell Genet 47: 106–107

Bikel I, Montano X, Agha ME, Brown M, McCormack M, Boltax J, Livingston DM (1987) SV40 small t antigen enhances the transformation activity of limiting concentrations of SV40 large T antigen. Cell 48: 321–330

Brown M, McCormack M, Zinn KG, Farrell MP, Bikel I, Livingston DM (1986) A recombinant murine retrovirus for simian virus 40 large T cDNA transforms mouse fibroblasts to anchorage-independent growth. J Virol 60: 290–293

Cao X, Mahendran R, Guy GR, Tan YH (1992) Protein phosphatase inhibitors induce the sustained expression of the Egr-1 gene and the hyperphosphorylation of its gene product J Biol Chem 267: 12991–12997

Chan WK, Chong T, Bernard HU, Klock G (1990) Transcription of the transforming genes of the oncogenic human papillomavirus-16 is stimulated by tumor promotors through AP1 binding sites. Nucleic Acids Res 18: 763–769

Cohen P (1989) The structure and regulation of protein phosphatases. Annu Rev Biochem 58: 453–508

de Ronde A, Sol CJA, van Strien A, ter Schegget J, van der Noordaa J (1989) The SV40 small t antigen is essential for the morphological transformation of human fibroblasts. Virology 171: 260–263

Fujiki H (1992) Is the inhibition of protein phosphatase 1 and 2A activities a general mechanism of tumor promotion in human cancer development? Mol Carcinog 5: 91–94

Griffin BE (1980) Structure and genetic organization of SV40 and polyoma virus. In: Tooze J (ed) Molecular biology of tumor viruses, part 2: DNA tumor viruses, 2nd edn. Cold Spring Harbor Laboratory Press, Cold Spring Harbor, pp 61–123

Hescheler J, Mieskes G, Ruegg JG, Takai A, Trautwein W (1988) Effects of a protein phosphatase inhibitor, okadaic acid, on membrane currents of isolated guinea-pig cardiac myocytes. Pflugers Arch 412: 248–252
Hurlin PJ, Kaur P, Smith PP, Perez Reyes N, Blanton RA, McDougall JK (1991) Progression of human papillomavirus type 18-immortalized human keratinocytes to a malignant phenotype Proc Natl Acad Sci USA 88: 570–574
Kim SJ, Lafyatis R, Kim KY, Angel P, Fujiki H, Karin M, Sporn MB, Roberts AB (1990) Regulation of collagenase gene expression by okadaic acid, an inhibitor of protein phosphatases. Cell Regul 1: 269–278
Koi M, Morita H, Yamada H, Satoh H, Barrett JC, Oshimura M (1989) Normal human chromosome 11 suppresses tumorigenicity of human cervical tumor cell line SiHa. Mol Carcinog 2: 12–21
Kriegler M, Perez CF, Hardy C, Botchan G (1984) Transformation mediated by the SV40 T antigens: separation of the overlapping SV40 early genes with a retroviral vector. Cell 38: 483–491
Loeken M, Bikel I, Livingston DM, Brady J (1988) Trans-activation of RNA polymerase II and III promoters by SV40 small t antigen Cell 55: 1171–1177
Mayer RE, Hendrix P, Cron P, Matthies R, Stone SR, Goris J, Merlevede W, Hofsteenge J, Hemmings BA (1991) Structure of the 55-kDa regulatory subunit of protein phosphatase 2A: Evidence for a neuronal-specific isoform. Biochemistry 30: 3589–3597
Pallas DC, Sharik LK, Martin BL, Jaspers S, Miller TB, Brautigan DL, Roberts TM (1990) Polyoma small and middle T antigens and SV40 small t antigen form stable complexes with protein phosphatase 2A. Cell 60: 167–178
Pecoraro G, Lee M, Morgan D, Defendi V (1991) Evolution of in vitro transformation and tumorigenesis of HPV16 and HPV18 immortalized primary cervical epithelial cells. Am J Pathol 138: 1–8
Pirisi L, Creek KE, Doniger J, DiPaolo JA (1988) Continuous cell lines with altered growth and differentiation properties originate after transfection of human keratinocytes with human papillomavirus type 16 DNA. Carcinogenesis 9: 1573–1579
Rauscher III FJ, Morris JF, Tournay OE, Cook DM, Curran T (1990) Binding of the Wilms'tumor locus zinc finger protein to the EGR-1 consensus sequence. Science 250: 1259–1262
Riccardi VM, Hittner HM, Francke U, Yunis JJ, Ledbetter D, Borges W (1980) The aniridia-Wilms tumor association: the critical of chromosome 11p13. Cancer Genet Cytogenet 2: 131–137
Rösl F, Durst M, zur Hausen H (1988) Selective suppression of human papillomavirus transcription in non-tumorigenic cells by 5-azacytidine. EMBO J 7: 1321–1328
Saxon PJ, Srivatsan ES, Stanbridge EJ (1986) Introduction of human chromosome 11 via microcell transfer controls tumorigenic expression of HeLa cells. EMBO J 5: 3461–3466
Scheidtmann KH, Mumby MC, Rundell K, Walter G (1991) Dephosphorylation of Simian virus 40 large-T antigen and p53 protein by protein phosphatase 2A: inhibition by small-t antigen. Mol Cell Biol 11: 1996–2003
Smits HL, Raadsheer E, Rood I, Mehendale S, Slater RM, van der Noordaa J and ter Schegget J (1988) Induction of anchorage-independent growth of human embryonic fibroblasts with a deletion in the short arm of chromosome 11 by human papillomavirus type 16 DNA. J Virol 62: 4538–4543
Smits PH, Smits HL, Jebbink MF, ter Schegget J (1990) The short arm of chromosome 11 likely is involved in the regulation of the human papillomavirus type 16 early enhancer–promoter and in the suppression of the transforming activity of the viral DNA. Virology 176: 158–165
Smits PHM, de Ronde A, Smits HL, Minnaar RP, van der Noordaa J, ter Schegget J (1992a) Modulation of the human papillomavirus type 16 induced transformation and transcription by deletion of loci on the short arm of human chromosome 11 can be mimicked by SV40 small t. Virology 190: 40–44
Smits PHM, Smits HL, Minnaar RP, Hemmings BA, Mayer-Jaekel RE, Schüurman R, van der Noordaa J, ter Schegget J (1992b) The 55 kDa regulatory subunit of protein phosphatase 2A plays a role in the activation of the HPV16 long control region in human cells with a deletion in the short arm of chromosome 11. EMBO J 11: 4601–4606
Smits PHM, Smits HL, Minnaar RP, ter Schegget J (1993) Regulation of human papillomavirus type 16 (HPV-16) transcription by loci on the short arm of chromosome 11 is mediated by the TATAAAA motif of the HPV-16 promoter. J Gen Virol 74: 121–124

Sreekantaiah C, de Braekeleer M, Haas O (1991) Cytogenetic findings in cervical carcinoma A statistical approach. Cancer Genet Cytogenet 51: 75–81
Stanbridge EJ (1990) Human tumor suppressor genes. Annu Rev Genet 24: 615–657
Van Heyningen V, Hastie ND (1992) Wilms' tumour: reconciling genetics and biology. TIG 8: 16–21
von Knebel Doeberitz M, Bauknecht T, Bartsch D, zur Hausen H (1991) Influence of chromosomal integration on glucocorticoid-regulated transcription of growth-stimulating papillomavirus genes E6 and E7 in cervical carcinoma cells. Proc Natl Acad Sci USA 88: 1411–1415
Vousden KH (1989) Human papillomaviruses and cervical carcinoma. Cancer Cells 1: 43–50
Walter G, Ruediger R, Slaughter C, Mumby M (1990) Association of protein phosphatase 2A with polyoma virus medium tumor antigen. Proc Natl Acad Sci 87: 2521–2525
Woodworth CD, Doniger J, DiPaolo JA (1989) Immortalization of human foreskin keratinocytes by various human papillomavirus DNAs corresponds to their association with cervical carcinoma. J Virol 63: 159–164
Yang SI, Lickteig RL, Estes R, Rundell K, Walter G, Mumby MC (1991) Control of protein phosphatase 2A by Simian Virus 40 small-t antigen. Mol Cell Biol 11: 1988–1995
zur Hausen H (1986) Intracellular surveillance of persisting viral infections Human genital cancer results from deficient cellular control of papillomavirus gene expression. Lancet 2: 489–491
zur Hausen H (1989) Papillomaviruses in anogenital cancer as a model to understand the role of viruses in human cancers. Cancer Res 49: 4677–4681
zur Hausen H (1991a) Viruses in human cancers. Science 254: 1167–1173
zur Hausen H (1991b) Human papillomaviruses in the pathogenesis of anogenital cancer. Virology 184: 9–13

Molecular Pathogenesis of Cancer of the Cervix and Its Causation by Specific Human Papillomavirus Types

H. ZUR HAUSEN

1 Introduction

Less than 10 years after publication of the hypothesis of a papillomavirus etiology of cancer of the cervix (ZUR HAUSEN 1975, 1976, 1977), the DNAs of the first cervical cancer-associated human papillomavirus (HPV) types were cloned and characterized. These DNAs are regularly and frequently found in biopsies obtained from cervical cancer patients throughout the world (DÜRST et al. 1983; BOSHART et al. 1984).

The last decade initiated the revival and broader acceptance of an old concept: a possibly important role for virus infections in human cancers (reviewed in GROSS et al. 1983). The now fast growing number of data on papillomaviruses as potential human carcinogens was even preceded by seroepidemiological data on a role for hepatitis B virus infections in hepatocellular carcinoma, for human T cell leukemia type 1 (HTLV-1) in adult T cell leukemia, and already in 1964 on Epstein-Barr virus in Burkitt's

Deutsches Krebsforschungszentrum, Im Neuenheimer Feld 280, 69120 Heidelberg, Germany

Current Topics in Microbiology and Immunology, Vol. 186

lymphoma, nasopharyngeal carcinoma, B cell lymphomas, and possibly Hodgkin's disease (reviewed in ZUR HAUSEN 1991a).

An etiological relationship of all these infections with the respective tumors has been questioned up to today, even though, as in the case of HPV infections in cervical cancer, experimental as well as epidemiological data leave scarcely any room to doubt causality.

As pointed out previously by EVANS (1976), the specific mode of virus–host cell interaction in carcinogenesis precludes the application of postulates originally phrased by KOCH (1881) for the establishment of causal relationships between infections and a specific disease. This is in part due to our inability to propagate several of these agents in tissue culture systems, even more so, however, to the remarkable species specificity of most of these infections, permitting no experimentation in animal models, and to their frequent ubiquity in non-tumor-bearing hosts.

An attempt has been made to rephrase criteria which should permit the establishment of causal relationships between infections and tumor development, at least under conditions of direct effects (*trans* or *cis*) contributed by the persisting agent (ZUR HAUSEN 1991b). These criteria were as follows:

1. Epidemiological evidence that the respective infections represent risk factors for the development of specific tumors.
2. Regular presence and persistence of nucleic acid of the respective infectious agent in cells of specific malignant tumors.
3. Stimulation of proliferation upon transfection of the respective genome or parts thereof in corresponding tissue culture cells.
4. Demonstration that the induction of proliferation and the malignant phenotype of specific tumor cells depend on effects or functions exerted by the persisting DNA of the infectious agent.

It was realized from the outset that these criteria are not applicable to indirect modes by which infections may contribute to tumor development (e.g., by causing immunosuppresion, or host cell genome modifications, or by modifying endocrine functions). Yet the four criteria appear to be useful in unambiguously assessing the role of ubiquitous potential tumor viruses with DNA persistence in cancer cells in their relationship to specific human tumors.

The following tries to summarize specifically three aspects of papillomavirus infections in cervical carcinogenesis: the role of viral genes as oncogenes, the role of viral DNA integration for the malignant conversion, and the differential regulation of HPV transcription in non-malignant and malignant cells. This will be combined in a model featuring the concept of molecular pathogenesis of HPV infections leading to cancer of the cervix.

2 Role of E6/E7 Gene Expression in Cervical Carcinogenesis

2.1 E6/E7 Genes Represent Oncogenes

The expression of E6 and E7 genes has been noted in all cervical cancer biopsies carefully examined and found to be positive for persisting HPV DNA. The same accounts for cell lines derived from cancer of the cervix (SCHWARZ et al. 1985; YEE et al. 1985; reviewed in SCHWARZ, 1987). In view of the high percentage (> 90%) of cervical carcinoma biopsies containing HPV DNA (reviewed in ZUR HAUSEN, 1989), the regular expression of E6/E7 genes in these tumors already hints at an essential function of the derived proteins for the proliferation of these cancer cells.

Support for this assumption originated from experiments revealing the immortalizing functions, initially of transfected HPV-16 DNA (PIRISI et al. 1987; DÜRST et al. 1987), subsequently of transfected E6/E7 genes derived from HPV-16 or -18 genomes (MÜNGER et al. 1989; HAWLEY-NELSON et al. 1989; WATANABE et al. 1989). Initial studies were performed in foreskin and cervical keratinocytes; later on, however, immortalization was also achieved in epithelial breast cells (BAND et al. 1990), bronchial cells (WILLEY et al. 1991), and cells derived from the oral and nasal mucosa (PARK et al. 1991; DEBYEC-RYCHTER et al. 1991), as well as in other cell types (see McDougall, this volume). The immortalized cells quickly become aneuploid or pseudodiploid, an observation also made in HPV-16- or HPV-18-containing early premalignant lesions, in contrast to lesions containing HPV-6 or -11 (FU et al. 1983; CRUM et al. 1985).

In organotypic cultures E6 + E7 immortalized human keratinocytes reveal intraepithelial neoplasia-like growth characteristics virtually indistinguishable from those found in cervical dysplastic lesions (MCCANCE et al. 1988; HUDSON et al. 1990; WOODWORTH et al. 1990).

It is interesting to note that immortalization of human keratinocytes has only been achieved with HPV types regularly found in cancers of the cervix but not with types regularly present in genital warts (HPV-6 and -11) which rarely convert into malignant tumors. The former are therefore considered as "high-risk," the latter as "low-risk" viruses (ZUR HAUSEN 1986a).

Long-term cultivation of HPV-16- or -18-immortalized cells which are initially non-tumorigenic when tested in immunoincompetent animals spontaneously leads to the appearance of malignant sublines (PECORARO et al. 1991; HURLIN et al. 1991; M. Dürst, personal communication). This stresses the potential oncogenicity of these viruses and most likely of the E6/E7 genes which are uniformly expressed in all immortalized lines. However, it shows in addition that the expression of the latter genes per se is not sufficient for the development of a malignant phenotype.

Table 1. Effects of papillomavirus oncogene expression in transgenic mice

Transgene	Effect	Latency period	Papillomavirus oncogene expression	Author
BPV-1 genome	initial mild fibromatosis aggressive fibroma fibrosarcoma	8 months	upregulation of E6/E7 transcription from mild to aggressive fibromatosis	SIPPOLA-THIELE et al. 1989
CRPV genome	skin papillomas	20 days	transcriptional upregulation in tumors	PENG et al. 1993
HPV 16 E6/E7 β actine promoter	neuroepithelial tumors	2.5 months	transcriptional upregulation in tumors	ARBEIT et al. 1993
HPV 16 E6/E7 MTV promoter	seminomas	7 months	transcriptional upregulation in tumors	KONDOH et al. 1991
HPV 16 E6/E7 bovine keratine 6 promoter	glandular stomach cancer	6 months	transcriptional upregulation in tumors	SEARLE et al. 1993
HPV 16 E6/E7 bovine keratine 10 promoter	hyperproliferation of skin epithelium with parakeratosis No malignant sequelae	—	transcriptional upregulation in the hyperproliferative epithelium	CID et al. 1993
HPV 16 E6/E7 αA crystallin promoter	lens tumors	6 months	transcriptional upregulation in tumors	GRIEP et al. 1993

Table 2. Evidence for an oncogene function of high risk HPV E6/E7 genes

1. E6/E7 genes are expressed in all HPV positive cancer cell lines and in HPV containing cancer biopsies.
2. E6/E7 antisense RNA reduces cell proliferation in the respective cell lines and abolishes their tumorigenicity.
3. E6/E7 genes upon transfection induce immortalization of human keratinocytes.
4. E6/E7 genes are tumorigenic in transgenic animals.
5. E6/E7 share functions with other well established viral oncogenes (adenovirus E1A and SV 40 T) in binding the cellular proteins Rb and p53.

The role of E6/E7 genes as oncogenes can be unequivocally established in E6/E7 transgenic mice. The introduction of E6/E7 genes of HPV-16 or -18 under the control of the HPV upstream regulatory region (URR) does not lead to significant transgene expression and pathological changes. This may underline the remarkable species specificity of these viruses. The use of heterologous tissue-specific promoters, however, results in tumor formation in the respective target organ after varying latency periods, usually after 3–8 months of age, with high levels of E6/E7 gene expression. Table 1 lists transgenic papillomavirus systems reported up to now.

The role of E6/E7 oncogene expression for cell proliferation can also be demonstrated by using inducible antisense constructs in clones derived from carcinoma cell lines carrying the respective viral genomes (VON KNEBEL DOEBERITZ et al. 1988, 1992) or of HPV-transformed rodent cells (CROOK et al. 1989). These studies document that not only the proliferative but also the malignant phenotype of the lines tested depend on E6/E7 oncogene expression.

Finally, in HPV DNA-containing cervical carcinoma cells permitting negative or positive modulation of HPV E6/E7 gene expression by treatment with glucocorticoids, the growth properties as well as tumorigenicity correlate well with E6/E7 gene activity (VON KNEBEL DOEBERITZ et al. 1991). Table 2 summarizes the present evidence for an oncogene function of E6/E7 genes of high-risk HPVs.

2.2 Role of Viral Oncoprotein Binding to Host Cell Proteins

The discovery of E7 protein binding to the cellular growth-regulating protein Rb (DYSON et al. 1989) and of E6 binding to p53 (WERNESS et al. 1990) provided a potential clue to the understanding of viral oncogene functions and their mechanism of action. This will be discussed in other contribution in this volume (see the chapters by SCHEFFNER et al. and by MCDOUGALL). It is assumed that a heteromeric complex of Rb protein and transcription factor E2F and other proteins found in nonproliferating cells (NEVINS et al. 1992a; KAELIN et al. 1991) is resolved by E7-Rb binding releasing active E2F (NEVINS et al. 1992b; PAGANO et al. 1992). This should result in the activation of several E2F-regulated genes involved in the G1-S phase transition of the cell cycle.

Similarly, binding of the E6 protein to the cellular protein p53 promotes degradation of the latter (SCHEFFNER et al. 1990) and probably results in exemption of the respective cells from p53-mediated control of specific transcriptional initiations and DNA repair (KASTAN et al. 1991; KESSIS et al. 1993).

A number of indications exist today, however, which do not support a direct role of E7 protein-Rb protein binding in regulating cell immortalization and early events in in vivo pathogenicity. They are summarized as follows:

1. Mutation of the E7-binding site for the Rb protein does not abolish its immortalizing function for human keratinocytes and affects the transforming activity in the ras-cooperation assay in rodents (JEWERS et al. 1992).
2. Mutations preventing Rb binding to the E7 protein of cottontail rabbit papillomavirus (CRPV) do not prevent pathogenicity of these mutants in rabbits (DEFEO-JONES et al. 1993). CRPV E7 binds rabbit as well as human Rb with similar efficiency as human high-risk viruses (HASKELL et al. 1993).
3. Deletion of amino acid residues 6–10 in HPV-16 E7 results in a loss of transforming activity by maintaining pRb binding and E2F transactivation (PHELPS et al. 1992).
4. Adenovirus type 2 transformed, but revertant, rat cell lines are resistant to retransformation by E1A (which also binds Rb), but can be retransformed by HPV-16 E7 oncogenes (SIRCAR et al. 1991).

Although other reports stress the importance of E7 protein Rb-binding sites for cell immortalization (BARBOSA et al. 1990; HECK et al. 1992), these studies were conducted in rodent cells and may be due to species-specific reactions to conformational changes of mutated E7 proteins.

Similarly to E7-Rb binding, arguments exist suggesting that p53-E6 binding is also not responsible for cell immortalization and early pathogenicity. This can already be suspected from observations showing that E6 expression is not sufficient for human cell immortalization and requires the addition of functioning E7 genes (MÜNGER et al. 1989). There are, in addition, at least three further sets of data supporting this view:

1. p53 knock-out mice develop normally although they are prone to early cancer development (DONEHOWER et al. 1992).
2. There is no correlation between the quantity of endogenous p53 and intracellular concentration of E6 protein (SCHEFFNER et al. 1991; HUBBERT et al. 1992; LECHNER et al. 1992).
3. E6 gene expression of HPVs presumably inducing exuberant in vivo stimulation of cell proliferation (e.g., HPV-6 or -11 in genital warts) (IFTNER et al. 1992) reveal at most a very low affinity binding of p53 (HOWLEY, 1991).

Obviously p53 functions are at least temporarily dispensable, although their absence results in an increased risk for cancer development. The example of genital warts shows in addition that at least in vivo growth stimulation by HPV gene functions does not seem to require specific interactions with cellular Rb and p53 proteins. The inability of viral DNA from low-risk types to induce similar growth stimulation under tissue culture conditions probably depends on their need for specific factors absent from conventional tissue culture media, but required for their transcriptional activity and DNA persistence.

2.3 E7-Rb and E6-p53 Binding in a Model for Tumor Development and Tumor Progression

Observations of p53 mutations in metastases of HPV-positive cervical cancers but not in the majority of primary tumors (CROOK and VOUSDEN 1992) points to a role of these modifications in tumor progression rather than in early events of cervical carcinogenesis. The latter interpretation can also be deduced from other results which had been combined into a model of cervical cancer progression (ZUR HAUSEN 1991):

Besides HPV-16 or -18 E6 and E7, simian virus 40 (SV40) large T and adenovirus E1A and E1B genes efficiently bind Rb and p53 proteins. Immortalization of cells by these transforming viral genes shows one common feature: the induction of chromosomal aberrations and mutations already in the early phase of immortalization. Chromosomal instability remains high in cells immortalized or transformed by these genes. The induction of chromosomal rearrangements by viral oncogenes has been shown directly (DREWS et al. 1992), and E7 has been identified as the gene responsible for chromosomal changes (HASHIDA and YASUMOTO 1991). In view of the involvement of p53 in DNA damage repair (KASTAN et al. 1991; reviewed in LANE 1992; HARTWELL 1992) and taking into account the negative regulation of DNA synthesis by Rb, it seems likely that the interaction of these proteins with the HPV oncoproteins introduces an element of continuous chromosomal instability in cells expressing the viral oncogenes at a significant level (see below). This should lead to an accumulation of mutational changes in the respective host cell DNA, providing an endogenous progression factor in cells latently infected by high-risk viruses.

Low-risk viruses, in spite of their high activity to stimulate host cell proliferation, are apparently unable to induce chromosomal changes in condolymata acuminata, the infected cells remain diploid or tetraploid (JAGELLA and STEGNER 1974; FU et al. 1983). Malignant conversion is rare and leads to verrucous carcinomas or Buschke-Löwenstein tumors which regularly contain HPV-6 or -11 DNA (BOSHART and ZUR HAUSEN 1986). Even these malignant tumors show little tendency towards progression, they grow invasively but rarely metastasize.

Thus, the frequently observed progression of HPV-16- or -18-positive premalignant lesions could be interpreted as a result of a cell endogenous function, possibly originating from interactions of viral oncogenes with Rb and p53. In low-risk infections, malignant conversion would be the rare consequence of additional mutational events induced by external carcinogenic factors.

As will be discussed subsequently, expression of E6/E7 genes of high-risk HPV is controlled by host cell factors. Modification of the respective host cell genes seems to result in increased E6/E7 expression and to further enhanced cell proliferation. In clinical lesions of low-risk viruses (HPV-6/11), the basal expression of E6/E7 appears to be high even in early stages (IFTNER et al. 1992).

Figure 1 depicts this scenario of tumor progression in infections by low- and high-risk viruses by assuming that host cell genes restricting HPV transcription in normal cells will be the most important targets for these probably random mutational events (ZUR HAUSEN 1977b, 1986b, 1981).

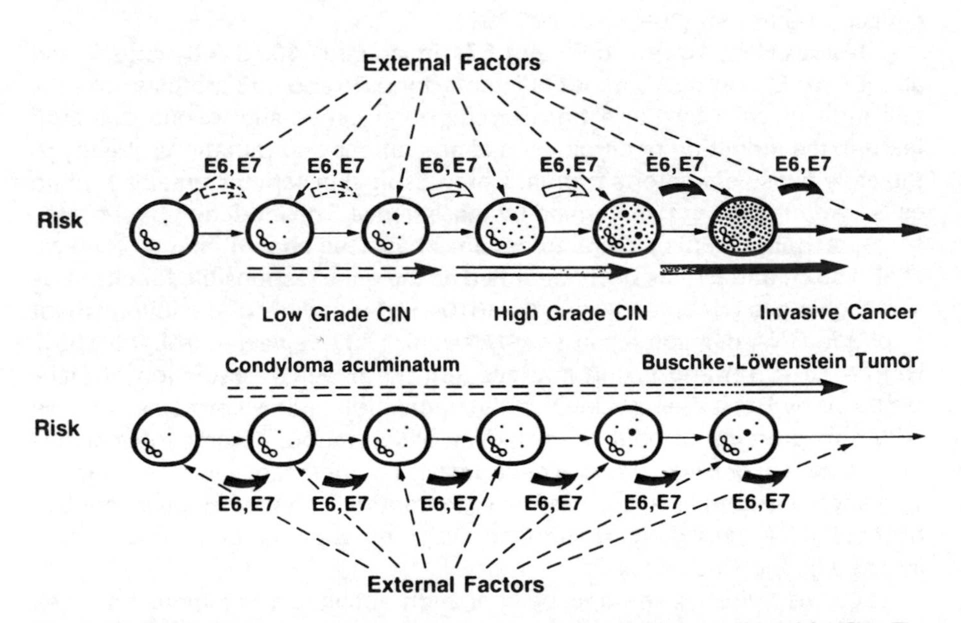

Fig. 1. Model of tumor development and tumor progression by high- and low-risk HPVs. The *circles* schematically show individual cells harboring latent viral DNA. The *small dots* outline random mutational events, the *big dots* specific modifications of genes involved in the suppression of viral E6/E7 transcription in high-risk infections (see also Figs. 2 and 3). The *arrows* indicate the relative intensity of E6/E7 expression. Cells with early stages of high-risk HPV infection efficiently control E6/E7 expression. Mutations induced preferentially by these oncogenes and less so by external factors lead to tumor progression. In low-risk viruses E6/E7 expression is less controlled but seems to be not significantly mutagenic. Progression appears to depend on rare externally induced mutagenic events

2.4 Cooperation Between E6 and E7 Genes in Immortalization and Transformation

Transfection studies established the co-operative action of E6/E7 genes in cell immortalization (MÜNGER et al. 1989; HAWLEY-NELSON et al. 1989). This is evidently the case after transfection of most human keratinocytes which appear to require both genes for effective immortalization. The E7 gene alone is able to immortalize cells when controlled by a strong foreign promoter (HALBERT et al. 1991). E6 does not seem to immortalize genital keratinocytes, although it appears to be the sole immortalizing factor for breast epithelial cells (BAND et al. 1991). Interestingly, even HPV-6 E6 immortalizes human breast epithelial cells, mediating here a rapid degradation of p53 similar to E6 of high-risk HPV E6 (BAND et al. 1993).

Recent studies have shown that HPV-induced cell immortalization requires genetic changes affecting the host cell genome since fusion of various HPV-immortalized clones leads in many instances to senescence and permits the establishment of complementation groups (CHEN et al. 1993). It is therefore suggestive that cooperation between E6/E7 genes more effectively mediates these host cell genetic changes in comparison to individual viral oncogenes.

Support for this interpretation stems from a recent publication by KESSIS et al. (1993) who showed that the E6 oncoprotein is able to override the G1 repair arrest following DNA damage. The G1 arrest appears to be a p53 function which is alleviated by E6 binding and the subsequent p53 degradation.

E7 alone appears to cause mutations in host cell DNA (HASHIDA and YASIMOTO 1991). It is therefore, tempting to speculate that the cooperation of E6 and E7 results from induction of mutations by E7 and faulty repair caused by E6-mediated p53 degradation, abolishing the G1 arrest required for efficient repair.

3 Role of Viral DNA Integration

Integration of HPV DNA is regularly observed in malignant tumors (BOSHART et al. 1984; SCHWARZ et al. 1985; YEE et al. 1985), although a small percentage of the biopsies appears to contain exclusively episomal sequences (MATSUKURA et al. 1989). This contrasts remarkably with the situation in premalignant lesions which regularly contain episomal DNA (DÜRST et al. 1985; MATSUKURA et al. 1989; CULLEN et al. 1991), although again a few exceptions have been noted here (LEHN et al. 1988; CULLEN et al. 1991). Integration therefore preferentially seems to occur at the borderline of the transition of still non-malignant to malignant tumors. This is suggestive for a specific role of integration in the process of malignant conversion. Fragile sites appear to be preferential loci for integration (SMITH et al. 1992).

In a number of tumors, deletions involving E2 (E4 and E5) and L2 have been noted (SCHWARZ et al. 1985, YEE et al. 1985). The integrate structures indicate that the disruption of the viral genome within a defined region represents a specific step in the integration of viral DNA. The viral ring molecule is commonly opened within the 3′ end of the E1 or the 5′ end of the E2 open reading frames. Since the E2 gene encodes a repressor for transcription of the E6 and E7 oncogenes, its inactivation due to DNA integration may play a role in the deregulation of E6/E7 gene expression in carcinoma cells (SPALHOLZ et al. 1985; LAMBERT et al. 1987; HUBBERT et al. 1988; BURNETT et al. 1990; ROMANCZUK et al. 1990; KARLEN and BEARD 1993).

In addition, however, some of the integrational events lead to transcriptional initiation within the 5′ flanking host cell region (RÖSL et al. 1989), apparently due to specific promoters at the integrational site. Thus, this mode of integration permits a more ready escape from intracellular control mechanisms down-regulating viral transcription in normal cells (see below).

This view is supported by experiments analyzing the response of HPV transcriptional activity to glucocorticoids in various HPV-positive cell lines. Cancer-linked anogenital HPV types contain hormone-responsive elements within their regulatory regions (GLOSS et al. 1987; PATER et al. 1988) reacting with progesterone or dexamethasone receptor complexes. In specific cell lines, dexamethasone treatment does not result in stimulation of HPV transcription (e.g., HeLa cells), in others (C4-1) it substantially enhances transcription, and in a third line (SW 756) it specifically inhibits HPV transcription (VON KNEBEL DOEBERITZ et al. 1991). This varying response to glucocorticoid exposure is not mediated by a transfunction and appears to be due to *cis*-regulation originating from flanking host cell sequences. The analysis of the viral regulatory region in C4-1 or SW 756 cells did not reveal modifications of the glucocorticoid response elements when compared to the HPV-18 prototype sequence.

It is of interest that, after grafting these cells into dexamethasone-treated nude mice, growth properties in vitro and tumorigenicity of the respective lines, correspond to the hormone effect. E6/E7 induction by dexamethasone leads to growth stimulation and increased tumorigenicity, E6/E7 inhibition in SW 756 cells by the same treatment results in spindle-shaped morphology, growth inhibition, and a drastic decrease in tumorigenicity (VON KNEBEL DOEBERITZ et al. 1991).

There is one cell line, Caski, harboring several hundred HPV-16 genome copies per cell (YEE et al. 1985). These cells reveal a specific type of transcriptional regulation of viral DNA: only a small minority of viral DNA molecules seems to be transcriptionally active (SMOTKIN and WETTSTEIN 1986), the vast majority of viral DNA molecules is heavily methylated (RÖSL et al. 1993) and remains transcriptionally silent. Fusion of Caski cells with cells derived from the HPV-16- positive SiHa line which express a CAT gene under the control of the HPV-18 promoter leads to a rapid extinction of CAT expression (F. Rösl and H. zur Hausen, unpublished observations), similar to fusion of SiHa-CAT cells with normal human keratinocytes (RÖSL et al. 1991). This suggests that the transcriptionally active HPV genes in Caski cells are probably controlled by flanking host cell sequences and thereby escape cellular repression, whereas the majority of latent HPV-16 genomes are either repressed by host cell genes or inactivated by heavy methylation. Caski cells are poorly tumorigenic when transplanted into nude mice. Interestingly, the arising tumors show a highly differentiated pattern (M. Dürst, personal communication). In HPV-16-immortalized partially differentiating keratinocytes, HPV E6/E7 gene expression is switched off (DÜRST et al. 1991). This is also the case in differentiating part of Caski tumors.

The data presently available point to the importance of viral DNA integration favoring malignant conversion. The malignant phenotype may result from dual effects: the interruption of viral intragenomic regulatory mechanisms and a partial or complete escape from intracellular control mechanisms suppressing HPV E6/E7 transcription in the absence of an E2-mediated control.

The data on HPV transcription in Caski and SiHa cells, together with various other observations, showing that fusion of cervical carcinoma cells containing integrated HPV DNA with normal cells (keratinocytes or fibroblasts) results in nontumorigenic hybrids, reveal that integrational dysregulation per se is not sufficient for the development of a malignant phenotype.

4 Cellular Control of Viral Oncogene Expression

The hypothesis of the existence of cellular control mechanisms down-regulating viral oncogene expression was first published in 1977 (ZUR HAUSEN 1977). It was based on the assumption that viral oncoproteins were required for the maintenance of the proliferative, but also malignant phenotype of latently tumour virus-infected cells. Subsequent modifications accounted specially for observations made in papillomavirus-containing cell systems (ZUR HAUSEN 1982, 1986b, 1989) and focused attention on transcriptional regulation.

During the past few years the postulated differential regulation of HPV transcription has been discovered in non-malignant cells when compared to malignant cells. The data originated in part from suppression of HPV transcription by 5-aza-cytidine in nonmalignant HeLa hybrid cells and immortalized keratinocytes, but not in their malignant progenitors or segregants (RÖSL et al. 1988). In part they resulted from a differential regulation of constitutively expressed reporter genes controlled by HPV promoters in malignant cells and their extinction in hybrids obtained after fusion with human keratinocytes (RÖSL et al. 1991.)

Although these experiments revealed the existence of different regulatory mechanisms for HPV oncogene expression in malignant and in nonmalignant cells, they did not solve the problem of an approximately equal transcription of E6/E7 RNA in various malignant and nonmalignant lines kept in tissue culture. There is no evidence for qualitative differences in E6/E7 expression correlating with the proliferative state of the cells as long as they are kept under optimal growth conditions.

The option remained open, however, that, due to different intercellular interactions, implantation of the cells into suitable hosts would differentiate between a non-malignant and a malignant state at the HPV transcript level. This expectation turned out to be correct: inoculation of HeLa normal cell hybrids into immunocompromised (nude) mice resulted in drastical reduction of HPV transcriptional activity within 3 days after transplantation (BOSCH et al. 1990). HPV-immortalized keratinocytes show only low levels of viral transcriptional activity after inoculation into nude mice (DÜRST et al. 1991). They form persisting epithelial cysts which differentiate into the inner part of the cyst. Malignant cells, in contrast, continue to transcribe E6/E7 genes under these conditions abundantly and grow invasively.

No detectable HPV transcription occurs in heterografted differentiating nonmalignant cells. This is contrary to observations made in clinical lesions. Particularly in low-grade cervical intraepithelial lesions (CIN) harboring HPV-16 DNA, on the other hand, most of the transcriptional activity is found in the differentiating cells (DÜRST et al. 1992; STOLER et al. 1992). This different pattern in differentiating cells probably results from the mode of viral DNA persistence. Contrary to the integrated state of viral DNA in immortalized cells, CIN lesions contain episomal HPV DNA (DÜRST et al. 1985; CULLEN et al. 1991).

In clinical lesions, there is frequently a remarkable difference in E6/E7 transcriptional activity between low- and high-grade CIN (DÜRST et al. 1992; STOLER et al. 1992; Auvinen et al., in press). Proliferating cells in low-grade CIN reveal only a low degree of transcriptional activity, barely detectable by standard in situ hybridization techniques. In contrast, high-grade lesions and invasive cancer exhibit abundant transcriptional activity throughout the proliferating layer. It thus appears that the low-grade/high-grade transition involves a switch in HPV E6/E7 gene regulation (ZUR HAUSEN 1991).

In conclusion, therefore, regulation of HPV transcription is different in non-malignant cells kept in tissue culture in comparison to the same cells implanted into an animal host. This difference is not detectable in malignant cells. It seems that the clinical correlate is the low E6/E7 transcriptional rate of low-grade CIN in comparison to the dysregulated E6/E7 transcription in high-grade lesions and invasive cancer.

During the past few years our laboratory (as well as others) has concentrated on the isolation and characterization of cellular factors regulating HPV E6/E7 gene expression. Today an increasing number of cellular proteins have emerged which negatively or positively influence E6/E7 gene expression. Superficially they can be subdivided into three groups:

1. Cytokines as intercellular mediators of viral gene regulation
2. Intracellular signal mediators
3. Transcription factors controlling the HPV promoter–enhancer region

4.1 Cytokines in HPV E6/E7 Gene Regulation

Whereas normal cervical epithelial cells constitutively secrete interleukin 1α (IL-1α), IL-1β, IL-6, IL-8, tumor necrosis factor (TNF)α and granulocyte macrophage colony-stimulating factor, synthesis of these lymphokines-appears to be significantly reduced in cell lines immortalized by HPV and in HPV-positive carcinoma cell lines (WOODWORTH and SIMPSON 1993). These observations led to the suggestion that decreased expression of lymphokines in HPV-carrying cells may influence immunity and inflammatory reactions in the cervical mucosa. TNFα expression of an HPV-16-immortalized keratinocyte line (SK-v) appeared to exert an autocrine loop for growth inhibition of these cells (MALEJCZYK et al. 1992). The constitutive release of IL-6 by the same cell line significantly augmented natural killer (NK) cell activity of human peripheral blood lymphocytes against these cells (MALEJCZYK et al. 1991).

Besides these observations pointing to immunological interactions resulting in cell lysis or in autocrine growth inhibition, several cytokines have been reported to suppress HPV E6/E7 expression selectively in nonmalignant cells:

1. TGFβ 1 and 2 (BRAUN et al. 1990; WOODWORTH et al. 1990),
2. Epidermal growth factor (YASUMOTO et al. 1991)
3. Leukoregulin and interferon gamma (WOODWORTH et al. 1992)
4. TNFα and interferon gamma (F. Rösl et al., in preparation).

Most of these observations extend to a limited set of HPV-immortalized lines and cell lines derived from cancer of the cervix.

Recently it was shown that activated human macrophages exert a dual effect on HPV-carrying non-malignant human cells: they effectively induce the macrophage-chemoattractant protein 1 (MCP-1) and repress HPV E6/E7 transcription (F. Rösl et al., in preparation). The excretion of MCP-1 leads to accumulation of activated macrophages in vivo and has been reported to result in tumor suppression (ROLLINS and SUNDAY 1991). In malignant cell lines neither MCP-1 induction nor HPV E6/E7 suppression has been observed after addition of activated human macrophages (F. Rösl et al., in preparation). MCP-1 induction and E6/E7 suppression are obviously mediated by specific cytokines, since cell-free supernatants obtained after co-cultivation of activated macrophages with non-malignant cells induce this effect. Although TNFα turns out to efficiently induce MCP-1, and a combination of TNFα and interferon gamma effectively blocks E6/E7 transcription in these cells, present evidence points to additional cytokines engaged in this reaction.

The process of HPV transcriptional inhibition by intercellular communication (specifically by non-HPV-harboring macrophages and possibly additional cell types) provides a convenient explanation for the growth of immortalized cells in tissue culture but not after heterografting. In the latter condition an intracellular cascade of regulatory events is triggered which is not initiated in homogeneous cell populations in tissue culture.

4.2 Signal Mediators in E6/E7 Gene Regulation

Early experiments performed by SAXON et al. (1986), by KAEBLING and KLINGER (1986), and by KOI et al. (1989) showed that the transfer of individual human chromosome 11 into HeLa or SiHa cervical carcinoma cells results in the suppression of the malignant phenotype. This suggested the existence of an important tumor suppressor gene in at least these cervical cancer cells in chromosome 11 and is in line with the observation of increased chromosomal aberrations involving chromosome 11 in cervical cancer biopsies (ATKIN et al. 1984, 1988, SREEKANTAIAH et al. 1991). Additional support originated from studies in human fibroblasts with a deletion on the short arm of one chromosome 11 (del 11 cells). These cells could be transformed by HPV-16 DNA whereas normal human embryonic fibroblasts were resistant to transformation (SMITS et al. 1988).

The data are extensively discussed in this volume (ter Schegget and van der Noordaa). The transformability of human fibroblasts correlated with a high level of HPV-16 early gene expression. SV40 small t antigen is essential for SV40-induced transformation of human diploid cells but dispensable for transformation of del 11 cells. This shows that SV40 small t can mimic the effect of the human chromosome 11 deletion (SMITS et al. 1992a). SV40 small t has been found to bind to protein phosphatase 2A (PP2A) and to inhibit the phosphatase activity (YANG et al. 1991). Since the B subunit of the holoenzyme PP2A regulates the activity of the catalytic subunit of this enzyme, the B subunit represented a possible candidate for a cellular factor mimicking SV40 small t activity and for inhibition of PP2A. It was indeed shown that the B subunit is highly expressed in del 11 cells but not in normal diploid fibroblasts (SMITS et al. 1992b). Moreover, specific inhibition of PP2A by ocadaic acid resulted in trans-activation of the HPV-16 control region in diploid normal cells. A common motif of four amino acids in the B subunit of PP2A and SV40 small t seems to be necessary for their ability to trans-activate the HPV-16 promoter.

The data available thus far point to at least two key events in the signal transfer regulating HPV E6/E7 gene expression: the activation of a gene, probably located on chromosome 11, and the activation or inactivation of protein phosphatases (at present mainly PP2A). The former appears to be important in regulating phosphatase activities influencing an unknown number of intracellular regulatory events. It should correspond to the previously postulated "CIF" (cellular interfering factor) gene (ZUR HAUSEN 1977, 1989). Its nature is still hypothetical. It seems to be functionally silent under tissue culture conditions, possibly due to methylation (RÖSL et al. 1988), and is re-activated in vivo, probably mediated by intercellular signals. Activation of phosphatases, here of PP2A, seems to correlate with an inhibition of HPV transcription.

4.3 Transcription Factors Controlling E6/E7 Expression

The HPV upstream regulatory region contains binding sites for numerous transcription factors resulting in a complex regulatory network for E6/E7 transcription. A number of factors activating HPV transcription have been identified including AP-1 (CHAN et al. 1990; CRIPE et al. 1990; OFFORD and Beard 1990) and here particularly the Jun B component (THIERRY et al. 1992), NF-II (GARCIA-CARRANCA et al. 1988; GLOSS et al. 1989), Sp-1 (GLOSS and BERNARD 1991; HOPPE-SEYLER and BUTZ 1992), a transcriptional enhancer factor, TEF-1 (ISHIJI et al. 1992), a keratinocyte specific transcription factor, KRF-1 (MACK and LAIMINS 1991), and activated glucocorticoid receptor complexes (GLOSS et al. 1987; CHAN et al. 1989).

Their absence, quantitative reduction, competitive binding, or functional modification could result in reduced HPV E6/E7 gene activity in non-malignant cells. There exists, however, no evidence in support of this view.

Similarly, an increasing number of negative cellular regulators have been identified: the Oct-1 transcription factor (HOPPE-SEYLER et al. 1991), the nuclear factor for interleukin 6 expression NF-IL (KYO et al. 1993), retinoic acid receptors (RAR) (BARTSCH et al. 1992; PIRISI et al. 1992), and the transcriptional repressor YY1 (BAUKNECHT et al. 1992).

It is of obvious interest to analyze a possible differential interaction of these negative regulators in malignant and non-malignant cells. Thus far neither structural nor functional modifications of the Oct-1 gene or its product have become apparent (HOPPE-SEYLER et al. 1991); NF-IL-6-repressed HPV transcriptional activity in HeLa and Caski cells, both representing carcinoma cell lines (KYO et al. 1993).

Two interesting exceptions are represented by the RAR complexes and the transcriptional repressor YY1 (SHI et al. 1991; review on RAR in LEID et al. 1992). Although retinoic acid addition results in repression of HPV transcription in both non-malignant and malignant cells, retinoic acid-mediated induction of the RARβ receptor was restricted to non-malignant cells (BARTSCH et al. 1992). No structural modifications of the RARβ gene or its regulatory region were noted in the non-reactive carcinoma cells. Transient transfection showed that the RARβ control region was more active in non-malignant than in malignant cells.

These observations point to an important dysregulation of the RARβ receptor in the malignant cells studied thus far, whereas other RARs (α, gamma, RXR) remained unaffected. This may gain additional importance following reports on a tumor-suppressive effect of RARβ in epidermoid lung cancer (HOULE et al. 1993), a high frequency of RARβ abnormalities in lung cancer (GEBERT et al. 1991), and the induction of TGFβ1 and 2 by retinoic acid in human keratinocytes (BATOVA et al. 1992). The latter could hint at an indirect mode of retinoic acid-mediated HPV transcriptional suppression.

Recently, BAUKNECHT et al. (1992) demonstrated that the transcriptional repressor YY1 plays a critical role in silencing HPV-18 promoter activity. In transient transfection assays using the proximal promoter region of HPV-18 containing the YY1-binding site, YY1 suppressed transcriptional activity in non-malignant and malignant cells. Interestingly, in the context of the whole promoter, mutations in the YY1-binding site led to increased transcriptional activity of the reporter plasmids in non-malignant cells. However, the higher basal activity of these constructs in some malignant lines was considerably reduced (T. Bauknecht et al., to be published). Overexpression of YY1 led to reduced activity of the reporter plasmid in HeLa cells, apparently due to squelching.

This seems to underline an important role of YY1 in HPV gene regulation. Whereas it acts as a silencer in non-malignant cells, it activates transcription in some malignant lines. It is likely that this activation results from interactions with protein complexes binding to the promoter–enhancer region. This binding may be profoundly affected by differences in protein phosphorylation (see also Fig. 4).

In cells containing a deletion in the short arm of chromosome 11, DNA–protein complexes formed with an HPV-16 promoter fragment are quantitatively different from those formed with nuclear cell extracts from diploid cells (SMITS et al. 1993). This study implicates TATA-binding proteins in this effects.

5 Model of HPV Pathogenesis

The data discussed in this review permit the development of a hypothetical model of the molecular pathogenesis of HPV-linked cervical cancer, stressing the role of intercellular and intracellular interactions. It probably represents a first functional model of individual steps involved in viral carcinogenesis in humans (Figs. 2, 3).

In normal cells a tight regulation of viral oncogene transcription, obviously coupled to the state of cellular differentiation, prevents the advance to dysplastic and invasive growth of the latently infected cells.

Activation of Macrophages
(Other Cells ?)

TNFα IFγ TGFβ Others ?

Activated Receptors

Activation of CIF
(Locus on Chrom. 11 ?)

Repression of PR 55β

Activation of PP 2A

Dephosphorylation of
yet Unidentified
Transcription Factor(s)

No Interaction with
YY-1 Transcriptional Silencer

Inhibition of HPV Transcription

Induction of
MCP-1 gene
Transcription

?

Fig. 2. Functional model of intra- and intercellular control of latent high-risk HPV oncogene expression. The individual steps are explained in the text

Activation of Macrophages
(Other Cells ?)

TNFα IFγ TGFβ Others ?

Activated Receptors

No Activation of CIF
due to Structural Modifications
of the CIF Genes

Activation of PR 55β

SV40 t

Ocadaic
Acid

Inhibition of PP 2A

No Dephosphorylation of
yet Unidentified
Transcription Factor(s)

Interaction of those Factor(s)
with YY-1 Results in Activation
of the HPV Promoter

Activation of HPV Transcription

No Induction of
MCP-1 gene

?

Fig. 3. Functional model of high-risk HPV oncogene transcriptional dysregulation. This scheme assumes an interruption of the intracellular surveillance due to structural modifications of the still hypothetical CIF gene, probably located on chromosome 11. SV40 small t and ocadaic acid mimic this pathway by directly inhibiting the catalytic subunit of PP2A

Early phases of carcinoma development may arise from a number of different events disrupting the cascade of intracellular regulations. Apparently the most important modification affects a yet unidentified gene most likely located on chromosome 11. This seems to have profound implications on subsequent steps in intracellular protein phosphorylation and dephosphorylation, and therefore probably on a broad spectrum of different genes. Dysregulation of PR55β (activation), the regulatory component of PP2A, by intragenic events (mutations, recombinations) would be another factor.

Inhibition of PP2A and possibly of other phosphatases by gene modifications of its catalytic subunit or competitive binding of other factors may further contribute to HPV activation.

The regulation at the transcriptional level is hypothetically depicted in Fig. 4. Modifications of the YY1-binding domain or associated transcriptional complexes could further contribute to transcriptional activation of HPV genomes. Indeed, in cervical carcinomas revealing exclusively episomal HPV DNA, three out of eight samples revealed mutations in the putative YY1-binding domain (H. Pfister, personal communication).

Obviously the present picture is still far from being complete, further intracellular control mechanisms are likely to exist, the function of E6/E7

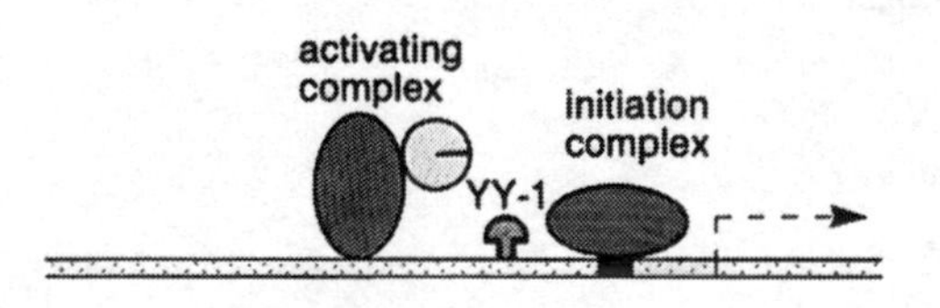

Suppression of HPV transcription due to non-interacting YY-1

Activation of HPV transcription due to interaction of YY-1 with putative phosphorylated transcription factor(s)

Inhibition of HPV transcription due to squelching

Fig. 4. A hypothetical simplified model of the critical role of the transcription factor YY1 in the regulation of E6/E7 transcription. The *crosses* schematically outline phosphorylation sites

oncoproteins is still poorly understood. Yet, cancer of the cervix, as the second most frequent cancer of women worldwide, has an excellent chance to represent the first human malignant tumor where molecular events starting at the level of intercellular communication up to the regulation of the actual tumor effector genes will be understood.

The discussion has omitted the existence of an immunological control of HPV infections which is presented in another chapter of this volume (Tindle and Frazer). It clearly plays an important role as indicated by the frequency of warts in immunocompromised individuals and the increased risk of carriers with certain HLA haplotypes and MHC class II gene expression for papillomavirus-caused cancers (WANK and THOMSSEN 1991; WANK et al. 1992; HELLAND et al. 1992; HAN et al. 1992). In evolutionary terms, immunological control may represent a third shield of protection, preceded by intracellular and intercellular growth control mechnisms. The development of this delicate and complex network protecting the host against potentially deleterious effects of papillomavirus infection represents an indirect hint to the ancient relationship between papillomavirus genotypes and their human hosts (see also BERNARD et al., this volume).

6 Conclusions

The oncogenes E6/E7 of high-risk papillomaviruses are decisive for the proliferative and malignant phenotype of HPV-linked cancers. In non-malignant cells intracellular and intercellular signals suppress these oncogenes at the transcriptional level. The available data permit the delineation of a functional model of interactive events controlling E6/E7 transcription. Modifications of host cell genes engaged in this control will interrupt this regulatory pathway and result in increased E6/E7 gene expression and, as a consequence, in stimulation of cell proliferation.

An intrinsic property of these viral gene products, i.e., to induce mutation and rearrangements of host cell DNA resulting in specific gene modifications and loss of gene function, becomes pronounced upon enhanced expression of the viral oncogenes. This leads to an accumulation of modifications of the host cell genome, in specific changes responsible for eventually resulting invasive and metastatic growth.

Cervical cancer, at present, appears to represent the first human cancer where effector genes (E6/E7) as well as the failure of their natural intracellular and intercellular suppressors can be analyzed in detail and therefore provides an excellent system to study carcinogenesis in humans.

Acknowledgment. Critical reading of this manuscript by Tobias Bauknecht, Ethel-Michele de Villiers, Matthias Dürst, Frank Rosl, Elisabeth Schwarz, and Magnus von Knebel Doeberitz is gratefully acknowledged.

References

Arbeit JM, Münger K, Howley PM, Hanahan D. Neuroepithelial carcinomas in mice transgenic with human papillomavirus type 16 E6/E7 ORF's. Am J Pathol (in press)

Atkin NB, Baker MC (1984) Non-random chromosome changes in carcinoma of the cervix uteri. I. Nine near-diploid tumors. Cancer Genet Cytogenet 7: 2209–2221

Atkin NB, Baker MC (1988) Deficiency of all or part of chromosome 11 in several types of cancer: significance of a reduction in the number of normal chromosomes 11. Cytogenet Cell Genet 47: 106–107

Auvinen E, Kujari H, Arstila P, Hukkanen V. Expression of α2 and E7 genes of the human papillomavirus type 16 in female genital dysplasias. Am J Pathol (in press)

Band V, Zaychowski D, Kulesa V, Sager R (1990) Human papillomavirus DNAs immortalize normal human mammary epithelial cells and reduce their growth factor requirements. Proc Nat Acad Sci USA 87: 463–467

Band V, Dalal S, Delmolino L, Androphy EJ (1993) Enhanced degradation of p53 protein in HPV-6 and BPV-1 immortalized human mammary epithelial cells. EMBO J 12: 1847–1852

Barbosa MS, Edmonds C, Fisher C, Schiller JT, Lowy DR, Vousden KH (1990) The region of the HPV E7 oncoprotein homologous to adenovirus E1A and SV40 large T antigen contains separate domains for Rb binding and casein kinase II phosphorylation. EMBO J 9: 153–160

Bartsch D, Boye B, Baust C, zur Hausen H, Schwarz E (1992) Retinoic acid-mediated repression of human papillomavirus 18 transcription and different ligand regulation of the retinoic acid receptor β gene in non-tumorigenic and tumorigenic HeLa hybrid cells. EMBO J 11: 2283–2291

Batova A, Danielpour D, Pirisi L, Creek KE (1992) Retinoic acid induces secretion of latent transforming growth factor beta 1 and beta 2 in normal and human papillomavirus type 16-immortalized human keratinocytes. Cell Growth Differ 3: 763–772

Bauknecht T, Angel P, Royer H-D, zur Hausen H (1992) Identification of a negative regulatory domain in the human papillomavirus type 18 promoter: interaction with the transcriptional repressor YY1. EMBO J 11: 4607–4617

Bosch F, Schwarz E, Boukamp P, Fusenig NE, Bartsch D, zur Hausen H (1990) Suppression in vivo of human papillomavirus type 18 E6–E7 gene expression in nontumorigenic HeLa-fibroblast hybrid cells. J Virol 64: 4743–4754

Boshart M, zur Hausen H (1986) Human papillomaviruses in Buschke-Löwenstein tumors: physical state of DNA and identification of a tandem duplication in the non-coding region of a human papillomavirus 6 subtype. J Virol 58: 963–966

Boshart M, Gissmann L, Ikenberg H, Kleinheinz A, Scheurlen W, zur Hausen H (1984) A new type of papillomavirus DNA, its presence in genital cancer biopsies and in cell lines derived from cervical cancer. EMBO J 3: 1151–1157

Braun L, Dürst M, Mikumo R, Guipposo P (1990) Differential response of nontumorigenic and tumorigenic human papillomavirus type 16-positive epithelial cells to transforming growth factor β. Cancer Res 50: 7324–7332

Burnett S, Ström AC, Jareborg N, Alderborn A, Dillner J, Moreno-Lopez J, Pettersson U, Kiessling N (1990) Induction of E2 gene expression and early region transcription by cell growth arrest: correlation with viral DNA amplification and evidence for differential promoter induction. J Virol 64: 5529–5541

Chan WK, Klock G, Bernard HU (1989) Progesterone and glucocorticoid control elements occur in the long control regions of several human papillomaviruses involved in anogenital neoplasia. J Virol 63: 3261–3269

Chan WK, Chong T, Bernard HU, Klock G (1990) Transcription of the transforming genes of the oncogenic human papillomavirus-16 is stimulated by tumor promoters through AP-I binding sites. Nucleic Acid Res 18: 763–769

Chellappan SP, Hiebert S, Mudryj M, Horowitz JM, Nevins JR (1991) The E2F transcription factor is a cellular target of the RB protein. Cell 65, 1053–1061

Chen T-M, Pecoraro G, Defendi V (1993) Genetic analysis of *in vitro* progression of human papillomavirus-transfected human cervical cells. Cancer Res 53: 1167–1171

Chittenden T, Livingston DM, Kaelin WG Jr (1991) The T/E1A-binding domain of the retinoblastoma product can interact selectively with a sequence-specific DNA binding protein. Cell 65: 1073–1082

Cid A, Auewarakul P, Garcia-Carranca A, Ovseiovich R, Gaissert H, Gissmann L (1993) Cell-type specific activity of the human papillomavirus (HPV) type 18 upstream regulatory region in transgenic mice and its modulation by TPA and glucocorticoids. J Virol 67: 6742–6752

Cripe TP, Alderborn A, Anderson RD, Pakkinen S, Bergman T, Haugen H, Petterson V, Turek LP (1990) Transcriptional activation of the human papillomavirus-16 P97 promoter by an 88 nucleotide enhancer containing distinct cell-dependent and AP-1 responsive modules. New Biol 2: 450–463

Crook T, Vousden KH (1992) Properties of p53 mutations detected in primary and secondary cervical cancers suggest mechanisms of metastasis and involvement of environmental carcinogens. EMBO J 11: 3935–3940

Crook T, Morgenstern JP, Crawford L, Banks L (1989) Continued expression of HPV16 E7 protein is required for maintenance of the transformed phenotype of cells co-transformed by HPV 16 plus EJ-ras. EMBO J 8: 513–519

Crum CP, Nagai N, Milao M, Levine RU, Silverstein SJ (1986) Histological and molecular analysis of early neoplasia. J Cell Biochem [Suppl] 9c: 70

Cullen AP, Reid R, Campion M, Lorincz AT (1991) Analysis of the physical state of different human papillomavirus DNA's in intraepithelial and invasive cervical neoplasms. J Virol 65: 606–612

Debiec-Rychter M, Zukowski K, Wang CY, Wen W-N (1991) Chromosomal characterizations of human nasal and nasopharyngeal cells immortalized by human papillomavirus type 16 DNA. Cancer Genet Cytogenet 52: 51–61

Defoe-Jones D, Vuocolo GA, Haskell KM, Hanobik MG, Kiefer DM, McAvoy EM, Ivey-Hoyle M, Brandsma JL, Oliff A, Jones RE (1993) Papillomavirus E7 protein binding to the retinoblastoma protein is not required for viral induction of warts. J Virol 67: 716–725

Donehower LA, Harvey M, Slagle BL, McArthur MJ, Montgomery CA, Butel JS, Bradley A (1992) Mice deficient for p53 are developmentally normal but susceptible to spontaneous tumours. Nature 356: 215–221

Drews RE, Chan VT-W, Schnipper LE (1992) Oncogenes result in genomic alterations that activate a transcriptionally silent dominantly selectable reporter gene (neo). Mol Cell Biol 12: 198–206

Dürst M, Gissmann L, Ikenberg H, zur Hausen H (1983) A new papillomavirus DNA from a cervical carcinoma and its prevalence in cancer biopsy samples from different geographic regions. Proc Natl Acad Sci USA 80: 3812–3815

Dürst M, Kleinheinz A, Hotz M, Gissmann L (1985) The physical state of human papillomavirus type 16 DNA in benign and malignant genital tumors. J Gen Virol 66: 1515–1522

Dürst M, Dzarlieva-Petrusevska RT, Boukamp P, Fusenig NE, Gissmann L (1987) Molecular and cytogenetic analysis of immortalized human primary keratinocytes obtained after transfection with human papillomavirus type 16 DNA. Oncogene 1: 251–256

Dürst M, Bosch F, Glitz D, Schneider A, zur Hausen H (1991) Inverse relationship between HPV 16 early gene expression and cell differentiation in nude mice epithelial cysts and tumors induced by HPV positive human cell lines. J Virol 65: 796–804

Dürst M, Glitz D, Schneider A, zur Hausen H (1992) Human papillomaviruses type 16 (HPV16) gene expression and DNA replication in cervical neoplasia: analysis by in situ hybridization. Virology 189: 132–140

Dyson N, Howley PM, Münger K, Harlow E (1989) The human papillomavirus 16 E7 oncoprotein is able to bind to the retinoblastoma gene product. Science 243: 934–937

Evans AS (1976) Epidemiological concepts and methods. In: Evans AS (ed) Viral infections of humans, epidemiology and control. Wiley, London, pp 1–32

Fu Y-S, Braun L, Shah KV, Lawrence WP, Robboy SJ (1983) Histologic, nuclear DNA and human papillomavirus studies of cervical condylomas. Cancer 52: 1705–1711

Garcia-Carranca A, Thierry F, Yaniv M (1988) Interplay of viral and cellular proteins along the long control region of human papillomavirus 18. J Virol 63: 4321–4330

Gebert JF, Moghal N, Frangioni JV, Sugarbaker DJ, Neel BG (1991) High frequency of retinoic acid receptor β abnormalities in human lung cancer. Oncogene 6: 1859–1868.

Gloss B, Bernard HU (1990) The E6/E7 promoter of human papillomavirus type 16 is activated in the absence of E2 proteins by a sequence-aberrant SP I distal element. J Virol 64: 5577–5584

Gloss B, Bernard HU, Seedorf K, Klock G (1987) The upstream regulatory region of human papillomavirus-16 contains an E2 protein-independent enhancer which is specific for cervical carcinoma cells and regulated by glucocorticoid hormones. EMBO J 6: 3735–3743

Gloss B, Chong T, Bernard HU (1989) Numerous nuclear factors bind the long control region of human papillomavirus type 16: a subset of 6 out of 23 DNase I-protected segments coincides with the location of the cell-type-specific enhancer. J Virol 63: 1142–1152
Griep AE, Herber R, Jeon S, Lohse JK, Dubielzig RR, Lambert P (1993) Tumorigenicity by HPV 16 E6 and E7 in transgenic mice correlates with alterations in epithelial cell growth and differentiation. J Virol 67: 1373–1384
Gross L (1983) Oncogenic viruses, 3rd edn. Pergamon, Oxford
Halbert CL, Demers GW, Galloway DA (1991) The E7 gene of human papillomavirus type 16 is sufficient for immortalization of human keratinocytes. J Virol 65: 473–478
Han R, Breitburd F, Marche PN, Orth G (1992) Linkage of regression and malignant conversion of rabbit viral papillomas to MHC class II genes. Nature 356: 66–68
Hartwell L (1992) Defects in a cell cycle check point may be responsible for the genomic instability of cancer cells. Cell 71: 543–546
Hashida T, Yasumoto S (1991) Induction of chromosome abnormalities in mouse and human epidermal keratinocytes by the human papillomavirus type 16 E7 oncogene. J Gen Virol 72: 1569–1577
Haskell KM, Vuocolo GA, Defoe-Jones D, Jones RE, Ivey-Hoyle M (1993) Comparison of the binding of the human papillomavirus type 16 and cottontail rabbit papillomavirus E7 proteins to retinoblastoma gene product. J Gen Virol 74: 115–119
Hawley-Nelson P, Vousden KH, Hubbert NL, Lowy DR, Schiller JT (1989) HPV 16 E6 and E7 proteins cooperate to immortalize human foreskin keratinocytes. EMBO J. 8: 3905–3910
Heck DV, Yee CL, Howley PM, Münger K (1992) Efficiency of binding the retinoblastoma protein correlates with the transforming capacity of the E7 oncoproteins of the human papillomaviruses. Proc Natl Acad Sci USA 89: 4442–4446
Helland A, Borresen AL, Kaern J, Ronningen KS, Thorsby E (1992) HLA antigens and cervical cancer. Nature 356: 23
Hoppe-Seyler F, Butz K (1992) Activation of human papillomavirus type 18 E6-E7 oncogene expression by transcription factor Sp1. Nucleic Acid Res 20: 6701–6706
Hoppe-Seyler F, Butz K, zur Hausen H (1991) Repression of the human papillomavirus type 18 enhancer by the cellular transcription factor Oct-1. J Virol 65: 5613–5618
Houle B, Rochette-Egly C, Bradley WEC (1993) Tumor-suppressive effect of the retinoic acid receptor β in human epidermoid lung cancer cells. Proc Natl Acad Sci USA 90: 985–989
Howley PM (1991) Role of human papillomaviruses in human cancer. Cancer Res 51: 5019–5022
Hubbert S, Schiller JT, Lowy DR, Androphy EJ (1988) Bovine papillomavirus transformed cells contain multiple E2 proteins. Proc Natl Acad Sci USA 85: 5864–5868
Hubbert NL, Sedman SA, Schiller JT (1992) Human papillomavirus type 16 E6 increases the degradation rate of p53 in human keratinocytes. J Virol 66: 6237–6241
Hudson JB, Bedell ML, McCance DJ, Laimins LA (1990) Immortalization and altered differentiation of human keratinocytes *in vitro* by the E6 and E7 open reading frames of human papillomavirus type 18. J Virol 64: 519–529
Hurlin PJ, Kaur P, Smith P, Perez-Reyes N, Blanton RA, McDougall JK (1991) Progression of human papillomavirus type 18 immortalized human keratinocytes to a malignant phenotype. Proc Natl Acad Sci USA 88: 570–574
Iftner T, Oft M, Böhm S, Wolczynski SP, Pfister H (1992) Transcription of the E6 and E7 genes of human papillomavirus type 6 in anogenital condylomata is restricted to undifferentiated cell layers of the epithelium. J Virol 66: 4639–4646
Ishiji T, Lace MJ, Parkhinen S, Anderson RD, Haugen TH, Cripe TP, Xiao J-H, Davidson I, Chambon P, Turek LP (1992) Transcriptional enhancer factor (TEF)-1 and its cell-specific co-activator activate human papillomavirus-16 E6 and E7 oncogene transcription in keratinocytes and cervical carcinoma cells. EMBO J 11: 2271–2281
Jagella HP, Stegner HE (1974) Zur Dignität der Condylomata acuminata. Klinische, histopathologische und cytophotometrische Befunde. Arch Gynaekol 216: 119–132
Jewers RJ, Hildebrandt P, Ludlow JW, Kell B, McCance DJ (1992) Regions of human papillomavirus type 16 E7 oncoprotein required for immortalization of human keratinocytes. J Virol 66: 1329–1335
Kaebling M, Klinger HP (1986) Suppression of tumorigenicity in somatic cell hybrids. III. Co-segregation of human chromosome 11 of a normal cell and suppression of tumorigenicity in intraspecies hybrids of normal diploid-malignant cells. Cytogenet Cell Genet 41: 65–70

Kaelin WG, Pallas DC, DeCaprio JA, Kaye FJ, Livingston DM (1991) Identification of cellular proteins that can interact specifically with the T/E1A-binding region of the retinoblastoma gene product. Cell 64, 521–532

Karlen S, Beard P (1993) Identification and characterization of novel promoters in the genome of human papillomavirus type 18. J Virol 67: 4296–4306

Kastan MB, Onyekwere O, Sidransky D, Vogelstein B, Craig RW (1991) Participation of p53 protein in the cellular response to DNA damage. Cancer Res. 51: 6304–6311

Kessis TD, Slebos RJ, Nelson WG, Kastan MB, Plunkett BS, Hau SM, Lorincz AT, Hedrick L, Cho KR (1993) Human papillomavirus 16 E6 expression disrupts the p53-mediated cellular response to DNA damage. Proc Natl Acad Sci USA 90: 3988–3992

Koch R (1891) Über bakteriologische Forschung. Verhandlungen des 10 Internationalen Medizinischen Congress, Berlin, Vol 1, p 35

Koi M, Morita H, Yamada H, Saboh H, Barrett JC, Oshimura H (1989) Normal human chromosome 11 suppresses tumorigenicity of human cervical tumor cell line SiHa. Mol Carcinog 2: 12–21

Kondoh G, Murata Y, Aozasa K, Yutsudo M, Hakura A (1991) Very high incidence of germ cell tumorigenesis (seminomagenesis) in human papillomavirus type 16 transgenic mice. J Virol 65: 3335–3339

Kyo S, Inoue M, Nishio Y, Nakanishi K, Akira S, Inoue H, Yutsudo M, Tanizawa O, Hakura A (1993) NF-IL 6 represses early gene expression of human papillomavirus type 16 through binding to the noncoding region. J Virol 67: 1058–1066

Lambert PF, Spalholz BA, Howley PM (1987) A transcriptional repressor encoded by BPV-1 shares common carboxy-terminal domain with the E2 transactivator. Cell 50: 69–78

Lane DP (1992) p53, guardian of the genome. Nature 358: 15–16

Lechner MS, Mack DH, Finicle AB, Crook T, Vousden KH, Laimins LA (1992) Human papillomavirus E6 proteins bind p53 *in vivo* and abrogate p53-mediated repression of transcription. EMBO J 11: 3045–3052

Lehn H, Villa LL, Marziona F, Hilgarth M, Hillemanns HG, Sauer G (1988) Physical state and biological activity of human papillomavirus genomes in precancerous lesions of the female genital tract. J Gen Virol 69: 187–196

Leid M, Kastner P, Chambon P (1992) Multiplicity generates diversity in the retinoic acid signalling pathways. TIBS 17: 427–433

Mack DH, Laimins LA (1991) A keratinocyte-specific transcription factor, KRE-1, interacts with AP-1 to activate expression of human papillomavirus type 18 in squamous epithelial cells. Proc Natl Acad Sci USA 88: 9102–9106

Malejczyk J, Malejczyk M, Urbanski A, Köck A, Jablonska S, Orth G, Luger T (1991) Constitutive release of IL 6 by human papillomavirus type 16 (HPV 16)-harboring keratinocytes: a mechanism augmenting the NK-cell-mediated lysis of HPV-bearing neoplastic cells. Cell Immunol 136: 155–164

Malejczyk J, Malejczyk M, Köck A, Urbanski A, Majewski S, Hunzelmann N, Jablonska S, Orth G, Luger TA (1992) Autocrine growth limitation of human papillomavirus type 16—harboring keratinocytes by constitutively released tumor necrosis factor-α. J Immunol 149: 2702–2708

Matsukura T, Koi S, Sugase M (1989) Both episomal and integrated forms of human papillomavirus type 16 are involved in invasive cervical cancers. Virology 172: 63–72

McCance DJ, Kopan R, Fuchs E, Laimins LA (1988) Human papillomavirus type 16 alters human epithelial cell differentiation *in vitro*. Proc Natl Acad Sci USA 85: 7169–7173

Münger K, Phelps WC, Bubb V, Howley PM, Schlegel R (1989) The E6 and E7 genes of human papillomavirus type 16 are necessary and sufficient for transformation of primary human keratinocytes. J Virol 63: 4417–4421

Nevins JR (1992a) E2F: a link between Rb tumor suppressor protein and viral oncoproteins. Science 258, 424–429

Nevins JR (1992b) Transcriptional regulation. A closer look at E2F. Nature 358, 375–376

Offord EA, Beard P (1990) A member of the activator protein 1 family found in keratinocytes but not in fibroblasts required for transcription from a human papillomavirus type 18 promoter. J Virol 64: 4792–4798

Pagano M, Dürst M, Joswig S, Draetta G, Jansen-Dürr P (1992) Binding of human E2F transcription factor to the retinoblastoma protein but not cyclin A is abolished in HPV 16-immortalized cells. Oncogene 7: 1681–1687

Park N-H, Min B-M, Li S-L, Huang MZ, Cherick HM, Doninger J (1991) Immortalization of normal human oral keratinocytes with type 16 human papillomavirus. Carcinogenesis 12: 1627–1631

Pater MM, Hughes GA, Hyslop DE, Nakshatri H, Pater A (1988) Glucocorticoid-dependent oncogenic transformation by type 16 and not type 11 human papillomavirus DNA. Nature 335: 832–835

Pecoraro G, Lee M, Morgan D, Defendi V (1991) Evolution of *in vitro* transformation and tumorigenesis of HPV 16 and HPV 18 immortalized primary cervical epithelial cells. Am J Pathol 138: 1–8

Peng X, Olson RO, Christian CB, Lang CM, Kreider JW (1993) Papillomas and carcinomas in transgenic rabbits carrying EJ-ras and cottontail rabbit papillomavirus DNA. J Virol 67: 1698–1701

Phelps WC, Baghi S, Barnes JA, Raychaudhuri P, Kraus V, Münger K, Howley PM, Nevius JR (1991) Analysis of transactivation by human papillomavirus type 16 E7 and adenovirus 12 S E1A suggest a common mechanism J Virol 65: 6922–8930

Phelps WC, Münger K, Yee CL, Barnes JA, Howley PM (1992) Structure-function analysis of the human papillomavirus type 16 E7 oncoprotein. J Virol 66: 2418–2427

Pirisi L, Yasumoto S, Fellery M, Doninger JK, DiPaolo JA (1987) Transformation of human fibroblasts and keratinocytes with human papillomavirus type 16 DNA. J Virol 61: 1061–1066

Pirisi L, Batova A, Jenkins GR, Hodam, JR, Creek KE (1992) Increased sensitivity of human keratinocytes immortalized by human papillomavirus type 16 DNA to growth control by retinoids. Cancer Res 52: 187–193

Rösl F, Dürst M, zur Hausen H (1988) Selective suppression of human papillomavirus transcription in non-tumorigenic cells by 5-aza-cytidine. EMBO J 7: 1321–1328

Rösl F, Westphal E-M, zur Hausen H (1989) Chromatin structure and transcriptional regulation of human papillomavirus type 18 DNA in HeLa cells. Mol Carcinog 2: 72–80

Rösl F, Achtstetter T, Hutter K-J, Bauknecht T, Futterman G, zur Hausen H (1991) Extinction of the HPV 18 upstream regulatory region in cervical carcinoma cells after fusion with non-tumorigenic human keratinocytes under non-selective conditions. EMBO J 10: 1337–1345

Rösl F, Arab A, Klevenz B, zur Hausen H (1993) The effect of DNA methylation on gene regulation of human papillomaviruses. J Gen Virol 74: 791–801

Rollins BJ, Sunday ME (1991) Suppression of tumor formation in vivo by expression of the JE gene in malignant cells. Mol Cell Biol 11: 3125–3131

Romanczuk H, Thierry F, Howley PM (1990) Mutational analysis of cis elements involved in E2 modulation of human papillomavirus type 16 P_{97} and 18 P_{105} promoters. J Virol 64: 2849–2859

Saxon PJ, Srivatsan ES, Stanbridge J (1986) Introduction of human chromosome 11 via microcell transfer controls tumorigenic expression in HeLa cells. EMBO J 5: 3461–3466

Scheffner M, Werness BA, Huibregtse JM, Levine JM, Howley PM (1990) The E6 oncoprotein encoded by human papillomavirus types 16 and 18 promotes the degradation of p53. Cell 63: 1129–1136

Scheffner M, Münger K, Byrne JC, Howley PM (1991) The state of the p53 and retinoblastoma genes in human cervical carcinoma cell lines. Proc Natl Acad Sci USA 88: 5523–5527

Schwarz E (1987) Transcription of papillomavirus genomes. In: Syrjänen K, Gissmann L, Koss LG (eds) Papillomaviruses and human disease. Springer, Berlin Heidelberg New York, pp 443–466

Schwarz E, Freese UK, Gissmann L, Mayer W, Roggenbuck B, Stremlau A, zur Hausen H (1985) Structure and transcription of human papillomavirus sequences in cervical carcinoma cells. Nature 314: 111–114

Searle PF, Thomas DP, Faulkner KB, Tinsley JM. Gastric carcinoma in transgenic mice expressing HPV 16 early region genes (submitted for publication)

Sippola-Thiele M, Hanahan D, Howley PM (1989) Cell-heritable stages of tumor progression in transgenic mice harboring the bovine papillomavirus type 1 genome. Mol Cell Biol 9: 925–934

Sircar S, Horvath J, Roberge D, Diouri M, Weber JM (1992) Adenovirus transformation revertant resistant to retransformation by E1 but not by SV40-T and HPV 16-E7 oncogenes. Virology 191: 187–192

Smith PP, Friedman CL, Bryand EM, McDougall JK (1992) Viral integration and fragile sites in human papillomavirus-immortalized human keratinocyte cell lines. Genes, Chrom Cancer 5: 150–157

Smits HL, Raadsheer E, Rood I, Mehendale S, Slater RM, van der Noordaa J, ter Schegget J (1988) Induction of anchorage-independent growth of human embryonic fibroblasts with a deletion in the short arm of chromosome 11 by human papillomavirus type 16 DNA. J Virol 62: 4538–4543

Smits PHM, deRonde A, Smits HL, Minaar RP, van der Noordaa J, ter Schegget J (1992a) Modulation of the human papillomavirus type 16 induced transformation and transcription by deletion of loci on the short arm of chromosome 11 can be mimicked by SV 40 small t. Virology 190: 40–44

Smits PHM, Smits HL, Minnaar R, Hemmings BA, Mayer-Jaekel RE, Schuurman R, van der Noordaa J, ter Schegget J (1992b) The trans-activation of the HPV 16 long control region in human cells with a deletion in the short arm chromosome 11 is mediated by the 55kDa regulatory subunit of protein phosphatase 2A. EMBO J 11: 4601–4606

Smits PHM, Smits HL, Minnaar RP, ter Schegget J (1993) Regulation of human papillomavirus type 16 (HPV-16) transcription by loci on the short arm of chromosome 11 is mediated by the TATAAAA motif of the HPV 16 promoter. J Gen Virol 74: 121–124

Smotkin D, Wettstein FO (1986) Transcription of human papillomavirus type 16 early genes in a cervical cancer and a cancer-derived cell line and identification of the E7 protein. Proc Natl Acad Sci USA 83: 4680–4684

Spalholz BA, Yang Y-C, Howley PM (1985) Transactivation of bovine papillomavirus transcriptional regulatory element by the E2 gene product. Cell 42: 183–191

Sreekantaiah C, de Braekeleer M, Haas O (1991) Cytogenetic findings in cervical carcinoma: a statistical approach. Cancer Genet Cytogenet 51: 75–81

Stoler MH, Rhodes CR, Whitbek A, Wolinsky SM, Chow LT, Broker TR (1992) Human papillomavirus type 16 and 18 gene expression in cervical neoplasias. Human Pathol 23: 117–128

Thierry F, Spyrou G, Yaniv M, Howley P (1992) Two AP-1 sites binding JunB are essential for human papillomavirus type 18 transcription in keratinocytes. J Virol 66: 3740–3748

von Knebel Doeberitz M, Oltersdorf T, Schwarz E, Gissmann L (1988) Correlation of modified human papillomavirus early gene expression with altered growth properties in C4-1 cervical carcinoma cells. Cancer Res. 48: 3780–3786

von Knebel Doeberitz M, Bauknecht T, Bartsch D, zur Hausen H (1991) Influence of chromosomal integration on glucocorticoid regulated transcription of growth-stimulating E6–E7 genes in cervical carcinoma cells. Proc Natl Acad Sci USA 88: 1411–1415

von Knebel Doeberitz M, Rittmüller C, zur Hausen H, Dürst M (1992) Inhibition of tumorigenicity of C4-1 cervical cancer cells in nude mice by HPV 18 E6–E7 antisense RNA. Int J Cancer 51: 831–834

Wank R, Thomssen C (1991) High risk of squamous cell carcinoma of the cervix for women with HLA-DQw3. Nature 352: 723–725

Wank R, Schendel DJ, Thomssen C (1992) HLA antigens and cervical carcinoma. Nature 356: 22–23

Watanabe S, Kanda T, Yoshiike K (1989) Human papillomavirus type 16 transformation of primary human embryonic fibroblasts requires expression of open reading frames E6 and E7. J Virol 63: 965–969

Werness BA, Levine AJ, Howley PM (1990) Association of human papillomavirus types 16 and 18 E6 proteins with p53. Science 248, 76–79

Willey JC, Broussoud A, Sleemi A, Bennett WP, Cerutti P, Harris, CC (1991) Immortalization of normal human bronchial epithelial cells by human papillomavirus 16 and 18. Cancer Res 51: 5370–5377

Woodworth CD, Simpson S (1993) Comperative lymphokine secretion by cultured normal human cervical keratinocytes, papillomavirus-immortalized, and carcinoma cell lines. Am J Pathol 142: 1544–1555

Woodworth CD, Waggoner S, Barnes W, Stoler MH, DiPaolo JA (1990a) Human cervical and foreskin epithelial cells immortalized by human papillomavirus DNAs exhibit dysplastic differentiation *in vivo*. Cancer Res 50: 3709–3715

Woodworth CD, Notario V, DiPaolo JA (1990b) Transforming growth factor beta 1 and 2 transcriptionally regulate human papillomavirus (HPV) type 16 early gene expression in HPV-immortalized human genital epithelial cells. J Virol 64: 4767–4775

Woodworth CD, Lichti U, Simpson S, Evans CH, DiPaolo JA (1992) Leukoregulin and gamma-interferon inhibit human papillomavirus type 16 gene transcription in human papillomavirus-immortalized human cervical cells. Cancer Res 52: 456–463
Yang S, Lickteig RL, Estes R, Rundell K, Walter G, Mumby M (1991) Control of phosphatase 2A by simian virus 40 small t antigen. Mol Cell Biol 11: 1988–1995
Yasumoto S, Taniguchi A, Sohma K (1991) Epidermal growth factor (EGF) elicits down-regulation of human papillomavirus-type 16 (HPV-16) E6/E7 mRNA at the transcriptional level in an EGF-stimulated human keratinocyte cell line: functional role of EGF-responsive silencer in the HPV-16 long control region. J Virol 65: 2000–2009
Yee C, Krishnan-Hewlatt I, Baker C, Schlegel R, Howley P (1985) Presence and expression of human papillomavirus sequences in human cervical carcinoma cell lines. Am J Pathol 119: 361–366
zur Hausen H (1975) Oncogenic herpesviruses. Biochem Biophys Acta 417: 25–53
zur Hausen H (1976) Condylomata acuminata and human genital cancer. Cancer Res 36: 530
zur Hausen H (1977a) Human papillomaviruses and their possible role in squamous cell carcinomas. In: Compans RW, Cooper M, Koprowski H et al. (eds) Current topics in microbiology and immunology, vol 78. Springer, Berlin Heidelberg New York, pp 1–30
zur Hausen H (1977b) Cell-virus gene balance hypothesis of carcinogenesis. Behring Inst Mitt 61: 23–30
zur Hausen H (1986) Human genital cancer: synergism between two virus infections or synergysm between a virus infection and initiating events. Lancet 2: 1370–1372
zur Hausen H (1986a) Genital papillomavirus infections. In: Rigby PWJ, Wilkie NM (eds) Viruses and Cancer. Cambridge University Press, Cambridge, pp 83–90
zur Hausen H (1989) Papillomaviruses in anogenital cancer as a model to understand the role of viruses in human cancers. Cancer Res 49: 4677–4681
zur Hausen H (1991a) Viruses in human cancers. Science 254: 1167–1173
zur Hausen H (1991b) Papillomavirus/host cell interactions in the pathogenesis of anogenital cancer. In: Brugge J, Curran T, Harlow E, McCormick F (eds) Origins of human cancer. Cold Spring Harbor Laboratory Press, Cold Spring Harbor, pp 685–705
zur Hausen H (1991c) Human papillomaviruses in the pathogenesis of anogenital cancer. Virology 184: 9–13

Epidermodysplasia Verruciformis: Immunological and Clinical Aspects

S. JABLONSKA and S. MAJEWSKI

1 Introduction

Epidermodysplasia verruciformis (EV) is a rare life-long genetic skin disorder associated with specific human papillomaviruses (EV HPVs) and characterized by disseminated wart-like or pityriasis versicolor (PV)-like lesions and skin cancers. The cases may be familial or sporadic, and the inheritance is believed to be autosomal recessive (LUTZNER 1978), although single cases of X-linked inheritance have been reported (ANDROPHY et al. 1985).

It can be speculated that in patients with EV, there is a defect of an unknown suppressor gene which is present in healthy individuals and

Department of Dermatology, Warsaw School of Medicine, Koszykowa 82A, 02–008 Warsaw, Poland prevents infection of humans with EV HPVs, or transfection of normal keratinocytes. It is conceivable that in a predisposed cell acquired from genetically affected parent, the single normal allele is sufficient for maintenance of normal phenotype. Second mutation inactivating the remaining normal allele is required to initiate malignant growth. In EV, however, there are no data on the defect of any characterized suppressor genes, e.g., p53 or pRb-105. A characteristic feature of the disease is immunotolerance, i.e., inability to recognize and reject own HPVs which are harmless for the general population. On the other hand, a genetically determined immunosurveillance appears to prevent invasive growth and metastases of skin tumors. Thus EV is a natural model for both cutaneous viral oncogenesis in humans and a model for studying factors involved in anti-tumor immune defense mechanisms.

Since the characterization of first EV HPVs in 1978 (ORTH et al.), over 20 specific HPVs have been detected but only a few of them, mainly HPV-5 and HPV-8 or related EV HPVs, have been found to be associated with carcinomas originating from benign lesions of EV.

2 Clinical Aspects

2.1 Epidemiology

Epidermodysplasia verruciformis is distributed worldwide, and benign lesions are associated with diverse EV HPVs, while the cancers usually harbor HPV-5 or HPV-8 DNA. Not infrequently EV patients are infected with more than one, sometimes with several specific HPVs and/or with HPV-3 (mixed infection).

Although EV HPVs do not infect normal individuals, in immunosuppressed patients wart-like and PV-like lesions containing EV HPV DNA may appear, they are frequently transitory, but are sometimes persistent and converting into full-blown EV (LUTZNER et al. 1980; GASSENMAIER et al. 1986; RUDLINGER et al. 1986; GROSS et al. 1988).

Epidermodysplasia verruciformis manifestations have also been reported in patients with human immunodeficiency virus infection (PROSE et al. 1990; BERGER et al. 1991), and in all these cases the diagnosis of acquired immunodeficiency syndrome (AIDS) preceded the appearance of characteristic EV lesions. Patients infected with human immunodeficiency virus or with various immune defects often have other viral infections, e.g., plantar or common warts, molluscum contagiosum, and others, whereas in EV patients infected exclusively with EV HPVs there is usually no susceptibility to other HPVs (JABLONSKA et al. 1972; JABLONSKA 1991).

Epidermodysplasia verruciformis HPVs have also been disclosed in the cutaneous plane warts of heavily immunosuppressed patients without any clinical and histological characteristics of EV lesions (VAN DER LEEIST et al. 1987; OBALEK et al. 1992). The warts harboring HPV-3 or HPV-10 may facilitate an EV HPV infection, possibly by providing in trans-viral proteins E1 and E2 (LAMBERT 1991), thus allowing the persistence of episomal EV HPV genomes. However, the host–cell restriction of the EV HPV genome, which does not allow full expression of EV morphology, appears to be responsible for the transitory character of this co-infection. It is to be presumed that the immunosuppressed population is infected with EV HPVs much more often than detected in single cases, in whom plaque lesions appear which resemble clinically and histologically those in EV patients. Thus, this population may constitute a reservoir for potentially oncogenic EV HPVs. Importantly, immunosuppressed patients may develop malignant tumors harboring EV HPVs (LUTZNER et al. 1983; BARR et al. 1989) and this favors the role of these viruses in oncogenesis.

2.2 Cutaneous Manifestations

2.2.1 Lesions Induced by EV HPVs

Cutaneous lesions associated with EV HPV are polymorphic. Most characteristic are red macules, achromic or brownish PV-like plaques, and plane wart-like lesions, usually somewhat flatter and more abundant than plane warts in the general population, sometimes confluent, usually widely dis-

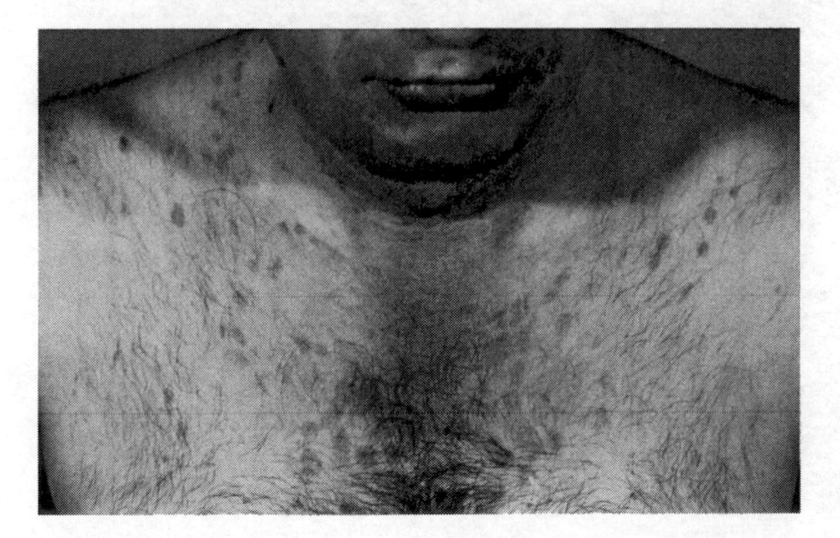

Fig. 1. Red plaques disseminated on the trunk, found to be induced by HPV-5 in a patient with widespread lesions associated with several EV HPVs

seminated all over the body (Fig. 1). On the face and extremities the lesions are wart-like, on the trunk red plaques and PV-like changes prevail.

2.2.2 Lesions Induced by HPV-3

Plane warts associated with HPV-3 are usually somewhat larger than plane wart-like lesions induced by EV HPVs and localized preferentially on the extremities and face (Fig. 2). Not infrequently they are pigmented and confluent, with uneven polycyclic outlines. Recognition of life-long generalized HPV-3 infection as a variety of EV is favored by cases of this type occurring in families with EV HPV infection and, not infrequently, mixed infection in EV patients (JABLONSKA 1991). Patients with disseminated HPV-3 infection have a severe defect of cell-mediated immunity (OBALEK et al. 1979; PFISTER et al. 1979; MAJEWSKI et al. 1986), and probably therefore, after a prolonged exposure, they may become sensitive to EV HPVs. We have followed such cases for 20–30 years and we have found that in the majority developed lesions harboring EV-HPVs after a long time. The mechanism here is probably similar to that of EV HPV infections in immunosuppressed persons, in whom, possibly by point mutations or deletions, the negative control becomes inoperative, leading to susceptibility to EV HPVs. A similar mechanism also appears to be responsible in EV patients for a not infrequent mixed infection with EV HPVs and HPV-3.

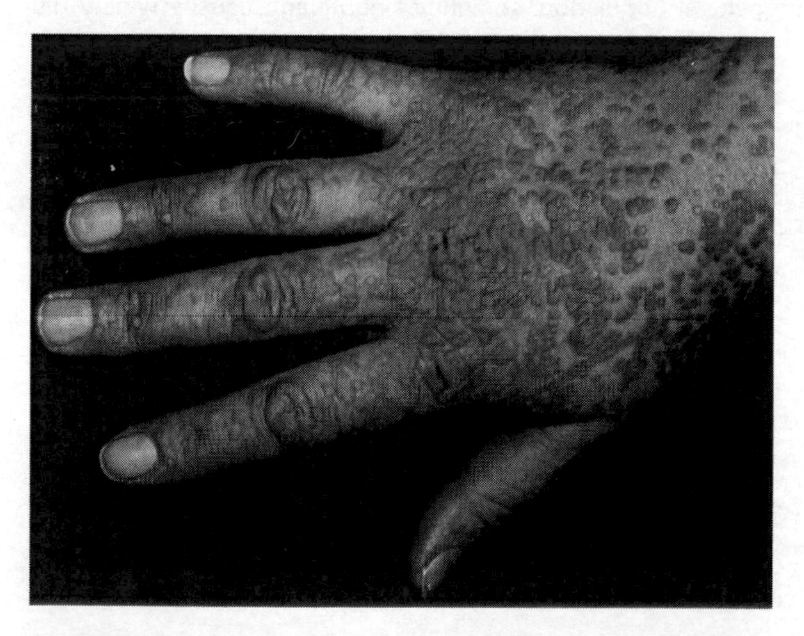

Fig. 2. Plane warts, very abundant and partially confluent, in a patient with mixed infection: HPV-3 on the hands and HPV-8 on the trunk and the face

2.3 Histopathology of Cutaneous Lesions

Large clear cells appear in the upper parts of the epidermis, often starting suprabasally. They are arranged in nests in the granular and spinous layers, displaying clear nucleoplasm and finally granular cytoplasm with prominent keratohyaline granules of various sizes and shapes (Fig. 3). This cytopathic

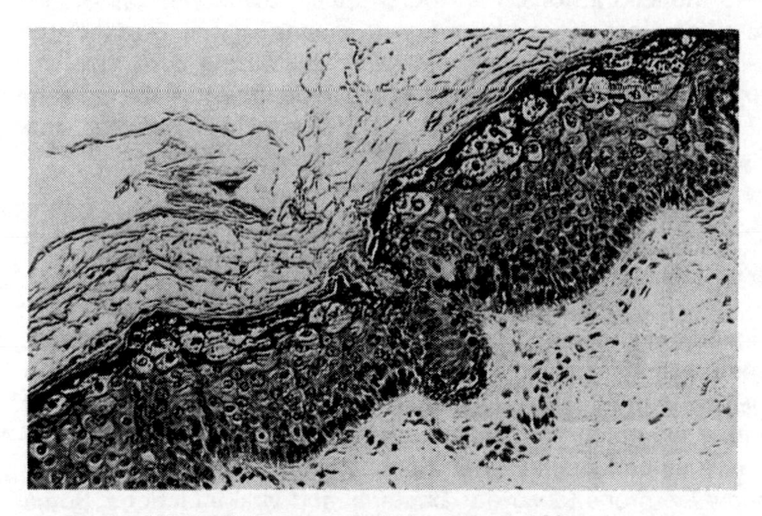

Fig. 3. Specific EV CPE. Hematoxylin and eosin; × 120

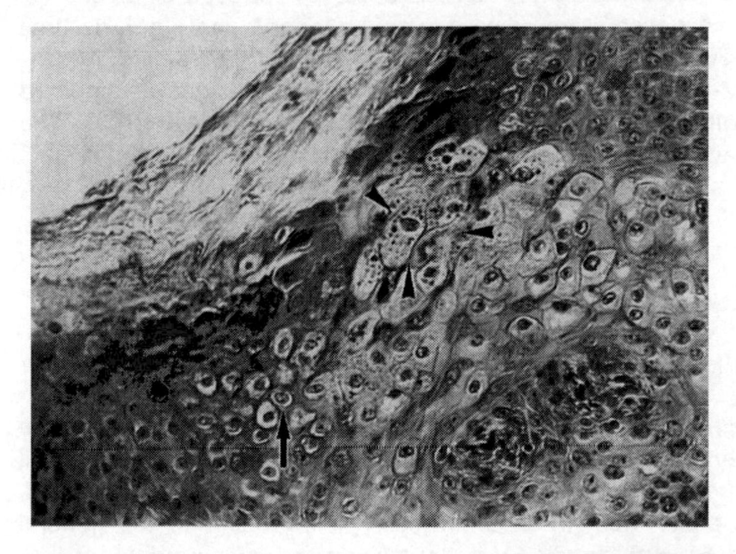

Fig. 4. Histology of elevated red plaque in a case of mixed infection. Typical CPE of HPV-3 (*arrow*): birds's eye-like round cells with pyknotic nuclei surrounded by clear halo. To the *right*, characteristic CPE of EV HPV (*arrowhead*) Hematoxylin and eosin; × 320

effect (CPE) is diagnostic for all EV HPVs irrespective of their types. In electron microscopy characteristic features are: clear nuclei filled with cristalline viral particles, margination of chromatine, and prominent nucleoli; the cytoplasm is devoid of any organelles except for ribosomes; and large keratohyaline granules are not associated with tonofilaments.

Lesions induced by HPV-3 have all the features of plane warts with a characteristic "bird's eye-like" appearance of the cells. In electron microscopy of HPV-3-induced lesions, the nuclei appear to be well preserved, without evident margination of chromatine in spite of the presence of cristalline viral particles. There is a perinuclear vacuolized zone and the displaced cytoplasmic organelles form a rim at the periphery of the cells. In patients with mixed infection, CPE is of the HPV-3 type in some areas, and characteristic of EV HPV in others (Fig. 4).

2.4 Course of Disease

The first cutaneous changes in EV are noticed at the age of 5–8 years, although in some patients red plaques on the trunk may be overlooked. If infected exclusively with EV HPVs, EV patients are usually in a good general condition; routine laboratory studies do not show any abnormalities. In general, they have no concomitant infections, whereas patients with HPV-3/EV HPV variety are prone to various bacterial and viral infections. Some patients have been reported to be mentally retarded (LUTZNER 1978).

We have not seen a complete regression of EV induced by EV HPVs, although in some patients the disease was stationary, and single lesions disappeared. However, we have observed a patient with a typical EV associated with HPV-3, in whom generalized plane wart-like lesions disappeared progressively following two deliveries (JABLONSKA and ORTH 1985). The mother of the patient, her sister, and her aunt had mixed EV HPVs and HPV-3 infection.

2.5 Malignant Conversion

2.5.1 Clinical Manifestations

Over half of the patients followed for 20–30 years developed cancers, some developed multiple actinic keratoses and Bowen's disease. The preferential localization of premalignant lesions, mainly actinic keratoses, is the forehead, i.e., this is the area where these changes typically occur in the general population (Fig. 5). However, in EV patients they are very abundant, appear at a younger age, and much more frequently convert into malignancies than actinic keratoses in the general population (the latter transform into carcino-

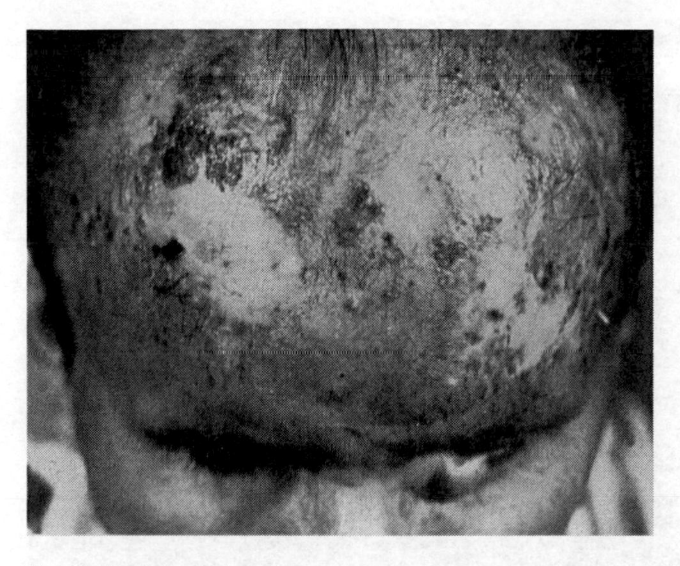

Fig. 5. Multiple carcinomas on the forehead and scalp in a patient with EV. The white scars were left after cryotherapy

mas in about 5% cases). Red plaques and PV-like lesions on the covered areas as well as seborrheic wart-like papillomas, preferentially located on the forehead, palpebra, and neck, rarely undergo malignant transformation. Cancers start to develop after the age of 30, usually in the fourth or fifth decades; they are mainly on the forehead and in traumatized areas. They are microinvasive or invasive and may be locally destructive, but, if not treated with X-rays, grow slowly and do not metastasize.

2.5.2 Histological Patterns

The characteristic histological feature of EV tumors is Bowen's atypia with monstruous dyskeratotic cells (Fig. 6a). The CPE is detectable exclusively in premalignant changes (Fig. 6b). In some tumors originating from actinic keratoses, CPE is still preserved in the superficial epidermis while the proliferating downwards rete ridges display atypical features.

2.5.3 Internal Malignancies

Visceral tumors are unusual for EV. We have seen lymphoma in one patient, tumor cerebri (astrocytoma) in one, and intestinal adenocarcinoma in another. There is no evidence for a direct relationship of internal malignancies with EV HPVs. It is conceivable that general immunosuppression is a predisposing factor here.

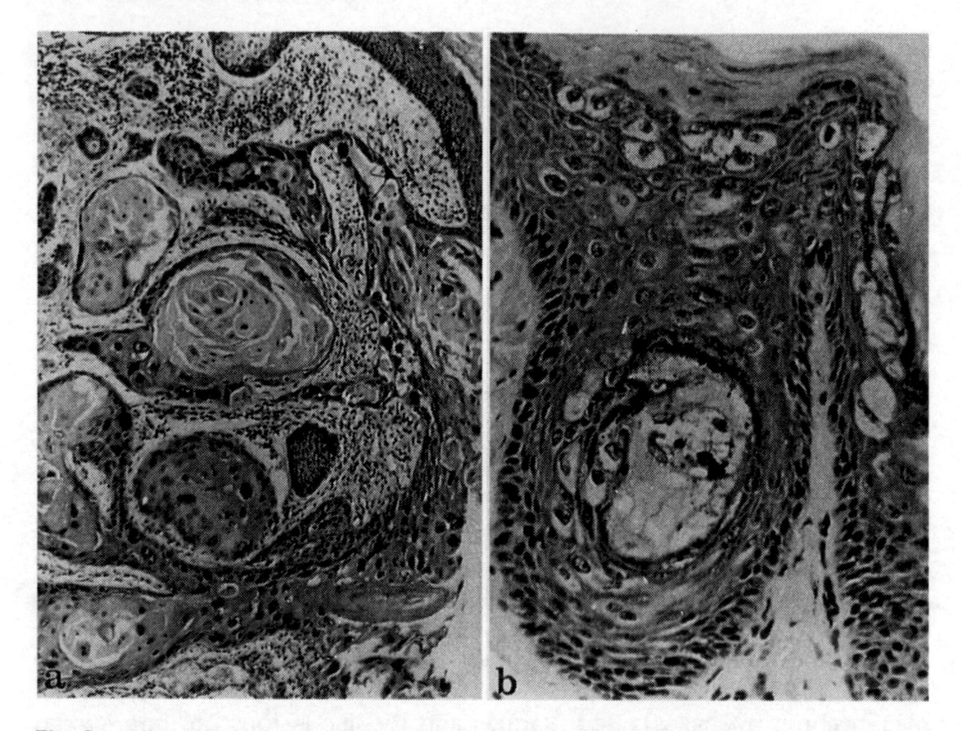

Fig. 6. a Histology of tumor displaying features of Bowen's atypia with pronounced dyskeratosis and formation of keratotic cysts. Such tumors are locally invasive and destructive, but in general do not metastasize. Hematoxylin and eosin; × 60. **b** Pronounced EV CPE in the epidermis and hair follicle in a premalignant lesion. Hematoxylin and eosin; × 320

2.6 Treatment

There is no specific therapy for EV. Promising results have been obtained in HPV-associated cervical cancers with combined therapy of 13-*cis*-retinoic acid and interferon α (IFNα) (LIPPMAN et al. 1992), probably due to a marked potentiation by IFNα of retinoid-induced differentiation (BOLLAG 1991). Retinoids and IFNα were found to have synergistic antiangiogenic and antiproliferative effect on HPV-harboring tumor cell lines (MAJEWSKI et al., 1993). This experimental therapy is now applied in EV patients. The tumors should be removed by surgery, cryotherapy, or laser. Healing is excellent, probably due to an increased activity of transforming growth factor β (TGFβ) in the lesional skin of EV patients. X-rays are strongly contraindicated, as in laryngeal papillomas, due to both the co-cancerogenic effect and the deleterious effect probably induced by the release of large amounts of tumor necrosis factor α (TNFα)

In patients with very numerous premalignant and malignant lesions involving almost the entire forehead, we replace the frontal skin with a graft taken from the uninvolved non-light-exposed skin of the interior aspect of

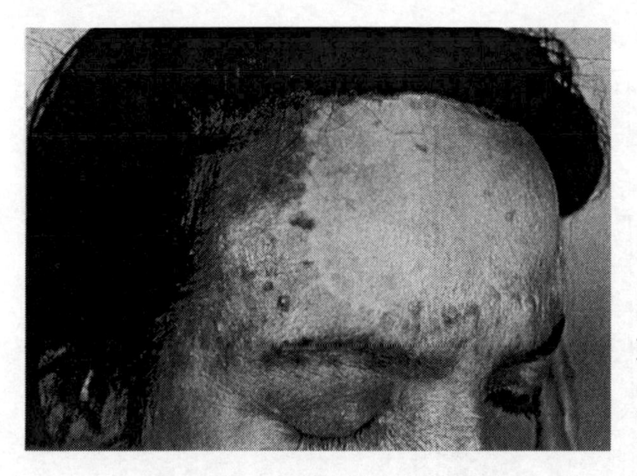

Fig. 7. Grafted skin from the internal aspect of the arm, 6 years after the transplantation. To the *left*, at the periphery multiple premalignant lesions, within the graft single small red plaques

the arm. Within 18 years of follow up we have not seen any malignancies in the grafted skin although grafts were surrounded by steadily developing premalignant and malignant lesions (Fig. 7). Benign red plaques started to appear in the grafted skin 10–14 years after the grafting. This indicates that the development of EV lesions is a slow process depending both on the local cutaneous and external factors, while carcinogenesis is a multistep process over several decades. This is in agreement with the long latency periods which elapse between infection and tumor development, as postulated by ZUR HAUSEN (1986, 1991).

3 Immunological Aspects

3.1 Specific Defect of Cell-Mediated Immunity

A highly characteristic feature of EV is the appearance of the first cutaneous lesions in early childhood, in parallel with the susceptibility of the skin to contact sensitizers. This would suggest a direct link with maturation of the immune system of the child. EV patients have deranged cell-mediated immunity (CMI); the specific defect is the inability to recognize EV HPVs and thus to reject the lesions which persist throughout life. This defect is manifested by the pronounced inhibition of natural killer (NK) cell activity (MAJEWSKI et al. 1990) and cytotoxic T cells (COOPER et al. 1990) against the specific target. The most important and constant abnormality is cutaneous anergy to locally applied strong contact sensitizers, e.g., dinitrochlorobenzene (DNCB) (Fig. 8). We found this defect most pronounced in mixed infections with EV HPVs and HPV-3 (GLINSKI et al. 1981). EV patients

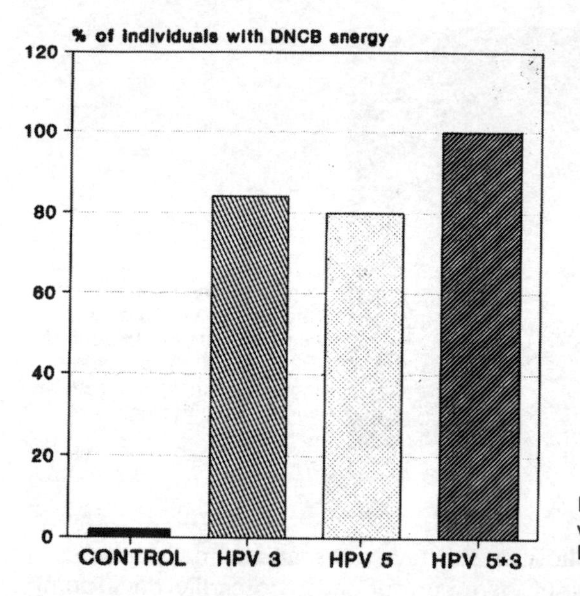

Fig. 8. Sensitization to DNCB in various forms of EV, as compared to healthy individuals

infected exclusively with HPV-3 had, in general, the most defective CMI, whereas patients infected only with EV HPVs had partially preserved cellular immunity: angiogenic activity, antibody-dependent cytotoxicity, etc. (MAJEWSKI et al. 1986).

3.2 Immunogenetic Control of EV HPV Infection

Since the immune response of T lymphocytes against various antigens, including HPV, requires their processing and presentation by proteins coded by the major histocompatibility complex (MHC) (BRACIALE and BRACIALE 1991; BRODSKY and GUAGLIARDI 1991), the polymorphism of these genes could provide a genetic basis determining the host immune reactions against EV HPVs. This is best illustrated by recent studies on RFLP of MHC in rabbits (HAN et al. 1992) which established in the Shope papilloma system a strong linkage between wart regression and DRα EcoRI fragment, and an increased relative risk of malignant transformation associated with a DQα PvuII fragment. This indicates a genetic control of wart evolution involving genes of the MHC. The hypothesis is further supported by studies in which a high risk of squamous cell carcinoma of the cervix was found to be associated with HLA-DQw3 (WANK and THOMSSEN 1991). The DQ differences were found to be due to an increase in DQw3 and decrease in DQw4. It is to be stressed that not only MHC class I and class II polymorphism but also variability of other candidate genes including TNFα, proteasome, and peptide transporter loci could exert both susceptibility and protective influences on the progression or regression of viral infection. Of special importance is TNFα polymorphism since the production of TNFα is genetically determined, as shown for both mice (JACOB and McDEVITT 1988) and humans (JACOB et al. 1990). Recently several genes encoding proteins responsible for antigen degradation (proteasomes, multicationic proteinases) and for peptide transportation were detected in the class II region of MHC (BROWN et al. 1991; GLYNNE et al. 1991; KELLY et al. 1992). A defect in either protein may result in the formation of the unstable class I molecules and a loss of presentation of antigens (KELLY et al. 1992). However, nothing is known about the polymorphism of proteasome and peptide-transporting genes in the human immune response against HPVs.

3.3 Abnormal Antigen Presentation in EV

A defect of specific immunity against EV HPVs (immunotolerance) appears to be related to the derangement of antigen presentation by antigen-presenting cells (APC) in the epidermis. The preserved number and function of Langerhans cells (LC) as well as co-stimulatory factors are required for optimal antigen presentation (BRACIALE and BRACIALE 1991). The nature of specific defect of antigen presentation in EV is unknown. It is possible that EV HPV antigens, similarly to other viral antigens, are presented in the context of both class I and class II MHC molecules, either by specialized APC (LC) or by "nonspecialist" cells, i.e., HPV-infected keratinocytes. In EV the function of LC appears to be preserved (HAFTEK et al. 1987; COOPER et al. 1990).

Therefore, some other cells might play a role in the induction of specific immunotolerance to EV HPVs, e.g., keratinocytes infected by these viruses. The heterogeneity of responses to virus of specific T cell clones suggests that the same antigenic peptide is bound to a specific MHC molecule in more than one conformation (KARR et al. 1990). Therefore, it is conceivable that the immunogenetic defect of EV HPV presentation by APC could be localized to the unknown allele of the MHC region, e.g., HLA-DQ, which serves as a preferred restricting element for suppression (HIRAYAMA et al. 1987; OLIVEIRA and MITCHINSON 1989).

There are several examples that virus-infected or tumor cells may act as their own APC when transfected by various cytokines, e.g., TNFα, interleukin 2 (IL-2), etc. (MURRAY and McMICHAEL 1992); and the presentation by "non-specialist" cells in the absence of co-stimulatory signals could lead to induction of immunotolerance. This is best illustrated by the finding that the epidermal keratinocytes $T6^-$ $HLA\text{-}DR^+$ present antigens in a tolerogenic manner (GASPARI et al. 1989; BAL et al. 1990). It is unknown which co-stimulatory signals are lacking in the EV epidermis. It could be presumed that this is not intercellular adhesion molecule 1 (ICAM-1) since this molecule is up-regulated on EV keratinocytes. The expression in EV epidermis of other accessory molecules (e.g., LFA-3, B7 ligand for CD28) is unknown. Various exogenous and endogenous factors can modify APC function by interfering

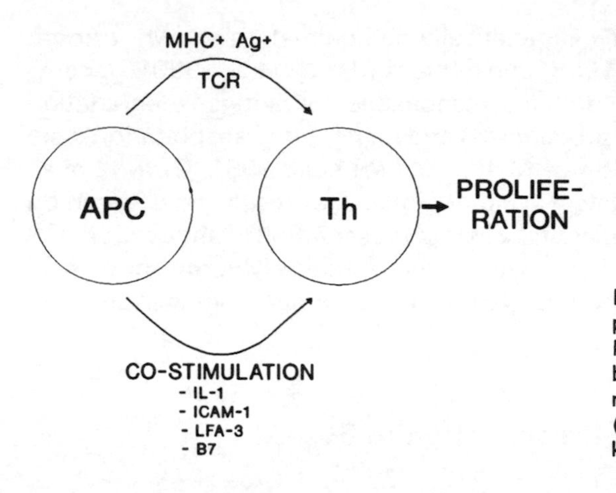

Fig. 9. Antigen presentation. The presentation of antigen in context of MHC to T helper cells (*Th*) requires both specific recognition by T cell receptor (*TCR*) and co-stimulatory (adhesion molecules and cytokines) signals

with accessory signal transduction (MURRAY and MCMICHAEL 1992) (Fig. 9). This could lead to an abrogation of proliferation of specific T helper 1 clones, probably due to the absence of autocrine stimulation by IL-2 (SIMON et al. 1992).

3.4 Role of Ultraviolet B in Generation of Specific Immune Defect

Ultraviolet B (UVB) may induce both nonspecific immunosuppression, e.g., by generation of immunosuppressive cytokines, and specific immunotolerance through alteration of antigen presentation.

3.4.1 *Cis*-Urocanic Acid as a Potent Immunosuppressant

One of the main factors affecting antigen presentation is the *cis*-isomer of urocanic acid (UCA) which is a major ultraviolet (UV)-absorbing component of the stratum corneum. It has been shown that UCA, when applied locally to the mouse skin, suppresses delayed-type hypersensitivity to herpes simplex virus type 1 (HSV-1) (ROSS et al. 1986) or to contact sensitizers with simultaneous generation of specific T suppressor cells (HARRIOTT-SMITH and HALLIDAY 1988; NOONAN et al. 1988). In our recent studies we found a high isomerization rate of UCA in the light-exposed skin of EV patients without additional UVB irradiation. The levels of *cis*-UCA were highest in the patients with most pronounced cutaneous malignancies. Of special interest is an increased isomerization of UCA also in healthy members of EV families, which suggests that this process is genetically controlled (Fig. 10). The exact mechanism of the immunosuppressive action of UCA is unknown; however, it has been suggested that it is mediated by TNFα (STREILEIN 1993).

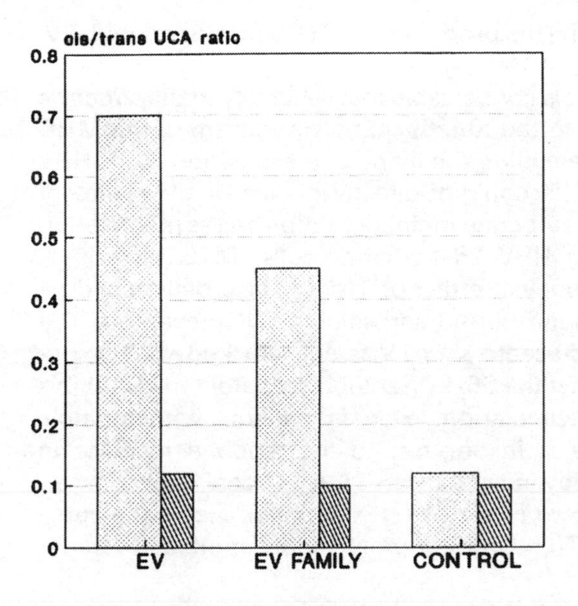

Fig. 10. *Cis/trans* ratio of UCA in light-exposed (*shaded columns*) and covered (*hatched columns*), not additionally UVB-irradiated skin in EV patients and in healthy family members.

3.4.2 Effects of Cytokines

It is well known that UVB exposure, in addition to UCA isomerization, decreases the number and function of $T6^+DR^+LC$, and causes a temporary decrease and then an increase in HLA DR^+ and ICAM-1 expression (NORRIS et al. 1990; SPENCER et al. 1993). UVB induces the generation of various cytokines, both stimulatory (IL-1, IL-6), and inhibitory (anti IL-1, TGFβ) (LUGER and SCHWARTZ 1990). Importantly, TNFα induced by UVB exposure has a double effect. On the one hand, it is responsible for the direct deleterious effect of UV irradiation, and on the other, it mediates a specific immunotolerance.

We found a significantly increased expression of TNFα mRNA in the lesional skin of EV patients, both by in situ hybridization and Northern blot techniques (MAJEWSKI et al. 1991). The mechanism of action of TNFα on afferent and efferent arms of immune responses is not known, but it appears to be of special importance for the generation of specific UVR-related T cell defects in patients with EV. Studies by YOSHIKAWA and STREILEIN (1990) strongly suggest that polymorphism of the TNFα and Lps loci governs susceptibility to the deleterious effect of UVB on the local cell-mediated hypersensitivity in mice. The authors showed that systemic administration of neutralizing TNFα-specific antibodies reconstituted the defect of induction of contact hypersensitivity produced by UVB in UVB-susceptible mice. This would suggest that UVB-induced overproduction of TNFα could mediate the local cutaneous immune defect. Studies in humans showed that susceptibility to effects of UVB radiation on induction of contact hypersensitivity constitutes a risk factor for skin cancer (YOSHIKAWA et al. 1990; STREILEIN

1993). We are presently studying the production of TNFα in patients with EV, as related to the polymorphism of TNFα gene and TNF$\alpha_{a/, b/, c/}$ microsatellites (JONGENEEL et al. 1991). It is conceivable that variability in the structure of TNFα gene may contribute to the functional polymorphism of the MHC gene complex, which could determine the mode of presentation of EV HPV antigens. On the other hand, TNFα could be directly involved in the control of tumor progression through the autocrine inhibitory action on the proliferation of keratinocytes, as shown for HPV-16-harboring cells (MALEJCZYK et al. 1992). However, the final biological effect of TNFα is also determined by expression of both cell membrane-bound and soluble TNFα receptors. The expression of cell membrane receptors in EV is still not known, but our preliminary findings suggest that the 55-kDa soluble receptor type I (inhibitor) is not increased in the circulation, even in patients with multiple cutaneous malignancies. This is in contrast to the reported increase in circulating TNFα receptors in advanced cancers (ADERKA et al. 1991). It can be speculated that non-increased levels of TNFα inhibitor enable the autocrine action of endogenous TNFα, limiting proliferation of infected, transformed EV keratinocytes.

Another mechanism responsible for UVB-related specific immunosuppression, especially by chronic exposure, could be due to the TGFβ, a potent immunosuppressive cytokine (MAJEWSKI et al. 1991). In the supernatants of short-term culture of EV keratinocytes we found TGFβ, detected by dot-blot analysis. Its production, increased by in vitro UVB irradiation, was correlated with IL-1-inhibitory activity of the supernatant (Fig. 11). The up-regulation of TGFβ could be of basic significance for HPV-associated tumor cell interactions with extracellular matrix (ECM) and for invasiveness (see below).

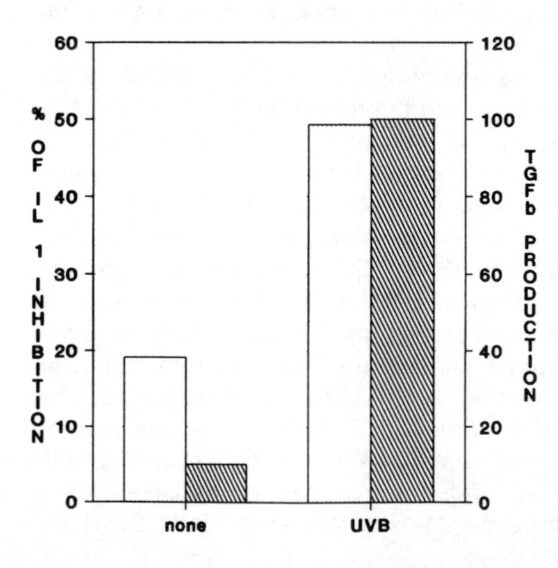

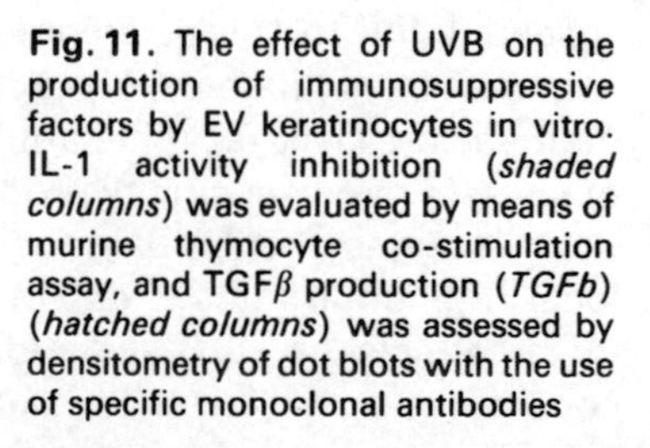

Fig. 11. The effect of UVB on the production of immunosuppressive factors by EV keratinocytes in vitro. IL-1 activity inhibition (*shaded columns*) was evaluated by means of murine thymocyte co-stimulation assay, and TGFβ production (*TGFb*) (*hatched columns*) was assessed by densitometry of dot blots with the use of specific monoclonal antibodies

3.5 Non-immunological Effects of UVB in Tumor Progression

Although the immunosuppressive effects of UVB appear to be of primary importance for EV HPV-associated carcinogenesis, there is also evidence for a direct mutagenic effect of UV light through the induction of pirymidine dimer promutagenic lesions. Specific for UVB C-T substitutions and CC-TT double-base mutations in sun-induced cancers of the skin were found to be the most frequent mutations of p53 (BRASH et al. 1991). The mutagenic effect of UVR in sun-induced cutaneous cancers is related to the inactivation of the wild form of antioncogen p53. The normal wild type of p53 may keep cell proliferation in check, while the mutated form acts as an oncogene. Overexpression of p53 generally reflects the presence of the mutant form because the altered protein accumulates in the tumor cells due to its longer half-life (HARRIS and HOLLSTEIN 1992). In our preliminary study in 12 patients with EV, we did not find accumulation of p53, with the exception of a slight up-regulation of this gene in one patient with multiple cutaneous malignancies and adenocarcinoma of the intestine. However, our study was performed by immunohistochemistry which might not be a sensitive enough method, and further studies with the use of molecular techniques are required. It is, however, possible that suppressor genes other than p53 are involved in EV tumor development and progression.

3.6 Role of Extracellular Matrix (ECM) in EV Tumor Progression

Interactions between tumor cells and ECM components, i.e., proteins and glycoproteins of the basement membrane, are important for cell proliferation, invasiveness, and metastasis formation. These interactions are mediated, inter alia, by cellular receptors of the VLA integrin family (HYNES 1992). The experimental studies of skin carcinogenesis in a mouse model have shown that altered expression of integrins is strongly linked to the appearance of early markers of tumor progression (TENNENBAUM et al. 1992). Several changes of integrin pattern were also disclosed in human cutaneous carcinomas (PELTONEN et al. 1989; CERRI et al. 1992).

We have studied the expression of $\alpha2$, $\alpha3$, $\alpha6$, and $\beta1$ chains of integrins in the skin of EV patients, including both benign lesions and advanced tumors. We found an up-regulation of $\alpha6$ (α chain of VLA6 = laminin receptor) in EV lesional skin, and this was associated with increased $\alpha2$ (α chain of VLA2 = collagen receptor) (S. Majewski et al., in preparation). Interestingly, an increased expression of $\alpha6$ was also detected in benign actinic keratoses in the general population but not in basal cell carcinomas. Recent studies by TENNENBAUM et al. (1992) in a mouse model of skin carcinogenesis, induced by chemical carcinogens or by oncogene transduction, revealed that the laminin receptor is predominantly expressed in proliferating tumor cells. DEMETER et al. (1992) have shown in humans that non-integrin high-affinity laminin receptor gene expression is associated with the development of malignancies of the cervical epithelia, but the increased expression appeared to correlate with the proliferative rather than the invasive properties of these cells. The authors found that the laminin receptor was dramatically induced in HPV-positive cervical glands, irrespective of HPV type, and this induction occurred before any evidence of invasion and progression to invasive carcinomas (DEMETER et al. 1992). It is conceivable that the low invasive and metastatic potential of EV tumors also depends on TGFβ, known to modulate VLA expression, and found to be overexpressed in EV (MAJEWSKI et al. 1991).

Besides its effect upon integrin expression, TGFβ is a potent suppressor of matrix-degrading enzymes via stimulation of metalloproteinase inhibitors (ROBERTS and SPORN 1990). Moreover, TGFβ, by enhancing tissue repair processes and new blood vessel formation, could contribute to the accelerated wound healing in EV patients, which is a surprising phenomenon in such a heavily immunosuppressed population (JABLONSKA 1991).

In Conclusion, EV is a model disease for genetic, HPV-associated, sun-induced cutaneous cancers that allows the study of complex interactions between the virus, extrinsic factors, and immunosurveillance mechanisms.

References

Aderka D, Engelman H, Hernik V, Skornick Y, Levo Y, Wallach D, Kushtal G (1991) Increased serum levels of soluble receptors for tumor necrosis factor in cancer patients. Cancer Res 51:5602–5607

Androphy EJ, Dvoretzky I, Lowy DR (1985) X-linked inheritance of epidermodysplasia verruciformis: genetic and virologic studies of a kindred. Arch Dermatol 121:864–868

Bal V, McIndoe A, Denton G, Hudson D, Lombardi G, Lamb J, Lechler R (1990) Antigen presentation by keratinocytes induces tolerance in human T cells. Eur J Immunol 20: 1893–1897

Barr BBB, Benton BC, McLaren K, Bunney MH, Smith IW, Blessing MH, Hunter JAA (1989) Human papillomavirus infection and skin cancer in renal allograft recipients. Lancet ii:124–129

Berger TG, Sawchunk WS, Leonardi C, Langenberg A, Tappero JT, Leboit PE (1991) Epidermodysplasia verruciformis-associated with human immunodeficiency virus disease. Br J Dermatol 124:79–83

Bollag W (1991) Retinoids and interferon: a new promising combination? Br J Haematol 79 [Suppl 1]:87–91

Braciale TJ, Braciale VL (1991) Antigen presentation: structural themes and functional variations. Immunol Today 12:124–129

Brash DE, Rudolph JA, Simon JA, Lin A, McKenna GJ, Baden HP, Halperin AJ, Ponten J (1991) A role of sunlight in skin cancer: UV-induced p53 mutations in squamous cell carcinoma. Proc Natl Acad Sci USA 88:10124–10128

Brodsky FM, Guagliardi LE (1991) The cell biology of antigen processing and presentation. Annu Rev Immunol 9:707–744

Brown MG, Driscoll J, Monaco JJ (1991) Structural and serological similarity of MHC-linked LMP and proteasome (multicatalytic proteinase) complexes. Nature 353: 355–357
Cerri A, Tadini G, Gitto R, Crosti L, Berti E (1992) Adhesion molecules and skin tumours: a peculiar pattern of integrin expression is present in carcinoma cells. Eur J Dermatol 4: 279–285
Cooper KD, Androphy EJ, Lowy DR, Katz SI (1990) Antigen presentation and T cell activation in epidermodysplasia verruciformis. J Invest Dermatol 94: 769–776
Demeter LM, Stoler MH, Sobel ME, Broker TR, Chow LT (1992) Expression of high-affinity laminin receptor mRNA correlates with cell proliferation rather than invasion in human papillomavirus-associated cervical neoplasms. Cancer Res 52: 1561–1567
Gaspari AA, Jenkins MK, Katz SI (1989) Class II MHC-bearing keratinocytes induce antigen-specific unresponsiveness in hapten-specific Th1 clones. J Immunol 141: 2216–2220
Gassenmeier A, Fuchs P, Schell H, Pfister H (1986) Papillomavirus DNA in warts of immunosuppressed renal allograft recipients. Arch Dermatol Res 278: 219–223
Glynne R, Powis SH, Beck S, Kelly A, Kerr L-A, Trowsdale J (1991) A proteasome-related gene between the two ABC transporter loci in the class II region of the human MHC. Nature 353: 357–360
Glinski W, Obalek S, Jablonska S, Orth G (1981) T cell defect in patients with epidermodysplasia verruciformis due to human papillomavirus type 3 and 5. Dermatologica 162: 141–147
Gross G, Ellinger K, Roussaki A, Fuchs PG, Peter HH, Pfister H (1988) Epidermodysplasia verruciformis in a patient with Hodgkins disease: characterization of a new papillomavirus type and interferon treatment. J Invest Dermatol 91: 43–48
Haftek M, Jablonska S, Szymanczyk J, Jarzabek-Chorzelska M (1987) Langerhans cells in epidermodysplasia verruciformis. Dermatologica 174: 173–179
Han R, Breitburd F, Marche PN, Orth G (1992) Linkage of regression and malignant conversion of rabbit viral papillomas to MHC class II genes. Nature 356: 66–68
Harriot-Smith TG, Halliday WJ (1988) Suppression of contact hypersensitivity by short-term ultraviolet irradiation. II. The role of urocanic acid. Clin Exp Immunol 72: 174–179
Harris CC, Hollstein M (1992) p53 tumor suppressor gene. Princip Pract Oncol 6: 1–12
Hirayama K, Matsushita S, Kikuchi I, Iuchi M, Ohta N, Sasazuki T (1987) HLA-DQ is epistatic to HLA-DR in controlling the immune response to schisosomal antigen in humans. Nature 327: 426–430
Hynes RO (1992) Integrins: versatility, modulation, and signaling in cell adhesion. Cell 69: 11–25
Jablonska S (1991) Epidermodysplasia verruciformis. In: Friedman RJ (ed) Cancer of the skin. Saunders, Philadelphia, pp 101–113
Jablonska S, Orth G (1985) Epidermodysplasia verruciformis. Clin Dermatol 3: 83–96
Jablonska S, Dabrowski J, Jakubowicz K (1972) Epidermodysplasia verruciformis as a model in studies on the role of papillomavirus in oncogenesis. Cancer Res 32: 585–589
Jacob CO, McDevitt HO (1988) Tumor necrosis factor-α in murine autoimmune "lupus" nephritis. Nature 331: 356–358
Jacob CO, Fronek Z, Lewis GD, Koo M, Hansen JA, McDevitt Ho (1990) Heritable major histocompatibility complex class II-associated differences in production of tumor necrosis factor: relevance to genetic predisposition to systemic lupus erythematosus. Proc Natl Acad Sci USA 87: 1233–1237
Jongeneel CV, Briant L, Udalovo IA, Sevin A, Nedospasov SA, Cambon-Thomsen A (1991) Extensive genetic polymorphism in the human tumor necrosis factor region and relation to extended HLA haplotypes. Proc Natl Acad Sci USA 88: 9717–9721
Karr RW, Yu W, Watts R, Evans KS, Celis E (1990) The role of polymorphic HLA-DRα chain residues in presentation of viral antigens to T cells. J Exp Med 172: 273–283
Kelly A, Powis SH, Kerr L-A, Mockridge I, Elliott T, Bastin J, Uchanska-Ziegler B, Ziegler A, Trowsdale J, Townsend A (1992) Assembly and function of the two ABC transporter proteins encoded in the human major histocompatibility complex. Nature 365: 641–644
Lambert PF (1991) Papillomavirus DNA replication. J Virol 65: 3417–3420
Lippman SM, Kavanagh JJ, Paredes-Epinosa M, Delgadillo-Madrueno F, Parades-Casillas P, Ki Hong Waun, Holdener E, Krakoff JH (1992) 13-cis retinoic acid plus interferonα-2a: High active systemic therapy for squamous cell carcinoma of the cervix. J. Natl Cancer Inst 84: 241–245
Luger TA, Schwartz T (1990) Epidermal cell-derived cytokines. In: Bos JD (ed) Skin immune system (SIS) CRC Press, Boca Raton, pp 257–291

Lutzner M (1978) Epidermodysplasia verruciformis: an autosomal recessive disease characterized by viral warts and skin cancer—a model for viral oncogenesis. Bull Cancer 65: 169–182

Lutzner M, Croissant O, Ducasse MF, Kreis H, Crosnier J, Orth G (1980) A potentially oncogenic human papillomavirus (HPV5) found in two renal allograft recipients. J Invest Dermatol 75: 353–356

Lutzner MA, Orth G, Dutronquay V, Ducasse MF, Kreis H, Crosnier J (1983) Detection of human papillomavirus type 5 DNA in skin cancers of an immunosuppressed renal allograft recipient. Lancet ii: 422–424

Majewski S, Skopinska-Rozewska E, Jablonska S, Wasik M, Misiewicz J, Orth G (1986) Partial defects of cell-mediated immunity in patients with epidermodysplasia verruciformis. J Am Acad Dermatol 15: 966–973

Majewski S, Malejczyk J, Jablonska S, Misiewicz J, Rudnicka L, Obalek S, Orth G (1990) Natural cell-mediated cytotoxicity against various target cells in patients with epidermodysplasia verruciformis. J Am Acad Dermatol 22: 423–427

Majewski S, Hunzelmann N, Nischt R, Eckes B, Rudnicka L, Orth G, Krieg T, Jablonska S (1991) TGFβ-1 and TNFα expression in the epidermis of patients with epidermodysplasia verruciformis. J Invest Dermatol 97: 862–867

Majewski S, Szmurlo A, Marczak M, Bollag W, Jablonska S (1993) Synergistic effects of retinoids and interferon α on tumor-induced angiogenesis: anti-angiogenic effect on HPV-harboring tumor cell lines. Int J Cancer (to be published)

Malejczyk J, Malejczyk M, Kock A, Urbanski A, Majewski S, Hunzelmann N, Jablonska S, Orth G, Luger TA (1992) Autocrine growth limitation of human papillomavirus type 16-harboring keratinocytes by constitutively released tumor necrosis factor-α. J Immunol 149: 2702–2708

Murray N, McMichael A (1992) Antigen presentation in virus infection. Curr Opin Immunol 4: 401–407

Noonan FP, DeFabo EC, Morrison H (1988) Cis-urocanic acid, a product formed by ultraviolet B irradiation of the skin, initiates an antigen presentation defect in splenic dendritic cells in vivo. J Invest Dermatol 90: 92–99

Norris DA, Lyons B, Middleton MH, Yohn JJ, Kashihara-Sawami M (1990) Ultraviolet radiation can either suppress or induce expression of intercellular adhesion molecule 1 (ICAM-1) on the surface of cultured human keratinocytes. J Invest Dermatol 95: 132–138

Obalek S, Glinski W, Haftek M, Orth G, Jablonska S (1979) Comparative studies on cell-mediated immunity in patients with different warts. Dermatologica 236: 1–12

Obalek S, Favre M, Szymanczyk J, Misiewicz J, Jablonska S, Orth G (1992) Human papillomavirus (HPV) types specific of epidermodysplasia verruciformis detected in warts induced by HPV3 or HPV3-related types in immunosuppressed patients. J Invest Dermatol 98: 936–941.

Oliveira DBG, Mitchison NA (1989) Immune suppression genes. Clin Exp Immunol 75: 167–177

Orth G, Jablonska S, Favre M, Jarzabek-Chorzelska M, Rzesa G (1978) Characterization of two new types of HPV from lesions of epidermodysplasia verruciformis. Proc Natl Acad Sci USA 75: 1537–1541

Peltonen J, Larjava H, Jaakkola S, Grainick H, Akiyama S, Yamada S, Yamada KM, Uitto J (1989) Localization of integrin receptors for fibronectin, collagen and laminin in human skin. Variable expression in basal and squamous cell carcinomas. J Clin Invest 84: 1916–1923

Pfister H, Gross G, Hagedorn M (1979) Characterization of human papilloma virus 3 in warts of a renal allograft patient. J Invest Dermatol 73: 349–353

Prose N, von Knebel-Doeberitz C, Miller S, Milburn PB, Heilman E (1990) Widespread flat warts associated with human papillomavirus type 5: a cutaneous manifestation of human immunodeficiency virus infection. J Am Acad Dermatol 23: 978–981

Roberts AB, Sporn MB (1990) The transforming growth factor-βs. In: Sporn MB, Roberts AB (eds) Peptide growth factors and their receptors I. Springer, Berlin Heidelberg New York (Handbook of experimental pharmacology, vol 95/I)

Ross JA, Howie SEM, Norval M, Maingay J, Simpson TJ (1986) Ultraviolet-irradiated urocanic acid suppresses delayed-type hypersensitivity to herpes simplex virus in mice. J Invest Dermatol 87: 630–633

Rudlinger R, Smith JW, Bunney MH, Hunter JAA (1986) Human papillomavirus infections in a group of renal transplant recipients. Br J Deramtol 115: 681–692

Simon JC, Krutmann J, Elmets CA, Bergstresser PR, Cruz PD (1992) Ultraviolet B-irradiated antigen-presenting cells display altered accessory signaling for T-cell activation: relevance to immune responses initiated in skin. J Invest Dermatol 98: 66S–69S

Spencer M-J, Vestey JP, Tidman MJ, McVittie E, Hunter JAA (1993) Major histocompatibility class II antigen expression on the surface of epidermal cells from normal and ultraviolet B irradiated subjects. J Invest Dermatol 100: 16–22

Streilein JW (1993) Sunlight and skin-associated lymphoid tissues (SALT): if UVB is the trigger and TNFα is its mediator, what is the message? J Invest Dermatol 100: 47S–52S

Tennenbaum T, Yuspa SH, Grover A, Castronovo V, Sobel ME, Yamada Y, DeLuca LM (1992) Extracellular matrix receptors and mouse skin cancerogenesis: altered expression linked to appearance of early markers of tumor progression. Cancer Res 52: 2966–2976

Van der Leest RJ, Zachow KR, Ostrow RS, Bender M, Pass F, Faras AJ (1987) Human papillomavirus heterogeneity in 36 renal transplant recipients. Arch Dermatol 123: 354–357

Wank R, Thomssen C (1991) High risk of squamous cell carcinoma of the cervix for women with HLA-DQw3. Nature 352: 723–725

Yoshikawa T, Streilein JW (1990) Genetic basis of the effects of ultraviolet light B on cutaneous immunity. Evidence that polymorphism at the TNFα and Lps loci governs susceptibility. Immunogenetics 32: 398–405

Yoshikawa T, Rae V, Bruins-Slot W, van den Berg J-W, Taylor JR, Streilein JW (1990) Susceptibility to effects of UVB radiation on induction of contact hypersensitivity as a risk factor for skin cancer in humans. J Invest Dermatol 95: 530–536

zur Hausen H (1986) Intracellular surveillance of persisting viral infections: human genital cancer results from deficient cellular control of papillomavirus gene expression. Lancet ii: 489–492

zur Hausen H (1991) Papillomavirus/host cell interactions in the pathogenesis of anogenital cancer. In: Brugge J, Curran T, Harlow E, McCormick F (eds) Origins of human cancer. Cold Spring Harbor Laboratory Press, Cold Spring Harbor, pp 685–705

Papillomaviruses and Cancer of the Upper Digestive and Respiratory Tracts

P.J.F. SNIJDERS, A.J.C. VAN DEN BRULE, C.J.L.M. MEIJER, and J.M.M. WALBOOMERS

1 Introduction

At present human papillomaviruses (HPVs) are linked to up to 10% of the worldwide cancer burden, mainly due to their involvement in anogenital cancer. Accumulating experimental data on cervical squamous cell carcinoma strongly support the assumption that certain HPV types have carcinogenic potential and are causally related to cervical cancer (ZUR HAUSEN 1991). HPV-16 and HPV-18, the major types associated with cervical cancer, exhibit transforming and immortalizing functions in primary rodent cells and human keratinocytes, respectively (MATLASHEWSKI et al. 1987; PECORARO et al. 1989; PIRISI et al. 1987), and the viral E6 and E7 open reading frames (ORFs) have been found to encode the HPV-transforming proteins (BARBOSA and SCHLEGEL 1989; HAWLEY-NELSON et al. 1989; MÜNGER et al. 1989a). The HPV-16 and HPV-18 E6 and E7 proteins can form complexes with the tumor suppressor gene products p53 and pRB, respectively (WERNESS et al. 1990; MÜNGER et al. 1989b), which at least in part may explain at what level these HPV types interfere with cell cycle control

Department of Pathology, Unit of Molecular Pathology, Free University Hospital, De Boelelaan 1117, 1081 HV Amsterdam, The Netherlands

mechanisms. However, viral E6–E7 functions are not sufficient for the development of malignant growth, and the additional modification of host cell genes by chemical factors has been considered to be important as well (ZUR HAUSEN 1982, 1989). Owing to this hypothesis, interest has been provoked for a role of HPV in the pathogenesis of carcinomas within the upper digestive and respiratory tracts since these tumors are known to be related to chemical factors such as tobacco and alcohol. Worldwide, squamous cell carcinoma of the ororespiratory tract, especially that of the lung, is the major form of cancer in humans. Therefore, efforts to study the relationship of specific HPV types with carcinomas of this tract are justified.

One of the first indications that HPV is involved in the development of head and neck tumors came from clinical observations that juvenile laryngeal papillomas originate from a perinatal infection of children from mothers with genital condylomatous lesions (QUICK et al. 1978). Indeed, HPV capsid antigens and DNA of HPV-6 and -11 have been detected in laryngeal papillomas by numerous groups (GISSMANN et al. 1983; MOUNTS et al. 1982; STEINBERG et al. 1983), and to date it is widely accepted that the low-risk HPV-6 and -11 are etiologically involved in laryngeal papillomas. In addition, the association of predominantly HPV-13 and HPV-32 with oral focal epithelial hyperplasia (Heck's disease), a benign lesion frequently found among Inuits and American Indians, has been documented (BEAUDENON et al. 1987; HENKE et al. 1989). Indications for an association between low-risk HPVs and inverted nasal papillomas have been obtained as well (WEBER et al. 1988; BRANDWEIN et al. 1988). HPV types have also been detected in benign oral lesions, either in association with human immunodeficiency virus (HIV) infection or not (ADLER-STORTHZ et al. 1986a; GREENSPAN et al. 1988; SYRJÄNEN et al. 1989; SNIJDERS et al. 1990b). The types identified thus far in benign lesions of the upper air and food passages include HPV types 2, 4, 6, 7, 11, 13, 32, and 57.

Human papillomavirus involvement in carcinomas of the upper digestive and respiratory tracts has initially been suggested on the basis of histologic and immunohistochemical studies. Histologic examination of laryngeal squamous cell carcinomas has revealed the presence of condylomatous changes, suggestive of HPV infections in a substantial proportion of cases (SYRJÄNEN and SYRJÄNEN 1981). Moreover, of 40 biopsy specimens of oral squamous cell carcinomas, 16 revealed histologic changes suggesting an HPV infection and eight of them showed positive staining with antiserum raised against the papillomavirus structural antigens (SYRJÄNEN et al. 1983). HPV-suggestive lesions have also been found next to bronchial squamous cell carcinomas in lung cancer patients (SYRJÄNEN and SYRJÄNEN 1987).

The involvement of HPV in these carcinomas, however, still needed further confirmatory evidence. Given the fact that the virus cannot be cultured in vitro, the introduction of newly developed molecular biologic methods has opened novel ways to examine the role of these viruses in the development of head and neck carcinomas as well as lung cancer.

2 Detection of HPV DNA in Carcinomas of the Upper Digestive and Respiratory Tracts

2.1 Methods for Detecting HPV DNA

Human papillomavirus DNA detection assays performed thus far include hybridization techniques such as dot blot, Southern blot, and in situ hybridization, and the polymerase chain reaction (PCR), the latter being the most sensitive (MELCHERS et al. 1989). Although the conventional use of these methods has been shown to be successful in the specific detection of a variety of HPV genotypes, problems arise when it is unknown which HPV type might be present in a certain lesion. Owing to the existence of more than 60 different HPV types (DE VILLIERS 1989), it is not feasible to perform successive type-specific screening assays. Consequently, HPV DNA detection assays have been developed that allow the detection of a broad spectrum of HPV genotypes at once. Among these, reverse blot hybridization (DE VILLIERS et al. 1986) uses cellular DNA as a probe for hybridization to a panel of cloned HPV DNAs, immobilized on a filter support. Moreover, based on sequence homologies between different HPVs, both Southern blot and PCR assays could be adapted to detect a broader spectrum of HPVs. Low-stringency Southern blot analysis (GISSMANN et al. 1982; KAHN et al. 1986), eventually using an HPV cocktail probe, makes use of the capability of HPV genotypes to cross-hybridize with more or less related HPV genotypes.

Universal PCR assays use general/consensus primers selected from highly conserved regions within the HPV genome (usually the E1 and L1 ORF), which anneal to target DNA of a broad spectrum of HPV genotypes (MANOS et al. 1989; GREGOIRE et al. 1989; SNIJDERS et al. 1990b; VAN DEN BRULE et al. 1990). Since sequenced HPV genotypes do not show nucleotide sequence identity over a length of more than 12 bases (GREGOIRE et al. 1989), the primers or the method have had to be adapted to allow efficient primer annealing to ensure the amplification of a broad spectrum of HPVs. These adaptations include primer degeneracy (MANOS et al. 1989; GREGOIRE et al. 1989; SNIJDERS et al. 1991), incorporation of inosine residues in the primers (GREGOIRE et al. 1989), and primer annealing under conditions that allow a certain degree of mismatch acceptance (SNIJDERS et al. 1990b; VAN DEN BRULE et al. 1990). Based on mismatch acceptance, a general primer-mediated PCR method using the primers GP5 and GP6 has been developed (SNIJDERS et al. 1990b). These primer sequences have been selected from a highly conserved region within the L1 ORF of mucosotropic HPVs using computer-assisted sequence analysis. The application of this method revealed a successful amplification of sequences from at least the HPV types 6, 11, 13, 16, 18, 30, 31, 32, 33, 35, 39, 40, 43, 44, 45, 51, 52, 56, 58 (VAN DEN BRULE et al. 1992; A.J.C. van den Brule et al., unpublished data). Based on sequence information of the L1 region flanked by GP5/GP6 of 22 different

```
T8 : T R S T N L S V C ASTTASIPNV Y T P T S F K E Y A R H V E E
     : : : : :   : : :            :         : : : :   : :   : :
HPV: T R S T N X T I C 8-13 aa    Y X X X X Y K E Y X R H X E E
         N   S   S L S            F         F R Q F
         G         V                        I N D
```

Fig. 1. Alignment of deduced amino acid sequences of GP5/GP6 flanked L1 sequences amplified from a tonsillar carcinoma (T8) with the HPV amino acid consensus sequence (HPV) as determined on basis of sequence information from 22 HPV genotypes (VAN DEN BRULE et al. 1992). Matched amino acids are indicated by *dots*. The middle part of the sequence, consisting of 8–13 amino acids is polymorphic and was not included in the alignment. *aa*, amino acid; *X*, variable amino acid (SNIJDERS et al. 1992a)

HPV genotypes, an amino acid consensus sequence could be determined (see lower part of Fig. 1). The consensus is built up from highly conserved amino acids and a more polymorphic region in between which together allow confirmation of HPV specificity as well as HPV genotype differentiation. In combination with sequence analysis, the GP5/GP6 PCR method allows the identification of putative novel HPV genotypes at high sensitivity (VAN DEN BRULE et al. 1992).

At present these PCR assays seem to be the most appropriate tools to search for novel HPV genotypes in carcinomas of the ororespiratory tract.

2.2 Detection of HPV DNA in Carcinomas of the Oral Cavity, Larynx, and Lung

Initial HPV DNA detection studies confirmed the presence of HPV in a proportion of oral carcinomas. Using the Southern blot method with HPV-11- and HPV-16-specific probes, LÖNING et al. (1985) detected HPV DNA in three out of six oral carcinomas. One of them contained HPV-11 DNA and one HPV-16 DNA, while typing of the third sample was not possible due to the limited amounts of DNA available. Reverse blot and standard Southern blot analysis of seven tongue carcinomas performed by DE VILLIERS et al. (1985) revealed the presence of HPV DNA in three cases. DNA from one tumor hybridized with the HPV-2-specific probe and the remaining two cases showed positivity for HPV-16 DNA. Several additional HPV DNA detection studies using either Southern blot or in situ hybridization techniques have been performed on oral carcinomas and the prevalence rates found, ranged from approximately 2.5% to 76%, as summarized in Table 1. A remarkably high prevalence of HPV-16 DNA was reported for oral carcinomas in Taiwanese patients (CHANG et al. 1989). Using genomic Southern blot analysis, 13 out of 17 of these carcinomas appeared to contain HPV-16

DNA. The different prevalence rates found by several groups may in part reflect the detection methods and probes used as well as geographic differences. However, HPV-16 is prominent among the positive cases, except for verrucous oral carcinomas which contain HPV-2. Remarkably, DNA of HPV-4, a type which is usually associated with benign cutaneous lesions, has been detected in an oral carcinoma by Southern blot hybridization (YEUDALL and CAMPO 1991).

Laryngeal carcinomas have also been studied for the presence of HPV DNA (Table 1). This led to the establishment of a clear association between HPV and verrucous carcinomas of the larynx, which are characterized by a low incidence (1%–2% of all carcinomas of the larynx). By Southern blot analysis at low stringency conditions, HPV-16-related DNA has been detected in all six verrucous carcinomas of the larynx studied by BRANDSMA et al. (1986). Squamous cell carcinomas of the larynx have shown HPV occurrence rates ranging from 2.7% to 12.9%, as determined by Southern blot or in situ analyses. KAHN et al. (1986) have isolated a novel HPV genotype (HPV-30) from a laryngeal carcinoma. However, the prevalence of this type appeared to be rather low since none of the 41 additional laryngeal carcinomas tested were found to contain HPV-30 DNA. HPV-6- and HPV-11-positive laryngeal squamous cell carcinomas have also been demonstrated; these are likely to have developed from preexisting laryngeal papillomas (ZAROD et al. 1988; LINDEBERG et al. 1989). Therefore, probably in conjunction with chemical or physical carcinogens, these low-risk types may occasionally contribute to malignant progression.

Human papillomavirus DNA has also been detected in lung carcinomas (Table 1). STREMLAU et al. (1985) have reported the presence of HPV-16-related DNA in an anaplastic carcinoma of the lung. Using in situ hybridization with an HPV cocktail probe, SYRJÄNEN and SYRJÄNEN (1987) could detect HPV DNA in five out of 99 lung carcinomas. By in situ hybridization using biotinylated probes, BEJUI-THIVOLET et al. (1990) found HPV DNA in six out of 33 lung carcinomas. Three of them contained HPV-18 DNA, one HPV-6, one HPV-11, and one HPV-16 DNA, while in one sample both HPV-16 and HPV-18 DNA could be detected. YOUSEM et al. (1991) detected HPV DNA by in situ hybridization in five out of 16 lung squamous cell carcinomas. None of the adenocarcinomas, bronchioloalveolar carcinomas, and small cell carcinomas examined showed HPV positivity. Only one out of six large cell undifferentiated carcinomas and one out of 17 squamous metaplasias tested also appeared to contain HPV DNA of types 6/11 and 16/18, respectively. As was the case for some laryngeal carcinomas, DNA of the apparently low-risk HPV types 6 and 11 has also been detected in lung carcinomas of patients with HPV-6- and HPV-11-positive chronic papillomatosis of the respiratory tract (DILORENZO et al. 1992; BYRNE et al. 1987). The newly developed PCR method has also been introduced to study HPV prevalence in carcinomas from the oral cavity and larynx (Table 1). In some studies an increased HPV positivity in carcinomas, but also in normal oral or

Table 1. Detection of HPV genotypes in carcinomas and normal tissue within the oral cavity, larynx, and lung

References	Site	HPV-positive carcinomas	HPV-positive normal tissue	HPV type(s)	Method
Löning et al. 1985	Oral cavity	3/6	–	11,16,X	SB
		3/6	–	–	IF
de Villiers et al. 1985	Tongue	3/7	–	2,16	RB, SB
Adler-Storthz et al. 1986b	Oral cavity	3/9[b]	–	2	SB
Maitland et al. 1987	Oral cavity	7/15	5/12	16	SB
Chang et al. 1989	Oral cavity	13/17	1/17	16	SB
Chang et al. 1990	Oral cavity	1/40	–	unknown	ISH
		11/40	–	6,16,18	PCR
Kashima et al. 1990	Oral cavity	7/74	–	6,16,X	SB, RB
Yeudall and Campo 1991	Oral cavity	19/39[c]	2/25	4,16,18	PCR, SB
Shindo et al. 1992	Tongue	8/24	–	16,16/18	PCR
Scheurlen et al. 1986	Larynx	1/36	–	16	SB
Brandsma et al. 1986	Larynx	6/6[b]	0/3	16–related	SB
Kahn et al. 1986	Larynx	1/42	–	30	SB
Syrjanen et al. 1987	Larynx	15/116	–	6,11,16	ISH
Pérez–Ayala et al. 1990	Larynx	26/48	3/6	16	PCR
		3/3[b]	3/6	16	PCR
Kiyabu et al. 1989	Larynx	4/10	–	16	PCR
Hoshikawa et al. 1990	Larynx	7/34	–	6,16	PCR
Bryan et al. 1990	Larynx/pharynx	7/8	9/14	6,11	PCR
Morgan et al. 1991	Larynx/pharynx	7/16	10/16	6/11,16/33	PCR
Syrjanen and Syrjänen 1987	Lung	5/99	–	unknown	ISH
Bejui–Thivolet et al. 1990	Lung	6/33	0/10	6,11,16,18,16/18	ISH
Yousem et al. 1991	Lung[a]	5/16	–	6/11, 16/18, 31/33/35	ISH

[a] Squamous cell carcinomas.
[b] Verrucous carcinomas.
[c] Adjacent dysplastic and normal epithelium also appeared HPV positive.
SB, Southern blot hybridization; IF, immunofluorescence; ISH, in situ hybridization; RB, reverse blot hybridization; PCR, polymerase chain reaction.

laryngeal epithelium, has been found (YEUDALL and CAMPO 1991; PEREZ-AYALA et al. 1990; MORGAN et al. 1991).

2.3 Comparison of HPV DNA Occurrence at Different Anatomic Sites: Preferential Association of HPV with Tonsillar Carcinomas

Owing to the different methods and probes/primers used for HPV DNA detection, it is difficult to compare results of HPV occurrence obtained by different groups. However, of particular interest are studies in which one method has been used to determine HPV DNA prevalence in carcinomas at different anatomic sites (Table 2). BRANDSMA and ABRAMSON (1989) have applied genomic Southern blot hybridization to examine carcinomas at different sites within the head and neck region for the presence of HPV DNA. Remarkable differences were observed and the highest occurrence rate was found in carcinomas of the tonsil (29%; two out of seven carcinomas were found to contain HPV-16 DNA), followed by the tongue (19%), pharynx (13%), and larynx (5%). No HPV DNA could be detected in carcinomas of the nose and mouth. In a study correlating HPV infection with occurrence of p53 mutations, BRACHMAN et al. (1992) examined 30 head and neck carcinomas for the presence of HPV DNA by PCR using HPV-16- and HPV-18-specific primers as well as the L1 region consensus primers MY11/MY06 (MANOS et al. 1989). HPV DNA was demonstrated in two out of three tonsillar carcinomas. Both of them were found to contain HPV-16 DNA. Of the remaining specimens tested, only one carcinoma originating from the hard palate was found to contain HPV DNA.

These findings suggest that the tonsil may be more susceptible to HPV infection or HPV-mediated carcinogenesis than other sites within the head and neck region, which was supported by results obtained in our laboratory. Using the GP5/GP6 general primer-mediated PCR method, biopsy specimens of squamous cell carcinomas from different sites within the upper digestive tract and respiratory tract were screened for the presence of HPV DNA (Table 2). GP5/GP6 PCR-positive specimens were analyzed by additional type-specific PCR for the HPV types 6, 11, 16, 18, 31, and 33. All ten tonsillar carcinomas examined appeared HPV positive, whereas all seven cases of tonsillitis were HPV negative SNIJDERS et al. 1992a). Four carcinomas contained HPV-16 DNA, four carcinomas contained HPV-33 DNA, and one carcinoma contained DNA of both HPV-16 and HPV-33. False positivity was excluded by additional genomic Southern blot analysis of type-specific PCR positive samples ($n = 4$). Two specimens showed evidence for the presence of an HPV type different from HPV-6, -11, -18, -31, or -33; one of them was negative by type-specific PCR and one carcinoma also contained HPV 33 DNA. Comparison of the sequences from the GP5/GP6 PCR products of these samples with L1 region sequences of HPV types with

Table 2. Comparison of HPV DNA occurrence in carcinomas of different anatomic sites within the upper digestive and respiratory tracts

References	Site	HPV positive (*n*)	(%)	HPV type(s)
Brandsma and Abramson 1989[a]	Nose	0/2		–
	Mouth	0/10		–
	Tongue	2/11	19	2x HPV-16 related
	Tonsil	2/7	29	2x HPV-16 related
	Pharynx	1/8	13	HPV-16 related
	Larynx	3/60	5	3x HPV-11 and HPV-16 related
Brachman et al. 1992[b]	Nasopharynx	0/1		–
	Oral cavity	1/6	17	HPV-18
	Tongue	0/5		–
	Tonsil	2/3	67	2x HPV-16 related
	Pyriform sinus	0/4		–
	Larynx	0/4		–
Snijders et al. 1992a; P.J.F. Snijders et al., unpublished data[c]	Nose	0/3		–
	Tongue	2/9	22	1x HPV-33, 1x HPV-X[e]
	Buccal mucosa	1/5	20	HPV-16
	Floor of mouth	1/6	17	HPV-16
	Tonsil[d]	10/10	100	4x HPV-16, 3x HPV-33, 1x HPV-16/33, 1x HPV-7, 1x HPV-X/33
	Larynx	2/12	17	1x HPV-11, 1x HPV-X
	Lung	6/31	19	1x HPV-6, 2x HPV-16, 3x HPV-X

[a] Determined by genomic Southern blot analysis at different stringencies using HPV-11, HPV-16, and HPV-18 DNA probes.
[b] Determined by HPV-16 and HPV-18 type-specific PCR and MY11/MY06 consensus primer PCR. Confirmation by genomic Southern blot analysis.
[c] Determined by GP5/GP6 general primer PCR followed by type-specific PCR for HPV types 6, 11, 16, 18, 31, and 33. Confirmation by genomic Southern blot analysis in several cases.
[d] HPV-7 positivity determined after sequence analysis of GP5/GP6 PCR product followed by type-specific PCR for HPV-7.
[e] HPV X = HPV type different from HPV types 6, 11, 16, 18, 31 and 33.

published sequence information (HPV types 1a, 2a, 5, 6, 8, 11, 13, 16, 18, 31, 33, 35, 39, 41, 42, 47, 51, 57, 58, ME180) as well as with preliminary L1 sequences which have been determined in a systematic sequencing effort at Heidelberg (HPV types 3, 4, 7, 9, 10, 12, 14, 15, 17, 19, 25, 26, 30, 32, 34, 40, 45, 49, 52, 53, 56, 63, 65; kindly provided by Dr. H. Delius, Heidelberg, FRG) revealed that one carcinoma contained HPV-7 DNA, whereas the other contained DNA of a still unknown HPV genotype (P.J.F. Snijders et al, unpublished data). HPV specificity of the GP5/GP6 PCR product of the latter sample (T8) was confirmed by fulfilment of the HPV amino acid consensus sequence, excluding a cellular origin of the PCR product (Fig 1).

To our knowledge, this is the first time that HPV-7, a type that has been associated with butcher's warts, has been found in a carcinoma. This type, however, has been found in benign oral lesions, often in association of HIV infection (GREENSPAN et al. 1988; SYRJÄNEN et al. 1989). The HPV-7-containing carcinoma was obtained from a 67-year-old male patient who did not show any sign of HIV infection. The presence of an apparently low-risk HPV type in a tonsillar carcinoma is not an isolated finding. BERCOVICH et al. (1991) reported the presence of HPV-6 DNA in a tonsillar carcinoma.

By DNA in situ analysis using probes specific for HPV-6, HPV-11, and HPV-16, NIEDOBITEK et al. (1990) have detected HPV-16 DNA in the neoplastic cells of six from 28 tonsillar carcinomas.

Using the same GP5/GP6 general primer-mediated PCR method in combination with type-specific PCR, HPV DNA could only be detected in about 20% of carcinomas originating from other sites within the upper digestive and respiratory tract (Table 2). The types found in these carcinomas include HPV-6, HPV-11, HPV-16, HPV-33, as well as types that reacted with general primers but failed to react with type-specific primers. Apart from the carcinomas tested, one out of four laryngeal hyperplasias with atypia was found to contain HPV-16 DNA (P.J.F. Snijders et al., unpublished data). Since these lesions are considered to be precursor lesions of laryngeal carcinomas, this argues for a role of HPV in the pathogenesis of at least some of these tumors.

The reason for the remarkably high prevalence of mainly "genital" HPV types in tonsillar carcinomas compared with carcinomas from neighbouring sites is presently unknown. Since initial HPV infection is suspected to take place at the basal cell layer of the mucosa, it is tempting to speculate that tonsillar epithelium is more often wounded due to bacterial infections, providing an entry which results in an increased exposure of basal cells to HPV particles. Alternatively, local differences (i.e., the release of specific cytokins) may render tonsillar epithelium more prone to HPV-mediated carcinogenesis than adjacent sites.

3 Detection of HPV Transcripts

In order to explore an etiologic role of HPV in the development of ororespiratory cancers, the detection of solely HPV DNA is not sufficient. Expression of virus-encoded functions, especially of the transforming genes E6 and E7, should also be considered. Unlike the detection of HPV genotypes, limited data are available about HPV expression patterns in carcinomas of the oral cavity and respiratory tract. DiLorenzo et al. (1992) have found transcriptionally active HPV-6 DNA in a lung carcinoma of a patient with recurrent HPV-6-positive laryngeal papillomatosis. Comparative Northern blot analysis of RNA from the carcinoma and a benign papilloma from the same patient revealed an increased amount of a 1.9-kb E6/E7 transcript relative to the other transcripts, in the carcinoma. In addition, a novel 7.3-kb transcript containing E6/E7 sequences was present in the carcinoma but absent in the papilloma. This pattern, probably resulting from rearrangements within the HPV-6 genome in the carcinoma, suggests that the viral E6 and E7 ORFs play an important role in these carcinomas, as is suspected for anogenital carcinomas.

We recently used RNA PCR and nonradioactive RNA is situ hybridization to study HPV expression in HPV-16- and HPV-33-containing carcinomas of the oral cavity and respiratory tract. The lack of data about the transcriptional pattern of HPV-33 forced us first to map HPV-33-specific early region transcripts in a tonsillar carcinoma. Using different HPV-33 primer-specific primer combinations in the RNA PCR, a total of five cDNA species could be identified including species which are spliced within the E6 ORF and having E6*I, E6*II, and E6*III coding potential (Fig. 2) (Snijders et al. 1992b). However, different splice acceptor sites were used for the HPV-33 E6*I and E6*II transcripts as compared with the HPV-16 and HPV-18 E6*I and II mRNAs, which had previously been mapped in cervical carcinoma cell lines (Smotkin and Wettstein 1986; Schwarz et al. 1987). In contrast to the HPV-16 and HPV-18 E6*II species, the HPV-33 E6*II mRNAs utilize a splice acceptor site located within the E7 ORF and consequently fail to encode a full-length E7 protein. Furthermore, HPV-33 E6* mRNAs were found to contain a short overlapping ORF, resulting in alternative coding potentials if translation were to start at an internal AUG codon within the E6 region. Interestingly, a second early region poly (A) site was mapped at the 3′ end of the E7 ORF, potentially allowing E6/E7 region expression independent of the poly (A) site localized downstream of the E5 ORF or eventually a cellular poly (A) site if viral DNA integration were to interrupt the early region of the HPV genome. Using RNA PCR with HPV-33-specific primers flanking the E6*I splice sites, E6*I transcripts could be detected in three HPV-33-containing tonsillar carcinomas and one HPV-33-positive cervical carcinoma that served as control (Snijders et al. 1992a). No full-length unspliced E6/E7 mRNAs could be detected, suggesting that these mRNAs were either absent or

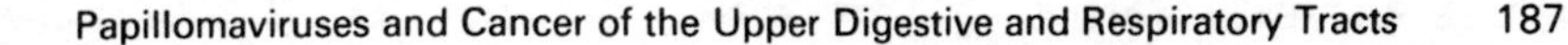

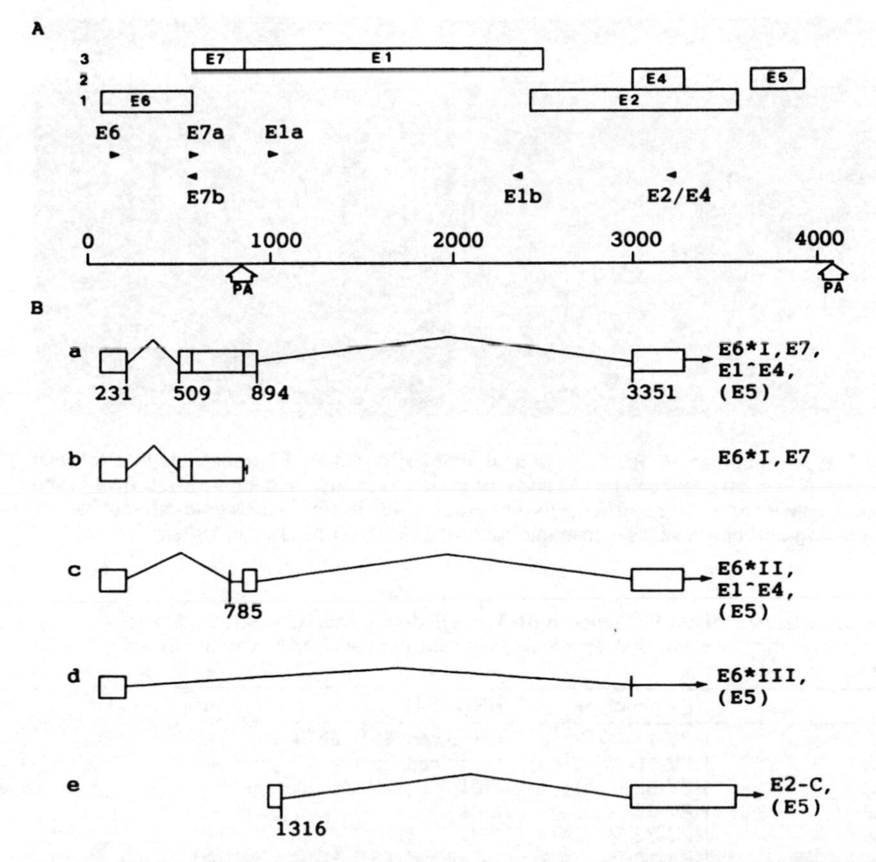

Fig. 2 A, B. Structures and coding potentials of HPV-33-specific cDNA species identified in a tonsillar carcinoma. **A** Primer positions in relation to the HPV-33 early region genome structure are indicated by *arrowheads*. *Arrowheads* point to the 5′ to 3′ direction. Putative poly (A) sites are indicated by *open arrows*. **B** Structures and probable coding potentials of the five cDNA species (*a–e*) identified depicted in relation to the genome structure shown in **A**. *Numbers* indicate the splice donor and acceptor sites. *Open boxes* represent potential coding regions. Coding potentials of the cDNA species are shown on the *right*. For species *a–d* the potential to encode E6* proteins depends on the utilization of the E6 AUG codon at nt position 109 (SNIJDERS et al. 1992b)

existed at very low levels within the tumors. RNA PCR with HPV-16-specific primers revealed the presence of unspliced and spliced E6*I mRNA in a tonsillar carcinoma. In addition, unspliced E6*I and E6*II mRNAs were detected in an HPV-16-positive carcinoma of the floor of the mouth and a lung carcinoma containing HPV-16 DNA (P.J.F. Snijders et al., unpublished data). By nonradioactive RNA in situ hybridization HPV-16- or HPV-33 specific E6/E7 region mRNAs could be demonstrated in the neoplastic cells of three tonsillar carcinomas examined (SNIJDERS et al. 1992a) and the HPV-16-containing carcinoma of the floor of the mouth (unpublished data). An example of RNA in situ hybridization on an HPV-33-containing tonsillar

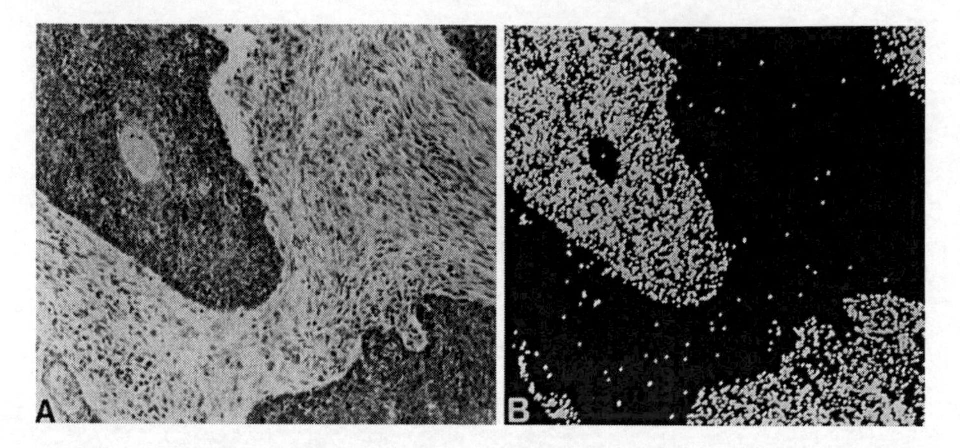

Fig. 3 A, B. Nonradioactive RNA in situ analysis of HPV-33 E6/E7 transcripts in a tonsillar carcinoma. **A** Section after hematoxylin and eosion staining and light microscopy. **B** RNA in situ hybridization with an HPV-33 E6/E7 antisense probe. Visualization was by gold/silver enhancement staining and confocal laser scan microscopy. (From SNIJDERS et al. 1992a).

Table 3. Detection of E6/E7 region mRNA in HPV-16- and HPV 33 containing carcinomas of the oral cavity and respiratory tract (SNIJDERS et al. 1992a, unpublished data)

Site	HPV genotype	RNA PCR[a]	RISH[b]
Floor of mouth	HPV-16	Unspliced, E6*I, E6*II	+
Tonsil	HPV-16	Unspliced, E6*I	+
Tonsil	HPV-33	E6*I	+
Tonsil	HPV-33	E6*I	+
Tonsil	HPV-33	E6*I	ND
Lung	HPV-16	Unspliced, E6*I, E6*II	ND

[a] Concerns RNA PCR detection of transcripts with E7 coding potential.
[b] Nonradioactive RNA in situ hybridisation (RISH) using biotinylated HPV-16- or HPV-33-specific E6/E7 region anti-sense RNA as probe. Positivity indicates cytoplasmic staining of the neoplastic cells exclusively.

carcinoma is shown in Fig. 3. The results of RNA PCR and RNA in situ hybridization performed on these carcinomas are summarized in Table 3.

4 Physical State and Intragenomic Rearrangements of HPVs in Carcinomas of the Upper Digestive and Respiratory Tracts

In cervical carcinomas the viral DNA is often found integrated into cellular DNA (DÜRST et al. 1985; SCHWARZ et al. 1985; CULLEN et al. 1991). In contrast, premalignant and benign cervical lesions predominantly contain episomal HPV DNA (DÜRST et al. 1985; CULLEN et al. 1991; LEHN et al. 1988). With respect to the opening and disruption of the HPV genome, integration regularly occurs within the viral E1–E2 region, whereas the E6 and E7 ORFs are invariably retained and transcriptionally active (SCHWARZ et al. 1985; CHOO et al. 1987). It has been suggested that HPV DNA integration is important in carcinogenesis by triggering the disruption of expression of E2-encoded transcriptional modulator proteins, which would result in the uncontrolled expression of the transforming genes E6 and E7.

Generally, an assessment of the viral physical state can be made by one-dimensional Southern blot analysis using different restriction enzymes as well as two-dimensional Southern blot analysis, allowing separation of episomal circular DNA from linear DNA (WETTSTEIN and STEVENS 1982).

DE VILLIERS et al. (1985) demonstrated the presence of HPV-16 DNA in two tongue carcinomas. One of them showed a PstI restriction enzyme pattern consistent with the presence of episomal HPV DNA. The other sample showed additional off-sized bands suggesting viral DNA integration, although a partial digestion could not completely be ruled out. MAITLAND et al. (1987) have found DNA of the HPV-16 prototype and an HPV-16 variant in a high proportion of normal mucosa, benign lesions, and carcinomas of the oral cavity. In all but one case of normal tissue, the digestion patterns suggested the presence of episomal HPV-16 DNA. CHANG et al. (1989) have detected HPV-16 DNA in 13 oral carcinomas from Taiwanese patients, all of which showed a Southern blot pattern suggesting the presence of episomal DNA. YEUDALL and CAMPO (1991) have detected episomal DNA of HPV-4, -16, and -18 in three individual oral carcinomas, respectively. In an attempt to find out whether HPV E6/E7 transcription in tonsillar carcinomas is correlated with viral DNA integration, we analyzed two HPV-16- and two HPV-33-containing carcinomas for both the viral physical state and E6/E7 transcription (SNIJDERS et al. 1992c). Southern blot analysis, DNA PCR, and two-dimensional gel electrophoresis revealed indications for the presence of only episomal DNA in the HPV-16-containing biopsies (Fig. 4A, B) and only integrated DNA in one HPV-33-containing biopsy. The second HPV-33-containing carcinoma, of which one biopsy and two resected tumor specimens were analysed, showed a rather complex physical state profile. One part of this tumor contained only episomal DNA (Fig. 4C), one part only contained integrated DNA, and the remaining tumor part contained both integrated and episomal HPV-33 DNA (Fig. 4D). Independent of the viral physical state, all biopsies and resected tumor parts tested showed the presence of E6/E7 transcripts, as determined by RNA PCR. Also and HPV-16-containing lung carcinoma tested showed a Southern blot pattern consistent with the presence of episomal HPV DNA (P.J.F. Snijders et al., unpublished data). Taking these data together, it seems that viral DNA integration, obviously less common in oral cancer and cancer of the respiratory tract than in cervical cancer, is not necessarily a prerequisite to ensure proper expression of the E6/E7 region. However, viral integration may take place, eventually as a late

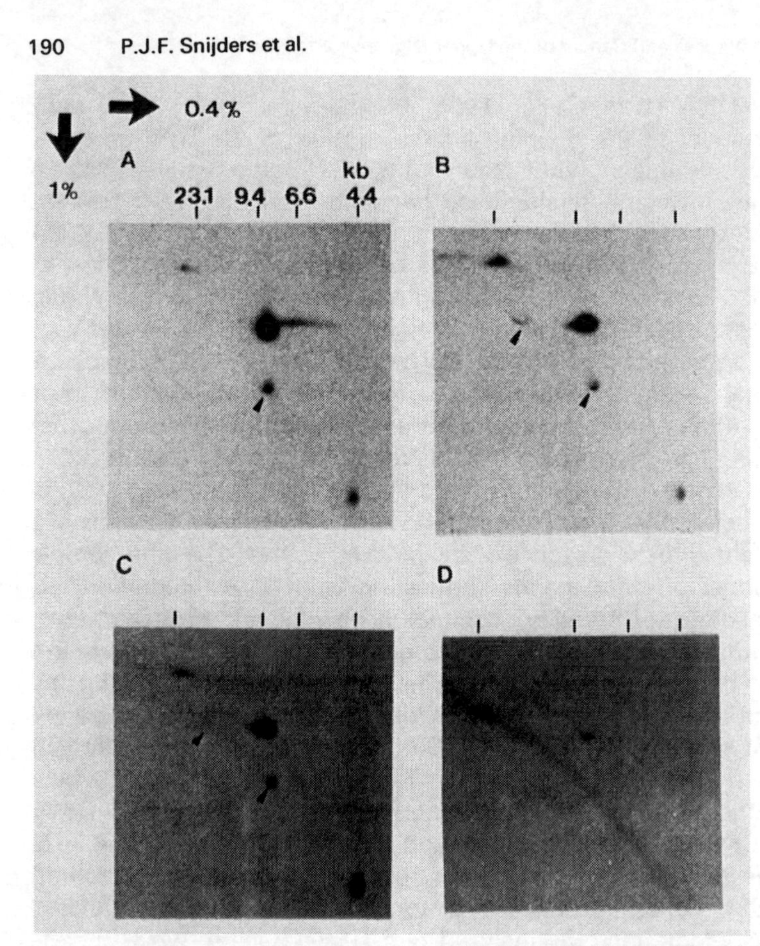

Fig. 4 A–D. Two-dimensional analysis of HPV-16-(**A, B**) and HPV-33 (**C, D**)-containing samples of tonsillar carcinomas. The HPV-16-containing samples were digested with HindIII and hybridized with HPV-16 DNA. The HPV-33-containing samples, which were derived from the same tonsillar carcinoma, were digested with BamHI and hybridized with HPV-33 DNA. $\lambda \times$ HindIII size markers are indicated in addition to directions of electrophoresis in the first (0.4%) and second (1%) dimension. Linear restriction fragments of 8 kb and 16 kb are indicated by *arrowheads* (SNIJDERS et al. 1992c)

event. BERCOVICH et al. (1991) have found integrated HPV-6 DNA in a tonsillar carcinoma. An HPV-16-containing laryngeal carcinoma studied by SCHEURLEN et al. (1986) was found to contain integrated as well as episomal HPV-16 DNA molecules.

Intragenomic modifications of the HPV genome have also been observed in ororespiratory cancers. DILORENZO et al. (1992) have analyzed HPV-6a DNA in a lung carcinoma of a patient with recurrent laryngeal papillomatosis. Detailed analysis of the viral DNA revealed that the carcinoma contained episomal HPV-6a genomes containing a duplication of the upstream regulatory region (URR), the late region, and a portion of the early region. Similarly, BYRNE et al. (1987) have found a duplication within the URR of HPV-11 found in a liver metastasis of a lung cancer patient with chronic HPV-11-positive laryngotracheobronchial papillomatosis. DE VILLIERS et al. (1985) reported the presence of HPV-2 DNA in a tongue carcinoma having a PstI restriction enzyme profile which differed from all known HPV-2 subtypes.

Rearranged high-risk HPV types have also been found in carcinomas of the larynx and oral cavity. In a laryngeal carcinoma, rearranged HPV-16 molecules have been detected with multiple duplications of the E4 ORF and part of the E2, E5, L1 and L2 ORFs (SCHEURLEN et al. 1986). These rearrangements resulted in the disruption of the viral L1 ORF. Altered episomal HPV-16 genomes showing the absence of a prototype PstI restriction fragment which maps to the L2-L1 region have been detected in oral carcinomas (MAITLAND et al. 1987; YEUDALL and CAMPO 1991). Further analysis of two of these HPV-16-containing oral carcinomas suggested the presence of repetitive cellular sequences within the viral genome (YEUDALL 1992). HPV-16 and HPV-33 genotypes found in tonsillar carcinomas did not contain major alterations affecting restriction enzyme patterns (SNIJDERS et al. 1992a). Also HPV-16 DNA from a lung carcinoma did not show major alterations (P.J.F. Snijders et al., unpublished data). Restriction enzyme analysis of a 7.3-kb EcoRI HPV 16 fragment which was cloned from this carcinoma has not revealed the presence of off-sized bands thus far. DNA sequence analysis of this fragment is necessary to find out whether minor alterations are present within this HPV-16 molecule.

5 Discussion and Perspectives

The data collected so far indicate that at least a proportion of carcinomas of the oral cavity and respiratory tract are associated with HPV DNA. HPV-16 appeared the most dominant type taking all data together. In addition, the mucosotropic high-risk HPV types 18, 30, and 33 have been detected, and recently HPV-31 has also been found in a cell line derived from a laryngeal carcinoma (BRADFORD et al. 1991). More than incidentally, DNA of the low-risk HPV types 2, 6, and 11 was present, and HPV-4 and HPV-7 have also been found in isolated cases.

These findings raise the question of what triggers these apparently low-risk HPV types to contribute to malignant progression. First, alterations within the viral genome affecting expression levels of the transforming genes could be of importance. This assumption may find support in the observed changes (often duplications) in regulatory sequences within HPV-6 and HPV-11 genomes in malignant lesions that have been extensively studied (BYRNE et al. 1987; DILORENZO et al. 1992). WU and MOUNTS (1988) have found that duplications within the URR of HPV-6 enhance viral gene expression. ROSEN and AUBORN (1991) demonstrated that the duplication of upstream regulatory sequences can increase the potential of HPV-11 to transform baby rat kidney cells in conjunction with an activated *ras* oncogene. However, more recently it has been reported that HPV-6 variants containing rearranged URR are commonly present in both genital benign condylomata acuminata and clinically malignant Buschke-Löwenstein tumors (RÜBEN et al. 1992). No clear correlation between genomic rearrangements and malignant potential of HPV-6 could be established. Minor alterations within the coding region of the viral transforming genes may also increase the oncogenic potential of low-risk HPVs. SANG and BARBOSA (1992) have demonstrated that subtle nucleotide changes resulting in single amino acid substitutions in the HPV-6 E7 protein give rise to biologic activities characteristic of the high-risk HPV E7 oncoproteins. Alternatively, changes within the cellular genome may contribute to the oncogenic potential of individual HPV types. Somatic mutations resulting in inactivation of wild-type functions of p53 and p*RB* may make infected cells independent of the viral functions responsible for inactivation of these tumor suppressor gene products. Indeed, mutant p53 could substitute for HPV-16 E6 in immortalizing human keratinocytes in cooperation with HPV-16 E7 (SEDMAN et al. 1992). Therefore, alterations of certain cellular genes may provide low-risk HPVs, expressing a low oncogenic variant of a certain transforming gene with full oncogenic potential. In the HPV-33-containing carcinomas we studied for the presence of E6/E7 region mRNA, only E6* transcripts could be detected while we failed to detect mRNA encoding full-length E6 (SNIJDERS et al. 1992a). Interestingly, in a survey correlating HPV status with the presence of p53 mutations in cervical carcinomas and cervical intraepithelial neoplasia (CIN), LO et al. (1992) only detected p53 mutations in two out of two HPV-33-containing carcinomas and not in HPV-16 (seven cases) and HPV-18 (one case) associated cervical carcinomas. This may argue for HPV-33 E6 being less competent in inactivating p53, although more samples have to be tested and in vitro studies are necessary to support this assumption.

Several studies revealed the presence of HPV related DNA which could only be visualized by low-stringency Southern blot hybridization. Together with the plurality of HPV types associated with lesions of the genital mucosa and skin, this may indicate that specific, still unidentified HPV types are also evolutionary adapted to the mucosa of the ororespiratory tract. General primer-based PCR methods can aid in the initial detection and subsequent isolation of putative novel HPV types. In addition to HPV DNA detection, these methods generate a type-specific DNA fragment that could be used as a probe for specific screening of genomic libraries. By applying GP5/GP6 PCR we did indeed find HPV genotypes different from the common genital types in carcinomas of the tongue, tonsil, larynx, and lung. Characterization of the PCR products followed by type-specific screening is necessary to find out whether these HPVs include novel putative ororespiratory tract-specific HPVs. However, HPV prevalences in these carcinomas do not reach the level of that found in tonsillar carcinomas, indicating that HPV-independent events have induced malignancy and/or that putative novel ororespiratory tract-specific HPVs were not recognized by these primers. Since the GP5 and GP6 primers were initially selected for mucosotropic genital HPVs, a significant proportion of cutaneous HPV types will fail to be amplified. Therefore, general primer PCR methods more specifically based on cutaneous types should also be applied to find out whether additional HPV positivity exists for these tumors. Primers selected on the basis of HPV types such as HPV-41, which do not show major homologies with any of the other HPV types indentified (DE VILLIERS 1989), should also be applied to allow more definitive statements to be made about HPV prevalence in these cancers.

It has been found that both oral and bronchial epithelial cells can be immortalized with HPV-16 and both HPV-16 and HPV-18, respectively (PARK et al. 1991; WILLEY et al. 1991), and HPV E6/E7 region transcripts appeared to be abundantly present in these cells. This, together with the transcriptional patterns observed in biopsy specimens, suggests a mechanism of malignant transformation in which the viral E6/E7 region plays a central role, as is the case for anogenital cancer. In fact, in a tonsillar carcinoma biopsy containing HPV-33 DNA integrated within the E2 region, the only transcript which could be detected was a short E6*I mRNA having a coding potential limited to E7 ORF (SNIJDERS et al. 1992c). This supports the idea that the viral E7 oncoprotein is the major HPV-transforming protein in these cells. More data have to be collected to find out whether additional HPV-encoded proteins may be involved in the malignant transformation of epithelium of the ororespiratory tract.

A significant proportion of ororespiratory cancers was found to contain extrachromosomal HPV DNA. Comparative analysis of E6/E7 transcription with viral physical state in tonsillar carcinomas revealed that E6/E7 transcription is not necessarily dependent on viral DNA integration. However, if integration were important in carcinogenesis, for example, by inactivating E2 transregulation, it is likely that viral intragenomic modifications or changes in host cell gene expression could also be alternative events to gain an equivalent effect from viral episomes. Gross alterations within episomal HPV-16 DNA have been detected in some studies. However, it is presently unknown what effect these changes have on viral E6/E7 expression. More detailed analysis by DNA sequencing will be necessary to find out whether HPV episomes which persist in carcinomas contain mutations which could affect E6/E7 expression.

The data collected thus far have clearly established the association of HPV genotypes with a proportion of carcinomas of the upper digestive and respiratory tract. In particular, the high prevalence of transcriptionally active HPVs in tonsillar carcinomas supports a viral etiology. For other carcinomas, established prevalence rates are most likely underestimated since none of the methods used to detect HPV DNA can claim to detect all HPV genotypes. Identification and subsequent isolation of novel HPVs from carcinomas of the respiratory tract has to be continued because new insight in a possible viral etiology may yield new immune therapeutic strategies in an area where treatment of these carcinomas is presently disappointing.

Acknowledgments. The authors would like to thank Yvonne Duiker for typing the manuscript. We are also very grateful to Prof. G.B. Snow and Dr. N. de Vries (Dept. of Otolaryngology, Free University Hospital, Amsterdam) for continuous interest in our work and Drs. L. Gissmann and H. Delius, DKFZ, Heidelberg, for providing unpublished HPV sequence information and helpful discussions. Part of the work presented in this chapter was supported by the Dutch Cancer Society "Koningin Wilhelmina Fonds," grant no. IKA 93-605. The research of P.J.F.S. has been made possible by a fellowship of the Royal Netherlands Academy of Arts and Sciences.

References

Adler-Storthz K, Newland JR, Tessin BA, Yeudall WA, Shillitoe EJ (1986a) Identification of human papillomavirus types in oral verruca vulgaris. J Oral Pathol 15: 230–233

Adler-Storthz K, Newland JR, Tessin BA, Yeudall WA, Shillitoe EJ (1986b) Human papillomavirus type 2 DNA in oral verrucous carcinoma. J Oral Pathol 15: 472–475

Barbosa MS, Schlegel R (1989) The E6 and E7 genes of HPV-18 are sufficient for inducing two-stage in vitro transformation of human keratinocytes. Oncogene 4: 1529–1532

Beaudenon S, Praetorius F, Kremsdorf D, Lutzner M, Worsaae N, Pehau-Arnaudet G, Orth G (1987). A new type of human papillomavirus associated with oral focal epithelial hyperplasia. J Invest Dermatol 88: 130–135

Béjui-Thivolet F, Liagre N, Chignol MC, Chardonnet Y, Patricot LM (1990) Detection of human papillomavirus DNA in squamous bronchial metaplasia and squamous cell carcinomas of the lung by in situ hybridization using biotinylated probes in paraffin-embedded specimens. Hum Pathol 21: 111–116.

Bercovich JA, Centeno CR, Aguilar OG, Grinstein S, Kahn T (1991) Presence and integration of human papillomavirus type 6 in a tonsillar carcinoma. J Gen Virol 72: 2569–2572

Brachman DG, Graves D, Vokes E, Beckett M, Haraf D, Montag A, Dunphy E, Mick R, Yandell D, Weichselbaum RR (1992) Occurrence of p53 gene deletions and human papilloma virus infection in human head and neck cancer. Cancer Res 52: 4832–4836

Bradford CR, Zacks SE, Androphy EJ, Gregoire L, Lancaster WD, Carey TE (1991) Human papillomavirus DNA sequences in cell lines derived from head and neck sqaumous cell carcinomas. Otolaryngol Head Neck Surg 104: 303–310

Brandsma JL, Abramson AL (1989) Association of papillomavirus with cancers of the head and neck. Arch Otolaryngol Head Neck Surg 115: 621–625

Brandsma JL, Steinberg BM, Abramson AL, Winkler B (1986) Presence of human papillomavirus type 16 related sequences in verrucous carcinoma of the larynx. Cancer Res 46: 2185–2188

Brandwein M, Steinberg B, Thung S, Biller H, Dilorenze T, Galli R (1988) Human papillomavirus 6/11 and 16/18 in Schneiderian inverted papillomas. Cancer 63: 1708–1713

Bryan RL, Bevan IS, Crocker J, Young LS (1990) Detection of HPV 6 and 11 in tumours of the upper respiratory tract using the polymerase chain reaction. Clin Otolaryngol 15: 177–188

Byrne JC, Tsao M-S, Fraser RS, Howley PM (1987) Human papillomavirus 11 DNA in a patient with chronic laryngotracheobronchial papillomatosis and metastatic squamous-cell carcinoma of the lung. N Engl J Med 317: 873–878

Chang F, Syrjänen S, Nuutinen J, Kärjä J, Syrjänen K (1990) Detection of human papillomavirus (HPV) DNA in oral squamous cell carcinomas by in situ hybridization and polymerase chain reaction. Arch Dermatol Res 282: 493–497

Chang K-W, Chang C-s, Lai K-S, Chou M-J, Choo K-B (1989) High prevalence of human papillomavirus infection and possible association with betel quid chewing and smoking in oral epidermoid carcinomas in Taiwan. J Med Virol 28: 57–61

Choo KB, Pau CC, Hau SH (1987) Integration of human papillomavirus type 16 into cellular DNA of cervical carcinoma: preferential deletion of the E2 gene and invariable retention of the long control region and the E6/E7 open reading frames. Virology 161: 259–261
Cullen AP, Reid R, Campion M, Lörincz AT (1991) Analysis of the physical state of different human papillomavirus DNAs in intraepithelial and invasive cervical neoplasms. J Virol 65: 606–612
de Villiers EM (1989) Heterogeneity of the human papillomavirus group. J Virol 63: 4898–4903
de Villiers EM, Weidauer H, Otto H, zur Hausen H (1985) Papillomavirus DNA in human tongue carcinomas. Int J Cancer 36: 575–578
de Villiers EM, Schneider A, Gross G, zur Hausen H (1986) Analysis of benign and malignant urogenital tumors for human papillomavirus infection by labelling cellular DNA. Med Microbiol Immunol (Berl) 174: 281–286
DiLorenzo TP, Tamsen A, Abramson AL, Steinberg BM (1992) Human papillomavirus type 6a DNA in the lung carcinoma of a patient with recurrent laryngeal papillomatosis is characterized by a partial duplication. J Gen Virol 73: 423–428
Dürst M, Kleinheinz A, Hotz M, Gissmann L (1985) The physical state of human papillomavirus type 16 DNA in benign and malignant genital tumors. J Gen Virol 66: 1515–1522
Gissmann L, Diehl V, Schultz-Coulon HJ, zur Hausen H (1982) Molecular cloning and characterization of human papillomavirus DNA derived from a laryngeal papilloma. J Virol 44: 393–400
Gissmann L, Wolnik L, Ikenberg H, Koldovsky U, Schnürck HG, zur Hausen H (1983) Human papillomavirus type 6 and 11 DNA sequences in genital and laryngeal papillomas and in some cervical cancers. Proc Natl Acad Sci USA 80: 560–563
Greenspan D, de Villiers EM, Greenspan JS, de Souza YG, zur Hausen H (1988) Unusual HPV types in oral warts in association with HIV infection. J Oral Pathol 17: 482–487
Gregoire L, Arella M, Campione-Piccardo J, Lancaster WD (1989) Amplification of human papillomavirus DNA sequences by using conserved primers. J Clin Microbiol 27: 2660–2665
Hawley-Nelson P, Vousden KH, Hubbert NL, Lowy DR, Schiller JT (1989) HPV 16 E6 and E7 proteins cooperate to immortalize primary human foreskin keratinocytes. EMBO J 8: 3905–3910
Henke RP, Guerin-Reverchon I, Milde-Langosch K, Strömme-Koppang H, Löning T (1989) In situ detection of human papillomavirus types 13 and 32 in focal epithelial hyperplasia of the oral mucosa. J Oral Pathol Med 18: 419–421
Hoshikawa T, Nakajima T, Uhara H, Gotok M, Shimosato Y, Tsutsumi K, Ono I, Ebihara S (1990) Detection of human papillomavirus DNA in laryngeal squamous cell carcinomas by polymerase chain reaction. Laryngoscope 100: 647–650
Kahn T, Schwarz E, zur Hausen H (1986) Molecular cloning and characterization of the DNA of a new human papillomavirus (HPV 30) from a laryngeal carcinoma. Int J Cancer 37: 61–65
Kashima HK, Kutcher M, Kessis T, Levin LS, de Villiers EM, Shah K (1990) Human papillomavirus in squamous cell carcinoma, leukoplakia, lichen planus, and clinically normal epithelium of the oral cavity. Ann Otol Rhinol Laryngol 99: 55–61
Kiyabu MT, Shibata D, Arnheim N, Martin WJ, Fitzgibbon PL (1989) Detection of human papillomavirus in formalin fixed invasive squamous carcinomas using the polymerase chain reaction. Am J Surg Pathol 13: 221–224
Lehn H, Villa L, Marziona F, Hillgarth M, Hillemans HG, Sauer G (1988) Physical state and biological activity of human papillomavirus genomes in precancerous lesions of the female genital tract. J Gen Virol 69: 187–196
Lindeberg H, Syrjänen S, Kärjä J, Syrjänen K (1989) Human papillomavirus type 11 DNA in squamous cell carcinomas and pre-existing multiple laryngeal papillomas. Acta Otolaryngol (Stockh) 107: 141–149
Lo K-W, Mok C-H, Chung G, Huang DP, Wong F, Chan M, Lee JCK, Tsao S-W (1992) Presence of p53 mutation in human cervical carcinomas associated with HPV-33 infection. Anticancer Res 12: 1989–1994
Löning T, Ikenberg H, Becker J, Gissmann L, Hoepfner I, zur Hausen H (1985) Analysis of oral papillomas, leukoplakias and invasive carcinomas for human papillomavirus type related DNA. J Invest Dermatol 84: 417–420
Maitland NJ, Cox MF, Lynas C, Prime SS, Meanwell CA, Scully C (1987) Detection of human papillomavirus DNA in biopsies of human oral tissue. Br J Cancer 56: 245–250

Manos MM, Ting Y, Wright DK, Lewis AJ, Broker TR (1989) Use of polymerase chain reaction amplification for the detection of genital human papillomaviruses. Cancer Cells 7: 209–214
Matlashewski GJ, Schneider J, Banks L, Jones N, Murray A, Crawford L (1987) Human papillomavirus type 16 DNA cooperates with activated ras in transforming primary cells. EMBO J 6: 1741–1746
Melchers W, van den Brule A, Walboomers J, de Bruin M, Burger M, Herbrink P, Meijer C, Lindeman J, Quint W (1989) Increased detection rate of human papillomavirus in cervical scrapes by the polymerase chain reaction as compared to modified FISH and Southern blot analysis. J Med Virol 27: 329–335
Morgan DW, Abdullah V, Quiney R, Myint S (1991) Human papilloma virus and carcinoma of the laryngopharynx. J Laryngol Otol 105: 288–290
Mounts Ph, Shah KV, Kashima H (1982) Viral etiology of juvenile- and adult-onset squamous papilloma of the larynx. Proc Natl Acad Sci USA 79: 5425–5429
Münger K, Phelps WC, Bubb V, Howley PM, Schlegel R (1989a) The E6 and E7 genes of the human papillomavirus type 16 together are necessary and sufficient for transformation of primary human keratinocytes. J Virol 63: 4417–4421
Münger K, Werness BA, Dyson N, Phelps WC, Howley (1989b) Complex formation of human papillomavirus E7 proteins with the retinoblastoma tumour suppressor gene product. EMBO J 8: 4099–4105
Niedobitek G, Pitteroff S, Herbst H, Shepherd P, Finn T, Anagnostopoulos I, Stein H (1990) Detection of human papillomavirus type 16 DNA in carcinomas of the palatine tonsil. J Clin Pathol 43: 918–921
Park NH, Min BM, Li SL, Cherrick HM, Doniger J (1991) Immortalization of normal human oral keratinocytes with type 16 human papillomavirus. Carcinogenesis 12: 1627–1631
Pecoraro G, Morgan D, Defendi V (1989) Differential effects of human papillomavirus type 6, 16 and 18 DNAs on immortalization and transformation of human cervical epithelial cells. Proc Natl Acad Sci USA 86: 563–567
Pérez-Ayala M, Ruiz-Cabello F, Esteban F, Concha A, Redondo M, Oliva MR, Cabrera T, Garrido F (1990) Presence of HPV 16 sequences in laryngeal carcinomas. Int J Cancer 46: 8–11
Pirisi L, Yasumoto S, Feller M, Doniger J, DiPaolo JA (1987) Transformation of human fibroblasts and keratinocytes with human papillomavirus type 16 DNA. J Virol 61: 1016–1066
Quick CA, Faras A, Krzyzek R (1978) The etiology of laryngeal papillomatosis. Laryngoscope 88: 1789–1795
Rosen M, Auborn K (1991) Duplication of the upstream regulatory sequences increases the transformation potential of human papillomavirus type 11. Virology 185: 484–487
Rüben A, Beaudenon S, Favre M, Schmitz W, Spelten B, Grussendorf-Conen E-I (1992) Rearrangements of the upstream regulatory region of human papillomavirus type 6 can be found in both Buschke-Löwenstein tumours and in condylomata acuminata. J Gen Virol 73: 3147–3153
Sang B-C, Barbosa MS (1992) Single amino acid substitutions in "low-risk" human papillomavirus (HPV) type 6 E7 protein enhance features characteristic of the "high-risk" HPV E7 oncoproteins. Proc Natl Acad Sci USA 89: 8063–8067
Scheurlen W, Stremlau A, Gissmann L, Höhn D, Zenner HP, zur Hausen H (1986) Rearranged HPV16 molecules in an anal and in a laryngeal carcinoma. Int J Cancer 38: 671–676
Schwarz E, Freese UK, Gissmann L, Mayer W, Roggenbuck B, Stremlau A, zur Hausen H (1985) Structure and transcription of human papillomavirus sequences in cervical carcinoma cells. Nature 314: 111–114
Schwarz E, Schneider-Gädicke A, zur Hausen H (1987) Human papillomavirus type 18 transcription in cervical carcinoma cell lines and in human cell hybrids. Cancer Cells 5: 47–53
Sedman SA, Hubbert NL, Vass WC, Lowy DR, Schiller JT (1992) Mutant p53 can substitute for human papillomavirus type 16 E6 in immortalization of human keratinocytes but does not have E6-associated trans-activation or transforming activity. J Virol 66: 4201–4208
Shindo M, Sawada Y, Kohgo T, Amemiya A, Fujinaga K (1992) Detection of human papillomavirus DNA sequences in tongue squamous-cell carcinomas using the polymerase chain reaction method. Int J Cancer 50: 167–171
Smotkin D, Wettstein FO (1986) Transcription of human papillomavirus type 16 early genes in a cervical cancer and a cancer-derived cell line and identification of the E7 protein. Proc Natl Acad Sci USA 83: 4680–4684

Snijders PJF, Schulten EAJM, Mullink H, ten Kate RW, Jiwa M, van der Waal I, Meijer CJLM, Walboomers JMM (1990a) Detection of human papillomavirus and Epstein-Barr virus DNA sequences in oral mucosa of HPV-infected patients by polymerase chain reaction. Am J Pathol 137:659–666

Snijders PJF, van den Brule AJC, Schrijnemakers HFJ, Snow G, Meijer CJLM, Walboomers JMM (1990b) The use of general primers in the polymerase chain reaction permits the detection of a broad spectrum of human papillomavirus genotypes. J Gen Virol 71:173–181

Snijders PJF, Meijer CJLM, Walboomers JMM (1991) Degenerate primers based on highly conserved regions of amino acid sequence in papillomaviruses can be used in a generalized polymerase chain reaction to detect productive human papilloma virus infections. J Gen Virol 72:2781–2786

Snijders PJF, Cromme FV, van den Brule AJC, Schrijnemakers HFJ, Snow GB, Meijer CJLM, Walboomers JMM (1992a) Prevalence and expression of human papillomavirus in tonsillar carcinomas, indicating a possible viral etiology. Int J Cancer 51:845–850

Snijders PJF, van den Brule AJC, Schrijnemakers HFJ, Raaphorst PMC, Meijer CJLM, Walboomers JMM (1992b) Human papillomavirus type 33 in a tonsillar carcinom generates its putative E7 mRNA via two E6* transcript species which are terminated at different early region poly (A) sites. J Virol 66:3172–3178

Snijders PJF, Meijer CJLM, van den Brule AJC, Schrijnemakers HFJ, Snow GB, Walboomers JMM (1992c) Human papillomavirus (HPV) type 16 and 33 E6/E7 region transcripts in tonsillar carcinomas can originate from integrated and episomal HPV DNA. J Gen Virol 73:2059–2066

Steinberg BM, Topp W, Schneider PS, Abramson A (1983) Laryngeal papillomavirus infection during clinical remission. N Engl J Med 308:1261–1265

Stremlau A, Gissmann L, Ikenberg H, Stark M, Bannasch P, zur Hausen H (1985) Human papilloma virus type 16 related DNA in an anaplastic carcinoma of the lung. Cancer 55:1737–1740

Syrjänen KJ, Syrjänen SM (1981) Histological evidence for the presence of condylomatous epithelial lesions in association with laryngeal squamous cell carcinoma. ORL 43:181–194

Syrjänen KJ, Syrjänen SM (1987) Human papillomavirus DNA in bronchial squamous cell carcinomas. Lancet 1:168–169

Syrjänen K, Syrjänen S, Lamberg M, Pyrkonen S, Nuutinen J (1983) Morphological and immunohistochemical evidence suggesting human papilloma virus (HPV) involvement in oral squamous cell carcinogenesis. Int J Oral Surg 12:418–424

Syrjänen S, Syrjänen K, Mäntyjärvi R, Collan Y, Kärjä J (1987) Human papillomavirus DNA in squamous cell carcinomas of the larynx demonstrated by *in situ* DNA hybridization. ORL 49:175–186

Syrjänen S, von Krogh G, Kellokoski J, Syrjänen K (1989) Two different human papillomavirus (HPV) types associated with oral mucosal lesions in an HIV-seropositive man. J Oral Pathol Med 18:366–370

van den Brule AJC, Snijders PJF, Gordjin RLJ, Bleker OP, Meijer CJLM, Walboomers JMM (1990) General primer-mediated polymerase chain reaction permits the detection of sequenced and still unsequenced human papillomavirus genotypes in cervical scrapes and carcinomas. Int J Cancer 45:644–649

van den Brule AJC, Snijders PJF, Raaphorst PMC, Schrijnemakers HFJ, Delius H, Gissmann L, Meijer CJLM, Walboomers JMM (1992) General primer PCR in combination with sequence analysis to identify potentially novel HPV genotypes in cervical lesions. J Clin Microbiol 30:1716–1721

Weber RS, Shillitoe EJ, Robbins KT, Luna MA, Batsakis JG, Donovan DT, Adler-Storthz K (1988) Prevalence of human papillomavirus in inverted nasal papillomas. Arch Otolaryngol Head Neck Surg 114:23–26

Werness BA, Levine AJ, Howley PM (1990) Association of human papillomavirus types 16 and 18 E6 proteins with p53. Science 248:76–79

Wettstein FO, Stevens JG (1982) Variable-sized free episomes of Shope papillomavirus DNA are present in all non-virus-producing neoplasms and integrated episomes are detected in some. Proc Natl Acad Sci USA 79:790–794

Willey JC, Broussoud A, Sleemi A, Bennett W, Cerutti P, Harris CC (1991) Immortalization of normal human bronchial epithelial cells by human papillomaviruses 16 and 18. Cancer Res 51:5370–5377

Wu T-C, Mounts P (1988) Transcriptional regulatory elements in the noncoding region of human papillomavirus type 6. J Virol 62: 4722–4729

Yeudall WA (1992) Human papillomaviruses and oral neoplasia. Oral Oncol Eur J Cancer 28B: 61–66

Yeudall WA, Campo MS (1991) Human papillomavirus DNA in biopsies of oral tissues. J Gen Virol 72: 173–176

Yousem SA, Ohori NP, Sonmez-Alpan E (1991) Occurrence of human papillomavirus DNA in primary lung neoplasms. Cancer 69: 693–697

Zarod AP, Rutherford JD, Corbitt G (1988) Malignant progression of laryngeal papilloma associated with human papilloma virus type 6 (HPV-6) DNA. J Clin Pathol 44: 280–283

Zur Hausen H (1982) Human genital cancer: synergism between two virus infections or synergism between a virus infection and initiating events. Lancet 2: 1370–1372

Zur Hausen H (1989) Papillomaviruses in anogenital cancer as a model to understand the role of viruses in human cancers. Cancer Res 49: 4677–4681

Zur Hausen H (1991) Human papillomaviruses in the pathogenesis of anogenital cancer. Virology 184: 9–13

In Vitro Systems for the Study and Propagation of Human Papillomaviruses

C. MEYERS and L.A. LAIMINS

1 Introduction

Papillomaviruses were first shown to be pathogenic agents by SHOPE and HURST (1933) who identified them as the causative agent of infectious papillomatosis of cottontail rabbits. Later, ROUS and BEARD (1935) observed that benign papillomas of rabbits induced by this virus could progress to carcinomas. Attempts to develop systems for the propagation of papillomaviruses in vitro were unsuccessful and resulted in a dormancy of papillomavirus research until the 1970s. The late 1970s brought the beginning of molecular biology technology and the cloning of papillomavirus genomes. This allowed for the isolation of sufficient quantities of material to begin a systematic study of papillomaviruses. In the 1980s a correlation between papillomaviruses and human neoplastic lesions of the anogenital area led to a significant increase in interest in the study of these viruses.

Today papillomaviruses are recognized as human pathogens of major importance involved in the pathology of benign and malignant lesions of cutaneous and mucosal epithelium. Until recently a significant impediment to the understanding of the biology of papillomaviruses has been the lack of a reproducible cell culture system permissive for vegetative viral replication.

Howard Hughes Medical Institute and Department of Molecular Genetics and Cell Biology, 920 E 58th Street, The University of Chicago, Chicago, IL 60637, USA

Current Topics in Microbiology and Immunology, Vol. 186

Recent success in the development of such systems will allow for a detailed study of papillomavirus biology and for development of treatments of papillomavirus-induced lesions. In this review, we will summarize the successful efforts to develop systems to propagate papillomaviruses in the laboratory.

Most of our understanding of the life cycle of papillomaviruses comes from observational studies of naturally infected epithelial tissue. The vegetative life cycle of papillomaviruses has been shown to have an intimate association with their host cell differentiation program (BROKER and BOTCHAN 1986; HOWLEY 1990; ZUR HAUSEN and SCHNEIDER 1987; PFISTER 1987). Infection by papillomaviruses probably occurs through microtraumas in the epithelial architecture, exposing the basal cells to entry by viruses. Following the infection of basal cells, the viral genome is established in an episomal state at approximately 20–100 copies per cell. As the infected epithelial cells migrate through the suprabasal strata and undergo differentiation, viral genomes are amplified to thousands of copies per cell (LAMBERT 1991). Concurrent with the viral DNA amplification is the activation of late gene expression and capsid assembly (Fig. 1). In the cervix, virion production occurs in lesions referred to as the condyloma acuminata or in cervical intraepithelial neoplasia type I (CIN I), which display a slightly altered pattern of differentiation from that seen in the normal ectocervical keratinocytes (KOSS 1987). In these low-grade lesions, the basal layers typically constitute the lower third of the epithelium, and cells with large vacuoles surrounding dense nuclear chromatin are often observed in suprabasal layers (KOSS 1987). These vacuolated cells have been termed koilocytes and are pathonomonic of a human papillomavirus (HPV) infection. In high-grade lesions such as CIN III and carcinomas, papillomavirus DNA is usually found integrated into cellular DNA, and the entire epithelium has an undifferentiated basal cell-like appearance. In these high-grade lesions no virus is produced.

Several characteristics of papillomaviruses have complicated the development of a cell culture system for the propagation of virion. In general,

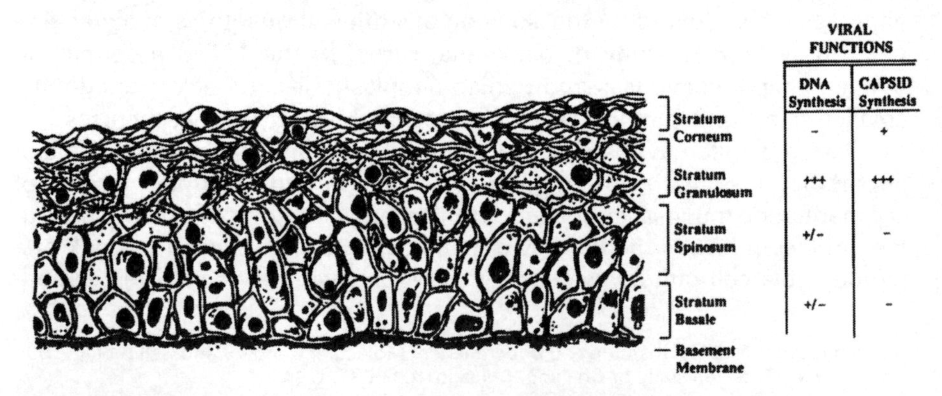

Fig. 1. Differentiating cells in a typical stratified epithelium with major viral functions indicated. Viral DNA amplification and late gene expression are restricted to highly differentiated suprabasal cells

papillomaviruses have a narrow host range, which is especially encumbering with the HPV, where no animal model has been developed to study infection of the human viruses. Papillomavirus are epitheliotropic viruses and, with the exception of bovine papillomavirus (BPV) and its related ungulate viruses that can nonproductively infect fibroblast cells, papillomaviruses have a strict tropism to epithelial cells. For a specific viral type, this tropism goes beyond merely host and tissue specificity as there is also an affinity for epithelium at specific anatomical sites. For example HPV-1 is commonly found on the dorsal aspects of the fingers and hands or on the soles of the foot, whereas HPV-16 is generally associated with the anogenital region. This preference for certain anatomical sites is most likely related to differences in replication requirements and differences in transcriptional requirements of different HPV types as well as the mechanisms of transmission.

Most experimental systems for the study of papillomavirus have utilized culture systems developed with cells of nonepithelial origins. Owing to the strict epitheliotropic nature of papillomaviruses, many of these studies in nonepithelial cell lines should be reexamined in epithelial cells to determine their relevance to the natural host cell. For the latter half of this century attempts have been made using numerous experimental techniques to culture keratinocytes for use in studies of papillomaviruses. Immortalized keratinocytes grown under standard cell culture conditions tend to quickly lose many of their epithelial characteristics, especially the ability to differentiate and undergo keratinization. These cell culture problems have restricted the study of papillomaviruses and in particular the understanding of the vegetative replication of papillomaviruses. Several descriptions of propagation of papillomaviruses in vitro have been reported (EISINGER et al. 1975; MENDELSON and KLIGMAN 1961; OROSZLAN and RICH 1964; NOYES 1965), but remain unconfirmed. Unsuccessful attempts to develop an in vitro culture system for the propagation of papillomaviruses have also been reported (BUTEL 1971; NIIMURA et al. 1975). The most significant obstacle in the propagation of papillomaviruses has been the inability to faithfully mimic the proliferation/differentiation cycles of the target tissue-squamous epithelium. Several reviews have been written in the past covering the topic of attempts to develop culture systems for papillomaviruses (TAICHMAN and LAPORTA 1986; TAICHMAN et al. 1984; BREITBURD 1987). This review will deal with new advances that have allowed for the vegetative viral replication of papillomaviruses in the laboratory outside of the natural host.

2 Laboratory Production of HPV Using Nude Mouse Xenografts

In an effort to study neoplastic transformation of normal human cervix, KREIDER et al. (1985) first incubated normal human cervical tissue with human condyloma acuminata extracts. Fragments of the human cervical tissue treated in this manner were then grafted beneath the renal capsules of athymic mice. Grafts which had been infected with virus extracted from condyloma acuminata tissue grew to significantly larger size than mock-infected grafts. When infected grafts were removed and examined histologically, they were found to be lined with an epithelium approximately 20–30 cells thick and displayed morphological transformation identical to that seen in cervical condyloma acuminata including koilocytosis, nuclear hyperchromasia, and binucleation. Immunohistochemical analysis with group-specific antigen to papillomaviruses exhibited specific staining only to the infected graft tissue. In addition, specific probes to HPV-11 and to a lesser extent HPV-6 hybridized only to DNA extracted from condyloma acuminata infected cervical grafts removed from the athymic mice.

Studies followed describing the neoplastic transformation of human epithelium obtained from numerous body sites by HPV-11 extracted from human condyloma acuminata material (KREIDER et al. 1986, 1987a). When grafted under the renal capsule of athymic mice, human epithelial tissue from the cervix, foreskin, vulva, urethra, penis, vocal cord, esophagus, and lower leg skin all underwent neoplastic transformation as defined by the morphological appearance. Epithelial tissues from the urinary bladder, ureter, abdominal skin, and breast were all negative for transformation.

When neonatal foreskin epithelium was infected with extracts of human vulvar condyloma acuminata and grafted under the renal capsule of athymic mice, the xenografts were able to support the vegetative replication of papillomaviruses (KREIDER et al. 1987b). The virions isolated from the xenografts had the morphology and density of HPV, and dot blot hybridization showed the DNA to be specifically HPV-11. Virions isolated from infected grafts were shown to be capable of infecting fresh foreskin grafts with subsequent isolation of HPV-11 virions. This was the first report of laboratory production of an HPV type and made it possible to study the characteristics of productive HPV-11 infection outside the natural host. A time course of a productive HPV-11 infection has been delineated by analyzing xenograft cyst development using in situ Southern and Northern techniques (STOLER et al. 1990). HPV E4 proteins of molecular weights 10 and 11 kDa have been identified in the condylomatous cyst walls and in the desquamated cells in the cavities of HPV-11-infected xenografts (BROWN et al. 1988). In additional studies HPV-11-specific polyclonal antiserum was shown to neutralize viral infection, as demonstrated by the prevention of transformation of human foreskin grafts (CHRISTENSEN and KREIDER 1990).

Interestingly, a specific isolate of HPV-11 (termed Hershey of HPV-11-H) is most amenable to propagation in the nude mouse xenograft system. HPV-11-H virions can be serially passaged using the xenograft system, but in general, extracts from other HPV lesions have failed to produce virions. Using the prototype HPV-11 (DARTMANN et al. 1986), physical and functional comparisons were performed between the prototype and HPV-11-H (DOLLARD et al. 1989), but restriction analysis, DNA sequencing of transcriptional control regions, and examination of the activity of E2 to transactivate the upstream regulatory region did not uncover any significant differences between the two HPV-11 subtypes. A genetic analysis of chimerics of these HPV-11 subtypes could define the genomic sequences required to produce virions in the xenograft system, but the ability to use purified viral DNA has not successfully been demonstrated in the xenograft system. In addition to HPV-11, HPV-1 has also been shown to be amenable to this xenograft system (KREIDER et al. 1990) with experimentally produced warts displaying the presence of HPV-1 DNA and papillomavirus capsid antigen.

A second system for propagation of HPV virus in nude mice involves the grafting of a cell line onto granuloma beds in the flanks of nude mice. A human CIN I-derived cell line, W12, maintains approximately 100 copies of HPV-16 DNA predominantly in an episomal state (STANLEY et al. 1989). When it was grafted onto richly vascularized sites on the flanks of nude mice, W12 cells formed a stratified differentiating epithelium and over time displayed the morphological features of a CIN I/H lesion in vivo. Terminally differentiating epithelium of the graft tissue contained amplified levels of HPV-16 DNA, HPV-16 major capsid protein, and viral particles (STERLING et al. 1990). Further description of the use of this system for the purification of infectious virus has not yet been reported.

3 Production of Virus-Like Particles

The lack of in vitro propagation systems and the inability to extract large amounts of virus from naturally infected tissue has restricted studies on viral morphogenesis and the production of suitable quantities of viral particles for immunological investigations. Compounding the problem has been the difficulty in identifying viral particles in naturally infected tissues of the "high-risk" HPV. One method to circumvent this problem has been the overexpression of the viral capsid proteins in heterologous systems. By using a recombinant vaccinia virus expression system, ZHOU et al. (1991) found that coexpression of both the HPV-16 L1 and L2 proteins resulted in the production of virus-like particles. These particles exhibited a density of 1.31 g/ml in agreement with known values of empty viral particles from other papillomaviruses. These experiments were soon corroborated by other laboratories using baculovirus expression vectors to produce empty viral particles of HPV-16 and BPV-1 (KIRNBAUER et al. 1992) or HPV 11 (ROSE et al. 1993), and simian virus 40 (SV-40) expression vectors (GHIM et al. 1992) or vaccinia virus vectors (HAGENSEE et al. 1993) to produce empty HPV-1 virus particles. XI and BANKS (1991) described the presence of protein complexes between HPV-16 L1 and L2 proteins when coexpressed in baculovirus-infected cells but did not demonstrate the presence of empty viral particles.

Interestingly, some of these reports demonstrated that expression of the L1 major capsid protein alone was sufficient for the production of virus-like particles (KIRNBAUER et al. 1992; ROSE et al. 1993; GHIM et al. 1992; HAGENSEE et al. 1993). However, capsids containing only the L1 protein were fewer in number and more variable in size and shape than capsids which contained both the L1 and L2 proteins (HAGENSEE et al. 1993). These data show that other gene products such as E4 of papillomaviruses are not required for the formation of capsids. The E4 protein had been suggested to be involved in morphogenesis because of its high expression in the superbasal cell layers of infected tissue. These studies, however, do not exclude the possibility that the E4 gene product(s) could still be involved in the proper packaging of viral DNA into the capsids or in some other unknown aspect of virion maturation.

An important aspect of virus-like particles is that they are immunologically similar to the native papillomavirus virions (GHIM et al. 1992; KIRNBAUER et al. 1992; ROSE et al. 1993). Antisera made against virus-like particles were able to inhibit infection of C127 cells in culture by BPV (KIRNBAUER et al. 1992) suggesting that this system might be useful for the eventual development of a viral vaccine. The preparation of virus-like particles with immunological similarities to the natural virions will also make these systems useful in seroepidemiology and pathogenesis studies.

4 Rafts: A Brief History

An assortment of techniques has been tried by various investigators to culture epithelial cells in tissue culture. The organotypic culture system (rafts) has proven to most accurately mimic the in vivo physiology of the epidermis. In this system epithelial cells are placed on top of a dermal equivalent component and raised to an air–liquid interface for growth and differentiation. The dermal equivalent is constructed by either the recombination of the epidermal cells with natural dermal elements or a synthetic collagen matrix maintained on a rigid support. Growing epithelial cells in either of these systems has allowed for a complete differentiation program, which is in contrast to that achieved in monolayer cultures. One of the first successful methods of epidermis organotypic culture was the placement of human skin explants onto the reticular aspect of split-thickness sections of pig skin (FREEMAN et al. 1976). Stratification and differentiation of the epithelial cells had occurred by day 14, displaying the characteristic basal, spinous, granular, and squamous layers. Proof of functional epithelium came from the successful grafting onto patients with third-degree burns. Similarly, deepidermized human skin flaps were used in place of pig skin with equivalent results (WOODLEY et al. 1982; RÉGNIER et al. 1981, 1984).

The use of collagen matrices as the dermal equivalent greatly simplified the use of organotypic cultures. The differentiation of liver epithelial cells was first observed on a collagen matrix which had been raised to the air–liquid interface by allowing it to float on the media surface (MICHALOPOULOS and PITOT 1975). The floating of the culture in this manner gave the system the commonly used name of "raft" culture. Liver epithelium grown by the raft system not only showed morphological features, but extended viability and functional features comparable to the in vivo state. Using raft technology, additional studies on mammary epithelium (EMERMAN and PITELKA 1977), rat lingual epithelial cells (FUSENIG et al. 1978), and mouse epidermal cells (LILLIE et al. 1980) all showed the ability to faithfully reproduce differentiation in vitro. The use of silicon chambers (FUSENIG et al. 1978) and stainless steel grids (LILLIE et al. 1980) to hold the rafts has made normal care of the cultures much easier. Furthermore, the accuracy of the differentiated state in vitro has been greatly improved by the introduction of fibroblasts into the collagen matrix (BELL et al. 1979, 1983; CHAMSON et al. 1982). Laboratories today use a form of a technique described by ASSELINEAU and PRUNIÉRAS (1984) (Fig. 2). Basically, collagen I is mixed with fibroblasts, media, and buffer at 0°C and then placed into tissue culture dishes and incubated at 37°C until the mixture has solidified. The collagen–fibroblast solution forms a lattice or raft which is used as the dermal equivalent. Epidermal cells of choice are seeded on top of the collagen–fibroblast matrix, allowed to attach, and grown to confluency while remaining submerged. The raft is then lifted to the liquid–air interface and fed from underneath so that the culture media only comes into contact with the bottom of the raft. The rafts are incubated for 2–3 weeks in which time the epidermal cells stratify and differentiate. Following stratification the rafts are harvested and examined by standard histological techniques.

It is important to ask how faithfully the raft system represents the situation seen in vivo. A comparison of artificial breast epithelium with the breast epithelium in vivo demonstrated that in general the "artificial epithelium" closely mimics its in vivo counterpart (ASSELINEAU et al. 1986). Only minor differences in the morphology of the basal keratinocytes and some differences in the distribution of a few keratinocyte-specific antigens, fibronectin, and the Ψ_3 antigen were detected in the raft culture epidermis, but are normally only seen in vivo in hyperproliferative epithelium such as during wound healing and psoriasis. Another report detected differences in the expression of keratins K1 and K2 between in vitro and in vivo grown epithelium (KOPAN et al. 1988). The ability of raft cultures to faithfully mimic the in vivo situation is not restricted to non-transformed epithelium, but rafts can also mimic tumoral architecture in vitro using cells derived from neoplasias (REGNIER et al. 1988).

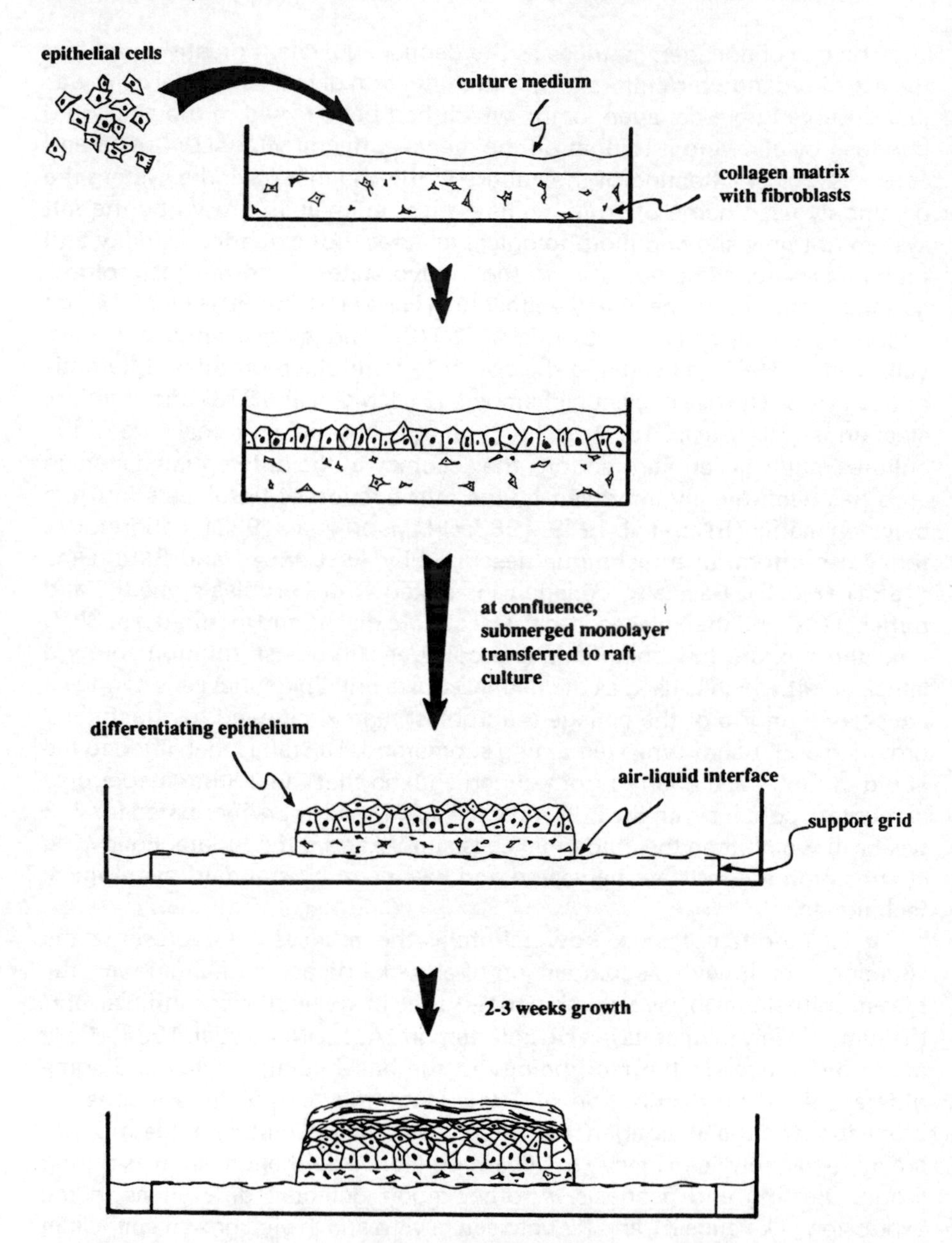

Fig. 2. Collagen raft system. Epithelial cells are plated onto a submerged collagen matrix containing fibroblast feeders in the presence of medium. When cells reach confluence the matrix is raised onto a metal grid at the air–liquid interface and subsequent feeding occurs from underneath. Cells stratify and differentiate over a 2-week period

5 Papillomavirus and Raft Cultures

The specific tropism of papillomaviruses for keratinizing epithelium and host specificity has been an impediment for developing a system to study virus-cell interaction in a natural situation. Research on papillomaviruses has greatly benefitted from the improvements made with raft culture system techniques. Using the raft culture system, the effects of HPV-16 transfected into primary human foreskin keratinocytes and SCC-13 cells have been examined in detail. Human foreskin keratinocytes containing transfected copies of HPV-16 maintained the ability to stratify and express differentiation-specific keratins but exhibited an altered pattern of differentiation (McCance et al. 1988). Gene products from HPV-16 induced differentiation abnormalities which were similar to those seen in anogenital intraepithelial neoplasias in vivo. In SCC-13 cells derived from a squamous cell carcinoma, the parental cells displayed some characteristics of differentiation on rafts but exhibited a total disruption of differentiation following the introduction of HPV-16. This work was extended to HPV-18 where the E6–E7 open reading frames (ORF) were identified as necessary and sufficient for alteration of normal differentiation patterns and for efficient immortalization of primary keratinocytes (Hudson et al. 1990). HPV-18-immortalized cell lines lost all ability to differentiate in raft culture following continuous passage in culture. The E7 ORF alone displayed a much diminished immortalization activity, whereas the E6 ORF by itself had no immortalizing activity. The ability of HPV-16 and -18 to immortalize keratinocyte cell lines and interfere with their differentiation program on raft cultures was identical to those effects seen in premalignant lesions in vivo. Similar results were demonstrated in studies using raft cultures by Pecoraro et al. (1989) and Blanton et al. (1991). Epithelial cells derived from low-grade CIN lesions as well as invasive cervical carcinomas grown in raft culture duplicated the histological features observed in vivo in cervical neoplasias (Rader et al. 1990). A common observation of these first papillomavirus-raft system investigations was that morphological and histological features of HPV-associated neoplasias were faithfully reproduced in the raft culture. Furthermore, these experiments suggested that changes in differentiation patterns seen during the progression of a lesion in vivo also develop in vitro in the raft system during the continued passage of HPV-immortalized cells. These studies collectively established the raft culture system as an important model system in which progression of HPV-induced lesions can be investigated in vitro.

Hurlin et al. (1991) examined HPV-18-immortalized cell lines in culture and described the progressive alteration in differentiation properties and the increase in chromosomal abnormalities that accompany passaging in culture. Biochemical studies on HPV-immortalized cell lines grown in raft cultures reported changes in the normal expression patterns of keratins and filaggrin (Merrick et al. 1992). Merrick et al. (1992) also showed a restriction of

PCNA expression to the basal layer in normal keratinocytes, whereas in HPV-immortalized keratinocytes expression of PCNA was seen throughout all the strata of the raft. This identifies a potentially useful marker of the effects of HPV on cellular proliferation in differentiating epithelium. Using PCNA as an indicator of cell cycle progression, BLANTON et al. (1992) demonstrated that the expression of HPV-16 E7 protein prevented withdrawal from the cell cycle but had no observable effect on terminal differentiation. Cell lines that coexpressed both E6 and E7 of HPV-16 were found to be similar to cell lines expressing E7 alone in their distribution of PCNA expression and ability to terminally differentiate in rafts. The expression of p53 in rafts expressing E6 was undetectable while in rafts expressing only E7, p53 was expressed throughout the basal and suprabasal layers of the rafts.

WILSON et al. (1992) undertook a characterization of a wide range of morphological and molecular changes that occur during the culturing of human foreskin keratinocytes on rafts. A comparison of cultures grown with mouse or human fibroblasts found that expression of differentiation specific markers varied both temporally and quantitatively. Interestingly, when human neonatal fibroblasts were used in the dermal equivalent, differentiation of the human foreskin keratinocytes was slower and did not obtain the same degree of differentiation by 32 days in culture as compared to cultures in which mouse fibroblasts were used. These studies have helped to define parameters important in controlling epithelial differentiation in vitro. More studies of this nature are needed to determine the parameters important for faithful duplication of the differentiation process in vitro. Furthermore, all aspects of raft culture systems need to be investigated in a systematic way, for example, source of collagen, dermal components, media supplements, incubation times and temperatures, and hormones commonly involved in the in vivo growth of the epidermis of choice.

A raft system has been used to examine differentiation-specific promoters of papillomaviruses. HUMMEL et al. (1992), using a cell line derived from a low-grade cervical lesion, characterized the differences in HPV gene expression in cells grown in monolayer and cells allowed to differentiate in raft culture. In terminally differentiating cultures there was a significant increase in the amount of expression of an mRNA encoding the E4 and E5 ORF. This increase was shown to be due to a transcript which initiated in the E7 ORF, defining a novel differentiation-dependent promoter. Since the activity of this promoter precedes the appearance of the other set of differentiation-specific transcripts for the late genes, L1 and L2, the E4/E5 p742 promoter has been termed an early–late promoter.

6 In Vitro Biosynthesis of Papillomavirus

Significant progress has recently been made in the ability to propagate papillomaviruses in culture. Using the xenograft system that KREIDER et al. (1985, 1987b) originally developed to propagate infectious HPV-11-H virions, DOLLARD et al. (1992) explanted infected xenograft tissue onto raft cultures and detected viral particles in outgrowths of the explant. In this study late gene expression was detected only in terminally differentiated tissue. This study demonstrated that the life cycle of HPV could be maintained in a coupled xenograft/raft system.

The raft system has recently been used to induce the production of virions from continuous cell lines. BEDELL et al. (1991) reported the isolation of a cell line derived from a CIN I lesion that maintains approximately 50 episomal copies of HPV-31 subtype b. When this cell line (CIN-612) was examined in the raft system, it was found to stratify and differentiate in a manner similar to a low-grade cervical lesion in vivo. In situ hybridization studies on CIN-612 raft culture serial sections demonstrated the amplification of HPV-31b genomes in distinct foci of the upper suprabasal regions. Although low levels of late gene transcription could be detected in stratified raft cultures, no capsid protein was demonstrated. This block to virion production was attributed to an incomplete differentiation program in raft culture of the CIN-612 cells. By the addition of the phorbol ester 12-*O*-tetradecanoylphorbol-13-acetate (TPA) to the raft culture media, MEYERS et al. (1992) observed a more complete differentiation program in CIN-612 raft cultures. In TPA-treated CIN-612 rafts the increased expression of the terminal differentiation markers keratin 10 and filaggrin was detected by immunohistochemical techniques. In addition, the major late capsid protein L1 of HPV was also detected by immunohistochemical techniques in the nuclei of suprabasal cells in CIN-612 rafts treated with TPA, but not in rafts grown in the absence of TPA.

When sections of the raft tissue were examined by electron microscopy, particles of approximately 50–55 nm were detected in the nuclei of suprabasal cells. These nuclear dense particles were of the correct size as papillomaviruses seen in naturally occurring infections in vivo and, as expected, were localized to the nucleus. Virus particles produced in this manner were purified by isopycnic centrifugation and were demonstrated to be complete virions by the presence of both the specific HPV DNA and virus particles sedimenting at the same fraction in the isopycnic gradient. This was the first system to describe the production of papillomavirus virions in vitro from a continuous cell line and demonstrated that latently infected keratinocytes can be reactivated into a productive infection with the biosynthesis of virions.

The increased expression of differentiation markers and the induction of production of virion by the introduction of TPA into the raft culture medium suggests a role for protein kinase C (PKC) in these processes. Phorbol esters have been demonstrated to induce virion production of other viruses such as Epstein-Barr virus (CRAWFORD and ANDO 1986; DAVIES et al. 1991; LI et al. 1992), Pichinde virus (POLYAK et al. 1991), Rift Valley fever virus (Lewis et al. 1989), cytomegalovirus (WEINSHENKER et al. 1988), human immunodeficiency virus (CULLEN and GREENE 1989; LAURENCE et al. 1990), and bovine leukemia virus (JENSEN et al. 1992). Both stratification at the air–liquid interface and PKC activation are required for the induction of papillomavirus biosynthesis. The induction of viral propagation by other known activators of the PKC signal transduction pathway such as 1,2-dioctanoyl-*sn*-glycerol supports this hypothesis (Fig. 3). The use of raft culture systems will allow for studies on the mechanisms of maintenance and termination of papillomavirus latency, as well as permit the systematic definition of the stages of vegetative viral propagation. Improved raft culture systems are being developed to expand the ability to propagate virions from all types of papillomaviruses and to begin genetic analysis of the complete papillomavirus life cycle.

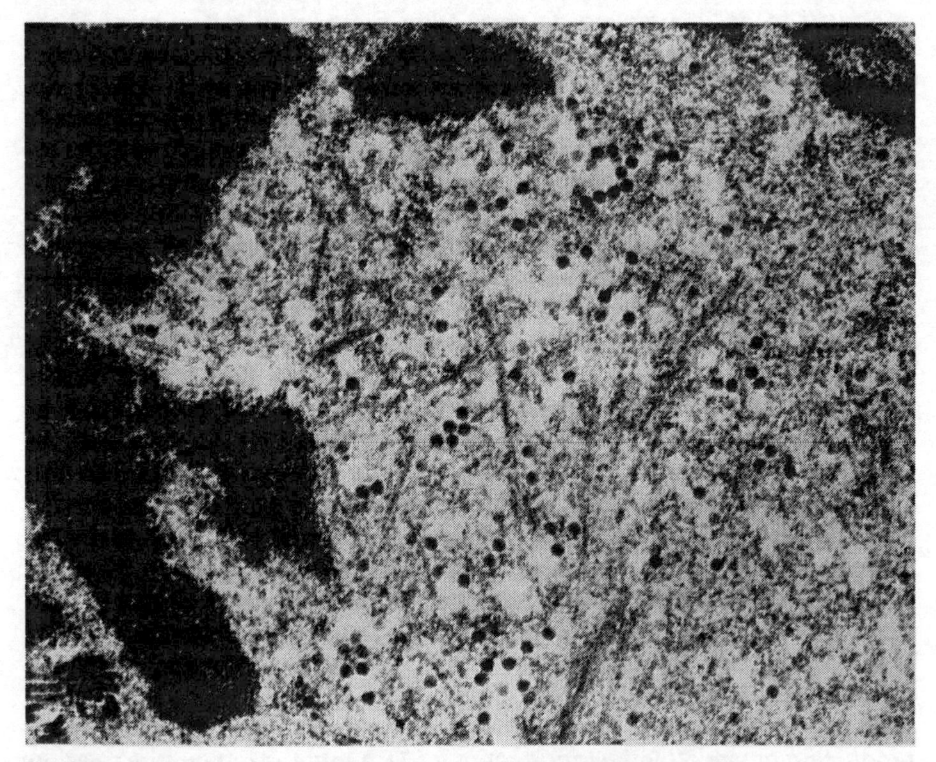

Fig. 3. Virions of approximately 50–55 nm in diameter detected by electron microscopy in stratified raft cultures of CIN-612 cells treated with 1,2-dioctanoyl-*sn*-glycerol

7 Conclusion

The ability of epithelial cells to undergo the process of differentiation is crucial for the replication of papillomaviruses. A small number of papillomaviruses such as BPV 1 are capable of infecting and inducing tumors in fibroblasts as well as epithelial cells, but infection of fibroblasts does not lead to a productive infection. Cell lines containing papillomavirus genomes that have been derived from carcinomas have all been unable to produce virus particles using current cell culture technology. Furthermore, cell lines derived following transfection of exogenous papillomavirus DNA have also been deficient in producing progeny virus since the transfected DNA usually integrating into host chromosomes.

In the last few years many advances have occurred in the development of several laboratory systems for the study of papillomaviruses. The nude mouse xenograft system was first shown to be capable of propagating papillomaviruses in the laboratory. The use of the xenograft system will allow for an understanding of the role of papillomavirus in the development of preneoplastic lesions and the identification of possible co-factors in the pathogenesis of human anogenital cancers. Candidate therapeutic agents for the treatment of anogenital cancers can now be tested for efficacy in the context of a whole animal using this system. This testing will also allow for the investigation of viral and host factors involved in papillomavirus regulation.

The production of virus-like particles in vitro using expression vectors for capsid proteins should greatly increase our knowledge of viral capsid morphogenesis and will be an important tool in understanding papillomavirus pathogenesis and seroepidemiology. Virus-like particles have already been shown to be highly immunogenic, displaying antigenic epitopes seen in a naturally occurring infection. These studies indicate that the empty capsid particles are good candidates for use in the development of a vaccine. It is also possible that these particles will be useful in studies to identify papillomavirus receptors. Ultimately, these studies will require the use of infectious virus to convincingly prove the identification of the receptor. Receptor studies using virus-like particles could be hampered by their inability to initiate a productive infection. The ability to express the capsid proteins in a cell line containing episomal copies of papillomavirus genomes might be an alternative way to overcome this problem.

The use of raft cultures is developing into a powerful system whereby the complete viral life cycle and the unique papillomavirus–keratinocyte interactions can be studied in a detailed manner in vitro. It is possible that the next few years will see the propagation of numerous papillomavirus types in vitro. Concomitantly with our gains in the understanding of the viral life cycle of papillomaviruses will come a greatly increased understanding of the mechanisms controlling the proliferation and differentiation of epithelium. The use of the raft system for the propagation of virions will allow for biochemical and molecular manipulations which are necessary for the eventual development of anti-viral treatments in the prevention of papillomavirus-associated neoplasias. The linking of the raft culture systems with the xenograft systems should also prove to be a useful tool in the identification of therapeutics.

We believe that the recent progress in the development of laboratory systems for the production of papillomaviruses has opened a new area of study. The next decade promises to be exciting as new investigations describe aspects of the vegetative replication of papillomaviruses and their role in pathogenesis.

References

Asselineau D, Pruniéras M (1984) Reconstitution of 'simplified' skin: control of fabrication. Br J Dermatol 111 [Suppl 27]: 219–222

Asselineau D, Bernard BA, Bailly C, Darmon D, Pruniéras M (1986) Human epidermis reconstructed by culture: is it normal? Soc Invest Dermatol 86: 181–186

Bedell MA, Hudson JB, Golub TR, Turyk ME, Hosken M, Wilbanks GD, Laimins LA (1991) Amplification of human papillomavirus genomes in vitro is dependent of epithelial differentiation. J Virol 65: 2254–2260

Bell E, Merrill C, Solomon D (1979) Characteristics of a tissueequivalent formed by fibroblasts cast in a collagen gel. J Cell Biol 83: 398a

Bell E, Sher S, Hull B, Merrill C, Rosen S, Chamson A, Asselineau D, Dubertret L, Coulomb B, Lapiere C, Nusgens B, Neveux Y (1983) The reconstitution of living skin. J Invest Dermatol 81: 2s–10s

Blanton RA, Perez-Reyes N, Merrick DT, McDougall JK (1991) Epithelial cells immortalized by human papillomaviruses have premalignant characteristics in organotypic culture. Am J Pathol 138: 673–685

Blanton RA, Coltrera MD, Gown AM, Halbert CL, McDougall JK (1992) Expression of the HPV 16 E7 gene generates proliferation in stratified squamous cell cultures which is independent of endogenous p53 levels. Cell Growth Differ 3: 791–802

Breitburd FV (1987) Cell culture systems for study of papillamaviruses. In: Syrjänen K, Gissmann L, Koss LG (eds) Papillomaviruses and human disease. Springer, Berlin Heidelberg New York, pp 371–392

Broker TR, Botchan M (1986) Papillomaviruses: retrospectives and prospectives. Cancer Cells 4: 17–36

Brown DR, Chin MT, Strike DG, (1988) Identification of human papillomavirus type 11 E4 gene products in human tissue implants from athymic mice. Virology 165: 262–267

Butel JS (1971) Studies with human papilloma virus modelled after known papovavirus systems. JNCI 48: 285–299

Chamson A, Finley J, Hull B, Bell E (1982) Differentiation and morphogenesis of keratinocytes grown of contracted collagen lattices. J Cell Biol 95: 59a

Christensen ND, Kreider JW (1990) Antibody-mediated neutralization in vivo of infectious papillomaviruses. J Virol 64: 3151–3156

Crawford DH, Ando I (1986) EB virus induction is associated with B-cell maturation. Immunology 59: 405–409

Cullen BR, Greene WC (1989) Regulatory pathways governing HIV-1 replication. Cell 58: 423–426

Dartmann K, Schwarz E, Gissmann L, zur Hausen H (1986) Virology 151: 124–130

Davies AH, Grand RJA, Evans FJ, Rickinson AB (1991) Induction of Epstein-Barr virus lytic cycle by tumor-promoting and non-tumor-promoting phorbol esters requires active protein kinase CJ Virol 65: 6838–6844

Dollard SC, Chow LT, Kreider JW, Broker TR, Lill NL, Howett MK (1989) Characterization of an HPV type 11 isolate propagated in human foreskin implants in nude mice. Virology 171: 294–297

Dollard SC, Wilson JL, Demeter LM, Bonnez W, Reichman RC, Broker TR, Chow LT (1992) Production of human papillomavirus and modulation of the infectious program in epithelial raft cultures. Genes Dev. 6: 1131–1142
Eisinger M, Kucarova O, Sarkar NH, Good RA (1975) Propagation of human wart virus in tissue culture. Nature 256: 432–434
Emerman JT, Pitelka DR (1977) Maintenance and induction of morphological differentiation in dissociated mammary epithelium on floating collagen membranes. In Vitro 5: 316–328
Freeman AE, Igel HJ, Herrman BJ, Kleinfeld KL (1976) Growth and characterization of human skin epithelial cell cultures. In Vitro 12: 352–362
Fusenig NE, Amer SM, Boukamp P, Worst PKM (1978) Characteristics of chemically transformed mouse epidermal cells in vitro and in vivo. Bull Cancer 65: 271–280
Ghim SJ, Jenson AB, Schlegel R (1992) HPV-1 L1 protein expressed in cos cells displays conformational epitopes found on intact virions. Virology 190: 548–552
Hagenesee ME, Yaegashi N, Galloway DA (1993) Self-assembly of human papillomavirus type 1 capsids by expression of the L1 protein alone or by coexpression of the L1 and L2 capsid proteins. J Virol 67: 315–322
Howley PM (1990) Papillomaviridae and their replication. In: Fields BN, Knipe DM, Chanock RM, Hirsch MS, Melnick JL, Monath TP, Roizman B (eds) Fields virology Raven, New York, pp 1625–1650
Hudson JB, Bedell MA, McCance DJ, Laimins LA (1990) Immortalization and altered differentiation of human keratinocytes in vitro by the E6 and E7 open reading frames of human papillomavirus type 18. J Virol 64: 519–526
Hummel M, Hudson JB, Laimins LA (1992) Differentiation-induced and constitutive transcription of human papillomavirus type 31b in cell lines containing viral episomes. J Virol 66: 6070–6080
Hurlin PJ, Kaur P, Smith PP, Perez-Reyes N, Blanton RA, McDougall JK (1991) Progression of human papillomavirus type 18-immortalized human keratinocytes to a malignant phenotype. Proc Natl Acad Sci USA 88: 570–574
Jensen WA, Wicks-Beard BJ, Cockerell GL (1992) Inhibition of protein kinase C results in decreased of bovine leukemia virus. J Virol 66: 4427–4433
Kirnbauer R, Booy F, Cheng N, Lowy DR, Schiller JT (1992) Papillomavirus L1 major capsid protein self-assembles into virus-like particles that are highly immunogenic. Proc Natl Acad Sci USA 89: 12180–12184
Kopan R, Traska G, Fuchs E (1988) Retinoids as important regulators of terminal differentiation: examining keratin expression in individual epidermal cells at various stages of keratinization. J Cell Biol 105: 427–440
Koss LG (1987) Cytologic and histologic manifestations of human papillomavirus infection of the female genital tract and their clinical significance. Cancer 60: 1942–1950
Kreider JW, Howett MK, Wolfe SA, Bartlett GL, Zaino RJ, Sedlacek TV, Mortel R (1985) Morphological transformation in vivo of human uterine cervix with papillomavirus from condyloma acuminata. Nature 317: 639–641
Kreider JW, Howett MK, Lill NL, Bartlett GL, Zaino RJ, Sedlacek TV, Mortel R (1986) In vivo transformation of human skin with human papillomavirus type 11 from condyloma acuminata. J Virol 59: 369–376
Kreider JW, Howett MK, Stoler MH, Zaino RJ, Welsh P (1987a) Susceptibility of varius human tissues to transformation in vivo with human papillomavirus type 11. Int J Cancer 39: 459–465
Kreider JW, Howett MK, Leure-Dupree AE, Zaino RJ, Weber JA (1987b) Laboratory production in vivo of infectious human papillomavirus type 11. J Virol 61: 590–593
Kreider JW, Patrick SD, Cladel, Welsh PA (1990) Experimental infection with human papillomavirus type 1 of human hand and foot skin. Virology 177: 415–417
Lambert PF (1991) Papillomavirus DNA replication. J Virol 65: 3417–3420
Laurence J, Cooke H, Sikder SK (1990) Effect of tamoxifen on regulation of viral replication and human immunodeficiency virus (HIV) long terminal repeat-directed transcription in cells chronically infected with HIV-1. Blood 75: 696–703
Lewis RM, Morrill JC, Jahrling PB, Cosgriff TM (1989) Replication of hemorrhagic fever viruses in monocytic cells. Rev Infect Dis 11: s736–s742
Li QX, Young LS, Niedobitek G, Dawson CW, Birkenbach M, Wang F, Rickinson AB (1992) Epstein-Barr virus infection and replication in a human epithelial cell system. Nature 356: 347–350

Lillie JH, MacCallum DK, Jepsen A (1980) Fine structure of subcultivated stratified squamous epithelium grown on collagen rafts. Exp Cell Res 125: 153–165
McCance DJ, Kopan R, Fuchs E, Laimins LA (1988) Human papillomavirus type 16 alters human epithelial cell differentiation in vitro. Proc Natl Acad Sci USA 85: 7169–7173
Mendelson CG, Kligman AM (1961) Isolation of wart virus in tissue culture. Successful re-inoculation into humans. Arch Dermatol 83: 559–562
Merrick DT, Blanton RA, Gown AM, McDougall JK (1992) Altered expression of proliferation and differentiation markers in HPV 16 and 18 immortalized cells grown in organotypic culture. Am J Pathol 140: 167–177
Meyers C, Frattini MG, Hudson JB, Laimins LA (1992) Biosynthesis of human papillomavirus from a continuous cell line upon epithelial differentiation. Science 257: 971–973
Michalopoulos G, Pitot HC (1975) Primary culture of parenchymal liver cells on collagen membranes. Exp Cell Res 94: 70–78
Niimura M, Pas F, Wooley R, Souter CA (1975) Primary tissue culture of human wart-derived epidermal cells (keratinocytes). JNCI 54: 563–569
Noyes WF (1965) Studies on the human wart virus II. Changes in primary human cell cultures. Virology 25: 358–363
Oroszlan S, Rich MA (1964) Human wart virus: in vitro cultivation. Science 146: 531–533
Pecoraro G, Morgan D, Defendi V (1989) Differential effects of human papillomavirus type 6, 16, and 18 DNAs on immortalization and transformation of human cervical epithelial cells. Proc Natl Acad Sci USA 86: 563–567
Pfister H (1987) Relationship of papillomaviruses to anogenital cancer. Obstet Gynecol Clin North Am 14: 349–361
Polyak SJ, Rawls WE, Harnish DG (1991) Characterization of Pichinde virus infection of cells of the monocytic lineage. J Virol 65: 3575–3582
Rader JS, Golub TR, Hudson JB, Patel D, Bedell MA, Laimins LA (1990) In vitro differentiation of epithelial cells from cervical neoplasias resembles in vivo lesions. Oncogene 5: 571–576
Régnier M, Pruniéras M, Woodley D (1981) Growth and differentiation of adult human epidermal cells on dermal substrates. Front Matrix Biol 9: 4–35
Régnier M, Pautrat G, Pauly G, Pruniéras M (1984) Natural substrates for the reconstruction of skin in vitro. Br J Dermatol 111 [Suppl 27]: 223–224
Régnier M, Desbas C, Bailly C, Darmon M (1988) Differentiation of normal and tumoral human keratinocytes cultured on dermis: reconstruction of either normal or tumoral architecture. In Vitro Cell Dev Biol 24: 625–632
Rose RC, Bonnez W, Reichman RC, Garcea RL (1993) Expression of human papillomavirus type 11 L1 protein in insect cells: in vivo and in vitro assembly of viruslike particles. J Virol 67: 1936–1944
Rous P, Beard JW (1935) The progression to carcinoma of virus-induced rabbit papillomas (Shope). J Exp Med 62: 523–548
Shope RE, Hurst EW (1933) Infectious papillomatosis of rabbits. J Exp Med 58: 607–623
Stanley MA, Browne HM, Appleby M, Minson AC (1989) Properties of a non-tumorigenic human cervical keratinocyte cell line. Int J Cancer 43: 672–676
Sterling J, Stanley M, Gatward G, Minson T (1990) Production of human papillomavirus type 16 virions in a keratinocyte cell line. J Virol 64: 6305–6307
Stoler MH, Whitebeck A, Wolinsky SM, Broker TR, Chow LT, Howett MK, Kreider JH (1990) Infectious cycle of human papillomavirus type 11 in human foreskin xenografts in nude mice. J Virol 64: 3310–3318
Taichman LB, Breitburd F, Croissant O, Orth G (1984) The search for a culture system for papillomavirus. J Invest Dermatol 83: 2s–6s
Taichman LP, LaPorta RF (1986) The expression of papillomaviruses in epithelial cells. In: Salzman NP, Howley PM (eds) The papovaviridae, vol 2: the papillomaviruses. Plenum, New York, pp 109–139
Weinshenker BG, Wilton S, Rice GPA (1988) Phorbol ester-induced differentiation permits productive human cytomegalovirus infection in a monocytic cell line. J Immunol 140: 1625–1631
Wilson JL, Dollard SC, Chow LT, Broker TR (1992) Epithelial-specific gene expression during differentiation of stratified primary human keratinocyte cultures. Cell Growth Differ 3: 471–483
Woodley D, Saurat JH, Pruniéras M, Régnier M (1982) Pemphoid, pemphigus and Pr antigens in adult human keratinocytes grown on nonviable substrates. J Invest Dermatol 79: 23–29

Xi SZ, Banks LM (1991) Baculovirus expression of the human papillomavirus type 16 capsid proteins: detection of L1–L2 protein complexes. J Gen Virol 72: 2981–2988
Zhou J, Sun XY, Stenzel DJ, Frazer IH (1991) Expression of vaccinia recombinant HPV 16 L1 and L2 ORF proteins in epithelial cells is sufficient for assembly of HPV virion-like particles. Virology 185: 251–257
zur Hausen H, Schneider A (1987) The role of papillomaviruses in human anogenital cancer. In: Salzman NP, Howley PM (eds) The papovaviridae, vol 2: the papillomaviruses. Plenum, New York, pp 245–263

Immune Response to Human Papillomaviruses and the Prospects for Human Papillomavirus-Specific Immunisation

R.W. TINDLE and I.H. FRAZER

Papillomavirus Research Unit, Lions Human Immunology Laboratories, University of Queensland, Princess Alexandra Hospital, Woolloongabba 4102, Queensland, Australia

Current Topics in Microbiology and Immunology. Vol. 186

1 Introduction

If you wished, as a machiavellian scientist or as an astute virus, to make study of the interaction between a virus and the human immune system difficult, you would start by making the infection chronic and the onset of the associated disease insidious, so that the epidemiology of the infection would be difficult to define. You would provide a range of immunologically distinct but morphologically similar viral types, each producing a similar disease, and a range of different diseases which can be produced by the same virus type. You would arrange that the virus couldn't be propagated easily in vitro or in a laboratory animal to ensure that authentic virus particles and viral proteins would be difficult to obtain. Finally you would allow the virus to become well adapted to its host's immune system through co-evolution, so that it would be only weakly immunogenic. This mixture provides a recipe for a successful virus and a frustrated immunologist and summarises the interaction between human papillomavirus (HPV) and the human host.

The immunological response to HPV infection and the prospects for immunological control have been the subject of a number of reviews to which the reader is referred (KIRSCHNER 1986; VARDY et al. 1990; GISSMANN and JOCHMUS-KUDIELKA 1989; DAVIES et al. 1991; CAMPO 1991; FRAZER and TINDLE 1992). Infection with the common skin and genital genotypes of HPV is generally chronic, even in late childhood when the immune system is at its peak. The chronicity of most HPV infections in an immunocompetent host suggests that there is a problem with PV antigen presentation to the immune system or, if a PV-specific immune response is induced, with this response recognising PV-infected cells. The increased incidence of clinically apparent HPV-associated disease in deliberately immunosuppressed transplant recipients (RUDLINGER et al. 1986), and patients immunosuppressed by human immunodeficiency virus (HIV) infection (FRAZER et al. 1986; KIVIAT et al. 1993), in each of whom the cellular arm of the immune system is abnormal but the humoral arm is relatively preserved, suggests that antigen-specific cell-mediated immunity is important in controlling HPV infection. In contrast, patients with common variable immunodeficiency without T cell deficit generally have normal skin and genital wart immunopathology, suggesting that antibody has little to do with prevention of PV infection or with maintenance of PV latency (I. Frazer, unpublished data).

2 Histological Evidence for Immune Responses to HPV Infection

Where an immune reaction is in progress against virally infected cells, there are generally increased numbers of lymphocytes, and particularly $CD8^+$ T cells if a cytotoxic T cell response is involved. There is also an inflammatory infiltrate around any necrotic or apoptotic cells.

2.1 Inflammatory Infiltrate in Genital HPV Infection

Histological examination shows little inflammatory infiltrate at the site of genital HPV infection. By comparison with normal cervical epithelium, there is rather a depletion of $CD4^+$ intra-epithelial lymphocytes (TAY et al. 1987b) in premalignant HPV-associated genital epithelial disorders and in genital warts. In cervical intra-epithelial neoplasia (CIN) there are increased numbers of cells with natural killer (NK) markers (TAY et al. 1987c; SYRJÄNEN et al. 1986) and $CD56^+$ large granular lymphocytes, (MCKENZIE et al. 1991) and in early micro-invasive carcinoma "cellular infiltrates" can be seen around the point where the carcinoma penetrates the basal lamina underlying the epidermis (DAVIES et al. 1991). Langerhans cell numbers are decreased in HPV-associated genital lesions (MORRIS et al. 1983; VIAC et al. 1990b; VARDY et al. 1990). Major histocompatibility complex (MHC) class 2 molecules and epidermal growth factor (EGF) receptors are not increased on keratinocytes (MCGLENNEN et al. 1991; CHEN et al. 1989; TAY et al. 1987a), and those adhesion molecules which have been studied are also unaltered. This suggests that there is little local cytokine release in response to PV infection, though many immunologically important adhesion molecules, including ICAMs and B7, have not been reported on.

2.2 Inflammatory Infiltrate in Skin Warts

There is generally a lack of inflammatory cells in skin warts (CHARDONNET et al. 1993). A chronic inflammatory infiltrate occurs in regressing human cutaneous warts (AIBA et al. 1986), but whether these are PV-specific T cells or a non-antigen-specific component of the chronic inflammatory reaction to a wart killed by other mechanisms is unknown. Langerhans cells are depleted from HPV-associated lesions (CHEN et al. 1989). Plane warts, associated mainly with HPV-3, often regress spontaneously. A mononuclear cell infiltrate, consisting chiefly of $CD4^+$ T cells, is observed in the dermis during this regression (ROGOZINSKI et al. 1988; VARDY et al. 1990), and Langerhans cells change in appearance, becoming enlarged with a vacuolated endoplasmic reticulum and golgi apparatus. Such an immune reaction suggests that HPV-3 is more immunogenic, or more accessible to the immune system, than the genital HPV genotypes. Spontaneous regression of common warts is accompanied by lymphocyte and macrophage infiltration and localised necrosis of keratinocytes (BENDER 1986).

3 Serological Evidence for Immune Responses to HPV Infection

3.1 Overview

Viral serology is concerned with:

1. Diagnosis and categorisation of viral infection for clinical and epidemiological purposes.
2. Definition of epitopes recognised by virus-neutralising antibodies as these could be used as the basis of a vaccine for prevention of infection.

3.1.1 Diagnosis of Infection

Papillomavirus is unusual amongst viruses in that, because it cannot be propagated in vitro and is present in small quantities in lesions in vivo, serological tests have not generally used virus particles or infected cells as antigenic substrates, but rather have employed recombinant viral proteins, or synthetic viral peptides. As acute infection with PV has been difficult to define or recognise, serological tests have been used to look for evidence of chronic infection in patients with clinically appropriate disease. They have also been employed in seroepidemiological studies where the link between HPV serology result and clinical problem is made by molecular tests for the presence of PV DNA and clinical and epidemiological history. As longitudinal studies of the natural history of genital HPV infection have not yet been completed, case–control studies have been the norm for establishing the clinical significance of antibody to PV proteins in human serum. One of the major problems has been to define control groups; the apparently ubiquitous nature of at least some genotypes of HPV infection, the possibility of genotype cross-reactive antibodies, and the possibility of non-sexual spread of the genital group of PVs have all hindered selection of control subjects who must be negative in any serological test. Even nuns and babies appear positive in some assays for antibody to genital HPV genotypes, an observation which does not match with the current paradigms of PV epidemiology.

3.1.2 Strategies for Detection of PV-Specific Antibody in Human Serum

3.1.2.1 Approaches Prior to the Molecular Era: Capsid Antibodies to Undefined Wart Types

Tests for PV antibodies prior to the recognition that there were many genotypes of wart virus showed, by a variety of techniques including immunodiffusion (PYRHONEN and PENTTINEN 1972) and immune electron microscopy (ALMEIDA and GOFFE 1965), that approximately 50% of patients with skin warts had antibody to wart capsid proteins and, in addition, to multiple proteins from wart-infected cells (VIAC et al. 1977). Antibodies to wart capsids were found with similar frequency in patients with warts and in healthy adolescents, and such antibodies were held to reflect continuing immunity resulting from prior infection (PFISTER and ZUR HAUSEN 1978). Distinct serotypes of PV were distinguished, and a distinction was also drawn between antibodies raised to denatured viral capsids, which appeared generally to be cross-reactive across different virus types, and antibodies raised to or recognising native PV virions, which were serotype specific. Sera used for immunotyping were raised in rabbits (PASS et al. 1977); naturally occurring human type specific sera were also used (PFISTER and ZUR HAUSEN 1978). Immunisation of patients with purified wart virus protein produced a humoral and cellular immune response and wart regression in some subjects, and no response and no regression in others (VIAC et al. 1978), confirming that PVs were immunogenic and that the different serotypes might reflect the existence of biologically different wart viruses. The multiplicity of different antigen preparations employed for antibody testing, coupled with the different sensitivity of the tests employed and limited clinical or microbiological categorisation of wart-associated diseases led to data showing a wide range of prevalence of PV capsid antibodies in human sera, and an unclear association between PV antibodies in serum and the outcome of PV infection, with or without treatment. Several reviews of early work in the area are recommended for further historical perspective (KIENZLER 1985; SPRADBROW 1988; PFISTER 1990).

3.1.2.2 Strategies for Measuring PV-Specific Humoral Immunity in the Molecular Era

The progressive cloning and sequencing of the various genotypes of HPVs through the 1980s allowed production, using recombinant expression systems in vitro, of purified PV viral antigens of specific genotype. The minor capsid protein L2 and the non-structural early proteins, which had previously been unavailable for immunological study, became the subject of a systematic study of immunogenicity following immunisation in animals, and of PV infection-induced immune response in humans. This work was carried out during a period in which there was a change in the immunological paradigms, such that it became accepted that it was possible for a host to mount both humoral and cellular immune responses to intracellular proteins, which had hitherto been regarded as immunologically inaccessible.

3.1.2.3 Potential Antigens for Diagnosis of PV Infection

3.1.2.3.1 Recombinant Proteins. Recombinant bacterial fusion proteins containing PV protein sequences can be produced relatively easily. The major problem with such proteins in serological assays has been the reactivity of human sera with the bacterial components of the protein preparation; despite purification of the proteins and absorption of the sera with bacterial cell lysates, such reactivity can still be demonstrated and has hindered use of enzyme-linked immunosorbent assays (ELISA) and complicated the use of immunoblots for the detection of PV protein-reactive antibodies. Immunoblots also bias results towards the detection of linear epitopes which, particularly for the capsid antigens, may be less relevant as markers of genotype-specific infection.

Recombinant protein produced in yeast and by baculovirus does not generally have the background reactivity problems of *Escherichia coli* fusion proteins, is more likely to assume a correct conformation and remain soluble, and can be produced in large amounts. Recombinant protein produced in mammalian cells, using either transient transfection assays or recombinant viral expression vectors, will have authentic post-translational modification, but yields of protein are smaller and purification correspondingly more difficult.

3.1.2.3.2 Synthetic Peptides. Synthetic peptides can be made in quantity and are relatively pure, though unexpected contaminants and synthesis problems mandate high-performance liquid chromatography (HPLC) and mass spectroscopy analysis of reagents if their use is the sole basis for a claim of specificity of an immune response. If a large protein is to be examined systematically for B cell epitopes the cost is high, and the method can only determine responses to short linear epitopes. Coupling of peptides to plates for ELISA reactivity is somewhat unpredictable and there have hitherto been no simple tests to confirm that a peptide has bound or that an epitope is displayed unless an antibody already exists to that epitope.

Use of peptides, at least in theory, allows for examination for immunoreactivity of a given serum with multiple epitopes from the same protein, which might indicate that the response to the epitopes was induced by the corresponding protein rather than cross-reactive; however, the concept of immunodominant epitopes has yet to be explored for the PV proteins, most of which are short and might therefore not have many reactive epitopes.

3.1.2.3.3 Whole Virions. Whole PV virions can be purified from clinical lesions. The yield is good from HPV-1-associated plantar warts and from some bovine papillomavirus (BPV) lesions. Smaller numbers of particles can be obtained from HPV-6 and HPV-11 lesions, and HPV-11 also can be grown in skin chips implanted under the renal capsule of nude mice, which provide a source of small numbers of viral particles (KREIDER et al. 1987). Recently, several systems have been developed for the production of PV capsids in vitro, using vaccinia (HPV-1) or baculovirus recombinant L1, or L1 and L2 proteins (ZHOU et al. 1991; KIRNBAUER et al. 1992; HAGENSEE et al. 1993), and also in raft cultures of human skin in vitro (MEYERS et al. 1992). Such particles will supplement naturally occurring virions for experiments using HPV-1 and will allow testing for antibodies to conformational as well as linear epitopes for the genital genotypes of HPV, as well as facilitating examination of sera for virus-neutralising antibodies.

3.2 Human Immune Responses to Specific Viral Proteins

The results obtained from serologic assays for antibody to HPV antigens in humans have varied markedly with the antigenic substrate employed, the assay conditions, the criteria for a positive result, and the patient and control groups studied. The results reported here are listed, for each group of viral proteins, by assay substrate and are documentary rather than interpretative. A subjective view of their significance is given in Sects. 3.2.1.7 and 3.2.2.3, while the conclusions drawn by the authors from the reported studies are given in Sect. 4.

3.2.1 Immune Responses to Viral Capsid Proteins

3.2.1.1 HPV-6 and HPV-11 L1

Recombinant Proteins. An early attempt to characterise human immune responses to HPV-6 L1 protein (Li et al. 1987) used an *E.coli* cII fusion protein; after absorption of the sera with non-recombinant bacterial cell lysate, the authors concluded that relatively weak immunoreactivity was observed by immunoblotting, with sera reactive to a dilution of 1:200 in 18 of 30 colposcopy clinic patients and two of 30 schoolchildren; they concluded that the observed reactivity was not likely to be the result of cross reactivity with the bacterial component of the fusion protein, but did not test this conclusion. A major study undertaken using partial length HPV-6 L1 TrpE fusion proteins and sera from patients with genital warts (Jenison et al. 1988) showed that reactivity specific to HPV-6 L1 protein, and not due to *E.coli* protein or fusion protein cross-reactive antibodies, was present in the sera of some of 75 patients with genital warts, and in none of 30 6-month-old children. After absorption with bacterial cell lysates, much reactivity with *E.coli* protein remained in the sera, including reactivity with proteins of similar molecular size to the HPV fusion proteins. A series of truncated and duplicated L1 constructs were therefore required to separate reactivity with the HPV-6 L1 fusion protein from "reactivity with *E.coli* proteins (which persisted after serum absorption and) that mimicked the anti p6L1XX1 (HPV-6 L1 fusion protein) pattern". The authors cautioned that the assays used were neither specific nor sensitive enough to give true prevalence figures for HPV L1-specific antibody. In subsequent papers using the same techniques, an epitope in HPV-6 L1 with which human sera reacted was mapped to aas 417–437, as distinct from the type common cross-reactive epitope which mapped to aas 193–220 (Jenison et al. 1989). Prevalence of antibody to the HPV-6 L1 TrpE fusion protein was estimated at over 50%, both amongst a sexually transmitted disease (STD) clinic population and in children (JENISON et al. 1990); interpretation of this assay was somewhat subjective, as "faint" bands in the immunoblot assay were scored as negative. Another study used β galactosidase fusion proteins and reported two positive tests for antibody to HPV-6 L1 amongst 46 patients with cervical cancer and four amongst 72 other adult subjects (KÖCHEL et al. 1991). At least one serum reactive with HPV-6 L1 was also reactive with a recombinant L1 protein expressed in yeast (CARTER et al. 1991).

3.2.1.2 HPV-6 and HPV-11 L2

Recombinant Proteins. Some of 75 sera tested by immunoblot were reactive with a HPV-6 L2 TrpE fusion protein (JENISON et al. 1988), and subsequent immunoblot studies showed a prevalence for antibodies to the recombinant HPV-6 L2 protein of 20% amongst STD clinic patients and children (JENISON et al. 1990), and of 48% for antibodies to the HPV-6 L2 or HPV-11 L2 TrpE fusion protein amongst 199 patients with cervical cancer and age-matched controls (YAEGASHI et al. 1992). Approximately one third of reactive sera in this study reacted with both HPV-6 and HPV-11 L2 proteins, while the remainder demonstrated genotype-specific reactivity. Reactivity was as common amongst cervical cancer patients as control subjects. Five separate epitopes reactive with human sera were defined in each L2 protein; two cross-reactive epitopes were defined.

Peptides. Human sera were examined for reactivity to four peptides selected from the L1 and L2 open reading frames (ORFs) of the capsid proteins of HPV (SUCHÁNKOVÁ et al. 1990). One peptide could be identified from the L2 ORF of HPV-6 and HPV-11 (aas 411–427), against which 11 of 21 sera from patients with condylomata but only one of 21 patients with cervical cancer were reactive. Reactivity was also found in 14% of healthy women, and there was some correlation, amongst this group, between reactivity and the lifetime number of sexual partners.

3.2.1.3 HPV-6 and HPV-11 Capsids

Antibodies to intact HPV-11 virions were sought by ELISA in sera from patients with juvenile laryngeal papillomatosis (JLP) (BONNEZ et al. 1992) and controls, using HPV-11 virions prepared using the mouse xenograft system. Of JLP patients, 47% had antibodies, whereas 10% of controls were positive—there was no correlation between detection of HPV-11 DNA in the patient's papillomas and a positive test for antibody to HPV-11 capsids. Patients with condylomatous disease and celibate controls (BONNEZ et al. 1991) were examined for HPV-11 antibodies with the same ELISA assay—in this study, the majority of patients had HPV-6-related disease, and 15 out of 46 patients were positive for HPV-11 antibodies; again no correlation with the HPV type of the genital wart disease was observed. Using the same method of preparing HPV-11 virions, another group (CHRISTENSEN et al. 1992) found no antibodies to intact virions by ELISA in a group of patients with condylomata, whereas four of eight patients with JLP had such antibodies—all patients with condylomata had antibodies to disrupted HPV-11 and BPV-1 virions, whereas only one of the JLP patients did. Antibodies to intact virions could neutralise PV particles, at least as measured by the inability of the particles to infect further skin grafts placed under the renal capsule of nude mice. Synthetic HPV-11 virus-like particles were made using recombinant baculovirus (ROSE et al. 1993) and, although the morphology of the particles was less typical than the HPV-1 and BPV-1 particles made by the same technique, the authors found that they expressed the major epitopes found on HPV-11 virions; 11 of 12 sera from patients with condylomata, and none of nine controls, were reactive with the synthetic particles, although none was reactive with denatured HPV-11 L1 protein by immunoblot.

3.2.1.4 HPV-16 and HPV-18 L1

Recombinant Proteins. Most studies (ABCARIAN and SHARON 1977; JENISON et al. 1990; CASON et al. 1992) have shown little or no reactivity by immunoblot with HPV-16 or -18 L1 fusion proteins in sera of patients with cervical cancer or HPV-related genital disease; one study using a TrpE fusion protein showed that several of the sera negative by immunoblot assay appeared positive by ELISA. Another study using HPV-16 L1 TrpE fusion protein (BARBER et al. 1992) showed 40% of women with CIN were reactive by immunoblot. We used an MS2 fusion protein, and with absorbed sera found 15% of sera from patients with genital warts and also age-matched controls were reactive with an L1 fusion protein, but not with MS2 control protein (SMILLIE et al. 1990). However, the sera positive in the immunoblot assay have subsequently proven negative when assayed in immunoblot and ELISA capture assay for antibody to HPV-16 L1 using baculovirus recombinant HPV-16 L1 (Park et al., submitted).

Synthetic Peptides. Peptides covering the entire L1 and L2 ORFs of HPV-16 were examined for reactivity with sera from 30 patients with CIN or invasive cancer (DILLNER et al. 1990); nine immunoreactive epitopes were identified to which at least 15% of the sera reacted, and the reactivity of control sera from patients with parotid tumours and healthy volunteers was then assessed. The peptide most frequently reactive with sera from the group of patients with cervical cancer, which was from HPV-16 L2, was as frequently reactive with control sera as with patient sera. Six peptides, all from HPV-16 L1, were more frequently reactive with sera from patients than control sera; however, all proved reactive to some extent with some control sera. The most discriminative peptide showed reactivity with 70% of sera from cervical cancer patients (mean OD 0.2), and with 5% of control sera. No comment was made by the investigators in this study as to whether sera could be divided into those which reacted with many peptides and those with none, or whether reactivity in each serum was commonly to a single peptide. One study in which no antibody to HPV-16 L1 protein was found by immunoblot using fusion protein examined the same sera for reactivity with L1 peptides by ELISA (CASON et al. 1992) and observed that reactivity to one peptide, corresponding to aas 473–492, was more common in patients with CIN than controls.

3.2.1.5 HPV-16 and HPV-18 L2

Recombinant Proteins. An initial study observed that two sera of 76 were reactive with HPV-16 L2 TrpE fusion protein by an immunospot assay (JENISON et al. 1988); the same investigators subsequently looked at 35 HPV-16 L2 TrpE fusion protein reactive human sera by immunoblot (JENISON et al. 1991) and found that 24 also reacted with HPV-18 L2 TrpE fusion protein; only one serum appeared to recognise an epitope common to both proteins. Further seroprevalence studies with the same TrpE fusion protein (JENISON et al. 1990) showed that 40% of STD patients, and of children, were reactive with HPV-16 L2, and 25% with HPV-18 L2. Epitopes reactive with human sera were mapped in HPV-16 L2 to aas 190–204 and 149–178. A similar prevalence of reactivity to HPV-16 L2 (48%) was found in another study (BARBER et al. 1992). A study using recombinant HPV-16 L2 prepared in yeast showed that eight sera recognised the yeast recombinant protein—one of eight reactive sera recognised native but not denatured L2, but seven recognised both. Another study using TrpE fusion proteins also found relatively frequent reactivity to HPV-16 L2 in human serum, with no disease specificity (BARBER et al. 1992)

Synthetic Peptides. Antibodies, of various isotypes, were detected to peptides of HPV-16 L2 in sera from patients with cervical cancer and controls (DILLNER et al. 1990), and none was more common in the patient than in the control group.

HPV-16 Capsids. When mice were immunised with recombinant HPV-16 virus-like particles, antisera were produced which recognised a range of epitopes distinct from those seen by human and animal sera (J. ZHOU et al. 1992); immunogenic epitopes of denatured L1 protein did not appear to be immunogenic in the context of synthetic viral particles. As yet no comparable data exist for human sera.

3.2.1.6 HPV-1 and Other Capsids/Capsid Proteins

Recombinant Proteins. Using a cleaved β galactosidase fusion protein, reactivity was observed (STEGER et al. 1990) to HPV-8 L1 in 20% of human sera, including 20% of children aged 6 years; at least two epitopes were recognised, but the majority of sera were only weakly reactive. Using L1 protein of HPV-1 expressed in a eukaryotic system (GHIM et al. 1992), it was shown that such HPV-1 L1 protein displayed epitopes not displayed by heat denatured L1—no human sera were tested in this study.

Particles. Purified HPV-1 virions were used in a comparison of immunoblot, ELISA and radioimmunoprecipitation assay (RIPA) for antibodies to HPV-1 capsid proteins (STEELE and GALLIMORE 1990); only eight of 83 sera were reactive using the immunoblot technique, whereas 74 were positive by ELISA and 61 with RIPA. Disrupting the virions abolished their reactivity with the majority of sera. There was a 100% correlation amongst this study population between a history of warts and presence of HPV-1 capsid antibodies. Similar results were found using purifed HPV-1 virions in another study (VIAC et al. 1990a); by ELISA, 15 of 21 patients with plantar warts of unspecified type had HPV-1 capsid-specific antibodies, whereas children under 2 years of age had no antibodies, and patients with other types of warts had a 20%–37% prevalence of antibodies. Immune electron microscopy was used to detect HPV-1 capsid antibody in another study (ANISIMOVÁ et al. 1990); more than 80% of patients who presented with warts had antibodies detectable by this technique against their own warts, and these antibodies were generally genotype specific; only three of 14 who had HPV-2 lesions, most of whom were over 10 years old, had any antibody against HPV-1.

Disrupted BPV-1 virions have also been used to look for antibodies to PV proteins in patients with HPV-related disease (DILLNER et al. 1989, 1990); a weak correlation between IgA antibodies to BPV 1 protein in serum and cervical secretions and a histological diagnosis of CIN were observed.

From cells infected with recombinant vaccinia virus expressing HPV-1 L1 and L2, synthetic HPV-1 virions could be purified (HAGENSEE et al. 1993), and such synthetic HPV-1 viral capsids are reactive with many human sera; reactivity correlated with a history of warts within the last 5 years (D. Galloway, personal communication).

3.2.1.7 Conclusions Concerning Reactivity with Capsid Proteins

Human papillomavirus type 1 disease is associated with genotype-specific antibodies to HPV-1 capsid proteins, and the majority of sera from patients with disease are reactive if a particle- or capsid-based assay is used: results obtained with capsid proteins prepared as bacterial fusion proteins do not correlate well with disease state. For the genital PV genotypes, there is little correlation in the epidemiological studies between the results for antibody prevalence obtained with the various different methods and substrates. There is also little correlation between the immunoreactive epitopes of capsid proteins defined by the different methods. Antibodies to the capsid proteins of HPV-6 are found commonly and with approximately equal frequency in sera from patients with HPV-related disease and control sera. Selected peptides, and synthetic capsids, may better discriminate between patients and controls but the prevalence of antibodies in different patient groups, as determined by these newer methods, is not yet clear. No significant disease-associated reactivity is seen to HPV-16 capsid protein by any method.

3.2.2 Early ORF Proteins

3.2.2.1 HPV-16 E7

Interest in immune responses to the early ORF proteins of HPV has centred on HPV-16 E7 because of an apparent association between serum antibodies to this protein and cervical cancer.

Recombinant Protein. An initial study (JOCHMUS-KUDIELKA et al. 1989) using two different preparations of E7 fusion protein showed a low incidence of IgG and IgM antibodies to E7 in control subjects (2% of 49 subjects), and a 20% incidence in patients with cervical cancer. There was a good correlation between reactivity by immunoblot with two different E7 fusion protein preparations used in this study. A further study (KÖCHEL et al. 1991) using a single β galactosidase fusion protein showed 18 reactive sera amongst 46 patients with cervical cancer, with two reactive of 72 controls. A study with TrpE E7 fusion protein (JENISON et al. 1990) showed 13% of patients at an STD clinic and 12% of children had antibodies to HPV-16 E7; three of the positive sera were able to precipitate "authentic" E7 from radiolabelled Caski cells, and some could precipitate E7 protein from recombinant E7 produced in yeast (CARTER et al. 1991). A similar study using ELISA with a β galactosidase–E7 fusion protein (HASHIDO et al. 1991) showed six of 54 cervical cancer patients had antibody to E7, while no control sera were positive; the four most strongly positive sera were able to precipitate E7 from cos cells transiently expressing HPV-16 E7. A further study from the same group (KANDA et al. 1992) with the same proteins, and including immunoblot analysis, extended the series of patients with cancer to 108, and 22% were positive for E7 antibody—again no controls were positive. Another study used a rabbit reticulocyte lysate transient expression system to produce E7 (MÜLLER et al. 1992), and 49 of 98 sera from patients with invasive cervical cancer had antibodies to HPV-16 E7, as opposed to two of 60 age-matched control sera. Reactivity by RIPA in this study correlated strongly with reactivity by ELISA with an E7 peptide spanning aa 5 to aa 35, but antibodies to this peptide not associated with a positive RIPA result were also seen in 9%–16% of control subjects, and 20% of patients with non-HPV-16 cervical cancer and CIN, as well as 37% of patients with invasive HPV-16-associated cancer.

Synthetic Peptides. A preliminary study using synthetic whole length E7 protein reported antibody of IgG class in one of 11 patients with early invasive cervical cancer and four of ten with late disease (REEVES et al. 1990); a higher prevalence of antibodies of IgA class was found (5/11, 6/10). Subsequent analysis of a more extensive series by the same group (MANN et al. 1990) failed to confirm the trend to higher prevalence of antibody with later stage of disease, or the specific association of cervical cancer with E7 specific IgA antibody, and showed a prevalence of antibody to the synthetic E7 peptide of about 20% amongst 186 women with cervical cancer, and about 9% amongst age-matched controls. The study mentioned earlier which compared RIPA with E7 peptide ELISA (MÜLLER et al. 1992) also examined reactivity to another E7 peptide from aas 29–52; less reactivity was seen to this peptide than to the aa 5 to 35 peptide, though patients with invasive HPV-16 cancer were more likely to be positive (20%) than controls. A study using six overlapping peptides from HPV-16 E7 (DILLNER 1990) showed significantly more reactivity to the first 20 aas of E7, and to aas 48–68, in patients with HPV-16-positive cervical cancer than in control subjects; only IgA reactivity was increased (23% compared to 2%). Overall, however, reactivity to all tested peptides was quite common both in the cancer patients and the controls. In another study using nine synthetic peptides spanning the E7 protein, sera pooled from patients with HPV-16-related disease were compared for IgG reactivity with sera from other control groups (KRCHNÁK et al. 1990). The most reactive peptide was aas 41–60; reactivity to this peptide was seen in 50% of patients with HPV-6/11-, HPV-16- and HPV-18-related disease, but not in control sera. HPV-16 infection-specific reactivity was seen to the peptides spanning the first 40 aas of HPV-16, particularly a peptide from aas 11–30; as pooled sera were used, no prevalence of antibody could be deduced. In a subsequent study from the same group, the peptide from aas 11–30 was compared for reactivity with peptides from aas 1–20 and 21–40 (SUCHÁNKOVÁ et al. 1992); background reactivity was considerable, and the cut-off for a positive result, based on sera from children, was an OD of 0.65–0.8, a problem common to all published assays for antibody to PV proteins employing peptide-based ELISAs. Of cervical cancer patients, 27.8% had antibody to at least one E7 peptide, compared with 8.6% of control patients; cervical cancer patients were more likely than control subjects to react with more than one of the E7 peptides. The same group compared ELISA reactivity to E7 aas 11–30 peptide with immunoblot reactivity to E7 fusion protein (SUCHANKOVÁ et al. 1991) and found good correlation between the two methods, with an overall prevalence of E7-specific antibodies for cervical cancer patients of 35.2%, compared with 5.7% for controls. We used a series of 11 overlapping peptides from E7 and compared patients with CIN and other HPV-related disease to nuns, infants, and laboratory staff (TINDLE et al. 1992). Antibodies were commonly found to peptides covering aas 10–80, though not to the N and C terminal peptides. Only the peptide from aas 12–30 was significantly more frequently reactive in patients than controls. Reactivity with a peptide frequently predicted reactivity with the adjacent overlapping peptide, suggesting shared epitopes, but reactivity with more than one of the E7 peptides in a single serum was no more common than would be expected by chance alone, either amongst the patients or the controls, providing no support for the hypothesis that the observed reactivity was induced by E7 protein.

IgA Antibodies. IgA antibodies to HPV-16 E7 peptides were found in cervical secretions of 18 of 29 patients with condylomas, some of whom were positive for HPV-16 DNA (DILLNER et al. 1993). One peptide from HPV-18 E7 (aas 14-33) was identified as more commonly reactive with sera from patients with cervical cancer (10/116) than with control sera (BLEUL et al. 1991).

3.2.2.2 Other Early ORF Proteins

3.2.2.2.1 HPV-16 E6

Recombinant Proteins. All of nine sera reactive by immunoblot to an E6–MS2 fusion protein, from a series of 26 patients with cervical cancer, were also reactive by RIPA with a recombinant E6 protein produced in baculovirus (STACEY et al. 1992), as was one serum from a control subject who was non-reactive with E6 by immunoblot.

Peptides. No increased frequency of E6 peptide reactivity was seen in two series of patients with cancer of the cervix, when compared with control subjects (BLEUL et al. 1991; DILLNER 1990). An increased incidence of E6 peptide specific antibody (16% versus 0%–6% in controls) was, however, seen in another series (MÜLLER et al. 1992).

3.2.2.2.2 HPV-16 E2

Peptides. An initial study (DILLNER et al. 1989) showed that IgA antibody to a C terminal peptide of HPV-16 E2 was more frequently seen in patients with cervical cancer (24/33 positive) than in control subjects (6/27 positive). Subsequent systematic analysis of overlapping HPV-16 E2 peptides defined three further N terminal peptides for which significantly more IgA reactivity was observed in cervical cancer sera than controls (DILLNER 1990). A further study of patients with cervical cancer showed 43% positive for E2 peptide IgA antibodies, with increasing prevalence of antibody with advancing disease (LEHTINEN et al. 1992a). A study of the HPV-18 E2 region corresponding to the HPV-16 E2 carboxy terminal peptide reactive in these studies showed IgA antibodies to this peptide more commonly in patients with adenocarcinoma of the cervix (LEHTINEN et al. 1992b) than in patients with squamous cancer or controls. Another group used the same E2 carboxy terminus peptide to study patients with cervical cancer and were unable to confirm either the finding of increased prevalence of IgA antibodies to the peptide, or the association with prognosis (MANN et al. 1990). A study of IgG antibodies to HPV-16 E2 fusion protein saw little reactivity in 92 patients with STDs (JENISON et al. 1990).

3.2.2.2.3 HPV-16 E4

Antibodies to HPV-16 E4 were rare amongst patients with cervical cancer (one of 72) in one study employing β galactosidase fusion proteins (KÖCHEL et al. 1991). They were more common (181 positive of 519 sera) and associated with CIN amongst samples in another study (JOCHMUS-KUDIELKA et al. 1989), although this study employed two fusion proteins as substrates and showed little correlation of reactivity with two different E4 fusion proteins. In a study employing TrpE fusion proteins (BARBER et al. 1992), antibodies to E4 were in 25% of subjects with HPV-16 DNA positive CIN, and in none of the controls.

3.2.2.3 Conclusions Concerning Reactivity with early ORF Proteins

There is an obvious lack of correlation between the results from the various assays which have been used for assessing HPV early (eORF) reactivity in serum. Nevertheless, antibodies to HPV-16 E7 are clearly found with increased frequency in patients with cervical cancer, and there may be a correlation with disease extent. The optimal assay system has yet to be defined and the prevalence of such antibodies and their utility as a disease marker must therefore be uncertain. Antibodies to other HPV eORF proteins have not yet been sought with sufficient rigour in large enough numbers of patients to determine their utility as disease markers or as indicators of PV protein immunogenicity following HPV infection.

3.3 Role of Antibodies in Protection Against PV Infection—Data from Animal Models

For epitheliotropic viruses which have no blood-borne phase of infection, serology is not much concerned with treatment, as neutralising antibodies in the blood would be of little relevance. However, PVs are potentially vulnerable to neutralising antibodies at mucosal and skin entry points.

Cottontail Rabbit Papillomavirus. In a study of rabbits exposed to cottontail rabbit papillomavirus (CRPV), which used fusion proteins as antigens for immunoblot, antibodies to E1, E2, E6 and E7 were seen throughout the course of infection, but antibodies to the L1 and L2 fusion proteins appeared only after cancer developed (LIN et al. 1993). Presumably the animal became exposed to degraded L1 and L2 proteins only when tumour cells were dying; antibodies to native L1 and L2 containing capsids might be elicited earlier in the course of the CRPV infection. Immunisation with bacterial recombinant L1 and L2 proteins can protect against CRPV infection; L1 or L2 recombinant vaccinia virus was better than L1 and L2 TrpE fusion proteins at eliciting antibodies against native as opposed to conformational epitopes, and also at eliciting protective immunity (LIN et al. 1992).

Bovine Papillomavirus Type 1. Synthetic BPV-1 particles, prepared using recombinant baculovirus protein, elicit rabbit antisera which neutralise the ability of BPV-1 to cause focus formation in C127 cells in vitro (KIRNBAUER et al. 1992). Immunisation of cattle with BPV-2 virus or BPV-2-associated tumour prevents subsequent infection with the BPV-2 virus (JARRETT et al. 1990a). This protection is type specific (JARRETT et al. 1990b) and can be induced more easily with synthetic L1 protein than with L2 protein (JARRETT et al. 1991), although the protection from recombinant protein is partial. Protection is associated with development of neutralising antibodies to BPV virions, though whether these are the means by which protection is effected is unclear.

4 PV Serology in the Future

4.1 Assay Systems and Substrates for Serology

The use of single viral proteins as substrates in immunoassays has not been well validated for diagnosis of viral infection; diagnostic serological assays are routinely based on viral preparations involving many or all of the proteins of the relevant virus, and even then are generally only held to be diagnostic of infection if a rise in titre with time is demonstrated. However, exceptions exist: past HBV infection can be diagnosed by antibody to the surface antigen, because the restricted number of epitopes available for reactivity and cross reactivity with human serum, and the multivalency of these antigenic epitopes in capsid preparations, has allowed reliable single protein assays to be developed. Some HIV-1 assays are also based on single capsid proteins because the immunogenicity of the capsid glycoprotein, coupled with sophisticated ELISA capture technology, has allowed the development of diagnostic assays based on reactivity to multiple epitopes of a single viral protein. The results of such single protein immunoassays are nevertheless confirmed by screening for reactivity with multiple HIV viral proteins by immunoblot, and a serum is not generally regarded as positive for HIV-1 antibody unless at least three viral proteins are simultaneously reactive. PV serology does not as yet meet the requirements for a single epitope diagnostic assay, and some genotypes of HPV might more appropriately be compared with hepatitis C virus where significant numbers of infected patients never appear to produce a measurable humoral immune response.

Viral particles, either natural or prepared synthetically using eukaryotic expression vectors, would appear to be the substrate of choice for future seroepidemiological studies of most human PV infections. Assays based on virions correlate with clinical status, appear to be genotype specific, and may in addition measure functionally important antibodies. There is little correlation between antibody to capsids and antibodies to capsid proteins recognised in immunoblot assays, and there are problems with contaminating *E. coli*-derived proteins when prokaryotic fusion proteins are used as substrate in immunoassay. It is possible that RIPA may overcome some of these difficulties if native sequence capsid protein is expressed in eukaryotic cells or yeast. Disrupted virions of the "wrong" type display group-specific rather than genotype-specific epitopes, and their use in epidemiological studies would therefore be difficult. Peptide-based assays have the same problem with loss of critical conformational epitopes as recombinant protein-based immunoblot assays, and in addition there is extensive reactivity against many peptides in "sticky" human serum. As synthetic capsids are now available for the common genital genotypes, future use of capsid peptides in epidemiological studies seems less than optimal.

4.2 Seroepidemiological Studies of HPV Infection

A corollary of the supposition that viruses have co-evolved with their hosts is that not all PVs will be equally immunogenic. In particular, it is unwise to assume that what applies to animal PVs also applies to HPV, and even that different HPV genotypes will elicit identical immunological scenarios in humans. Immunosuppressed transplant recipients have an increased relative risk (RR) for HPV-5 and HPV-8-associated squamous cell cancers (BUNNEY et al. 1987). These genotypes, unlike the genital strains, produce little disease in an immunocompetent host, indicating that they may be more immunogenic after natural infection and associated disease is controlled by an effective immune response. Other genotypes including HPV-16 are seemingly less well controlled by the immune system either because of a weaker immunogenicity or a better evasion strategy, and HPV-16 capsid antibodies have yet to be demonstrated.

The majority of common HPV genotypes, including HPV-6, HPV-11, HPV-1 and HPV-2, presumably lie somewhere between these extremes. HPV-1 capsid proteins are immunogenic. Patients with HPV-1-associated warts generally make antibody to capsid proteins; the majority of antibodies are directed against conformational determinants displayed only on intact virions. HPV-6 is possibly immunogenic; while antibodies to intact capsids are seen, the incidence in patients with HPV-6-associated disease is not sufficiently raised to convincingly demonstrate that these antibodies are associated with infection. The consequences of chronic infection, as exemplified by JLP, may differ from acute infection producing condylomata acuminata.

As the mucosal immune system is separate from the systemic immune system and characteristically secretes IgA locally in response to local immune stimuli, there is need to examine mucosally secreted IgA and blood IgA-neutralising antibody which to some extent reflects mucosal immunity. For genital wart disease, attention needs to be placed on measuring capsid-specific antibodies in cervical secretions as these would be the potentially virus-neutralising antibodies. For all genotypes of HPV, defining the natural history of development of capsid-specific antibodies through the course of acute PV infection will be crucial to evaluating the usefulness of serological testing. It will have to be borne in mind that for some virus types, especially HPV-16, little capsid-specific antibody may be made because of the low level of expression of capsid proteins in lesions.

The role in clinical and epidemiological studies of measuring antibodies to eORF proteins merits further study. In particular, study should be of E7, for which an immune response has some connection with HPV-associated cancer, and of E1 as the protein essential for maintenance of episomal HPV DNA. Current data at best indicate a 20% specificity and virtually no selectivity for E7 antibody assays as a test for cervical cancer, and the major conclusion from the work to date is that by the time invasive cancer has developed, but probably not before, the immune system can, but may not always, become aware of E7 as a non-self-protein.

5 Cell-Mediated Immunity to HPV Proteins

If antibodies to HPV proteins are elicited by infection, then cell-mediated immunity to the same proteins is certain to be present. What is not known presently is the nature and duration of that response, which protein epitopes are recognised, and the extent to which the response is effective in controlling HPV infection.

5.1 In Vitro Studies with Human Cells

Recombinant HPV fusion proteins contain T-helper (Tindle et al. 1991) and T-proliferative (Comerford et al. 1991; Shepherd et al. 1992) epitopes seen in the context of murine MHC and T cell repertoires, but this information gives little indication whether these proteins are immunogenic in the context of human MHC and T cell repertoires.

Early studies showed that peripheral blood lymphocytes from patients with a history of skin warts proliferate to wart antigen preparations in vitro (Eisinger and Lee 1976). More recent in vitro studies have shown proliferation of peripheral blood T cells from some normal individuals in response to relatively high concentrations of peptide from HPV E6 and L1 proteins

(STRANG et al. 1990), indicating the appropriate restriction element and a high frequency of T cell precursors with TcR recognising E6 and L1 epitopes. One or more of a set of overlapping peptides spanning the whole E7 molecule were recognised by 8% of normal controls, 50% of women with a history of CIN, and 56% of cancer patients (A. Kadish, personal communication). Studies under way in our laboratory have so far been unable to demonstrate consistent T proliferation in response to HPV-16 E7 protein or a range of overlapping peptides spanning the length of E7, in any of a panel of healthy subjects including most of the common MHC class 2 haplotypes, even after several rounds of in vitro stimulation of PBMC with antigen. T-proliferative epitopes have been defined (ALTMANN et al. 1992) in the HPV-16 E7 protein: three are recognised, in association with two HLA haplotypes, in E7-seropositive individuals. However, the "universal" HPV-16 E7 T epitope of mice (TINDLE et al. 1991) does not hold this property in humans, which exemplifies the limitations of looking for epitopes relevant to humans outside the natural host in which HPV has evolved.

The definition of cytotoxic T cell epitopes in HPV-infected subjects has proved difficult, as has attempting to raise HPV peptide-specific cytotoxic T lymphocytes (CTL) in mice. It has been possible to define an HPV-16 E7 CTL epitope in D^6 mice (FELTKAMP et al. 1993). Potential CTL epitopes have been sought in humans (STAUSS et al. 1992) by narrowing down the candidate peptides from HPV proteins to those which bind to MHC molecules, as determined by peptide-mediated stabilisation of "empty" MHC molecules at the surface of R-MAS (mouse) and HT-2 (human) cells, and by using predictive algorithms for peptides capable of being presented to the TcR by particular MHC molecules. Using these techniques, a nonamer CTL epitope functional in the HLA A2.1 background has been identified in the N terminal portion of HPV-16 E7 (KAST et al. 1993). Two of the sequences identified as T-proliferative epitopes in HPV-16 E7 functioned as CTL epitopes for CD4-restricted CTL from cervical carcinoma patients (ALTMANN et al. 1992). In addition to these antigen-specific HPV-directed responses, activated macrophages are reported to selectively recognise and destroy HPV-16-associated neoplastic cells with E7 as the target molecule (BANKS et al. 1991).

There are no data to indicate that the T epitopes to which human proliferative T cell responses have been defined in vitro using peptides would be presented in vivo by HPV-infected cells, or host antigen-presenting cells, or that the T cell response to any HPV protein or peptide is altered qualitatively or quantitatively following natural infection or immunisation with HPV proteins. We believe that this is the yardstick by which the T immunogenicity of HPV in humans must be measured (FRAZER and TINDLE 1992). Endogenously produced and naturally processed HPV peptides, preferably presented by keratinocytes, will need to be used to mimic the presentation of HPV molecules in the genital tract and skin. This should now be achievable through the use of appropriate transfection expression vectors for HPV genes to introduce HPV proteins into the cytosol compartment, on track for incorporation into MHC class I molecules at the keratinocyte surface.

5.2 Animal Models of HPV Peptide-Directed Tumour Immunity

A number of animal systems have demonstrated the induction of a cell-mediated immune response which will control the growth of cells expressing HPV ORFs either in vivo or in vitro. Immunisation of mice with non-tumorigenic syngeneic fibroblasts expressing recombinant HPV-16 E7 protein slowed the growth of recombinant E7-expressing melanoma cells in vivo, indicating an E7-associated rejection antigen (CHEN et al. 1991). In similar experiments the same group demonstrated the induction of E6-specific CTLs (CHEN et al. 1992b) in animals immunised with E6-expressing fibroblasts. CTLs from mice whose tumours had regressed could lyse, in an antigen- and MHC-restricted manner, E6-expressing target cells. In another model (MENEGUZZI et al. 1991), inoculation of rats with vaccinia recombinants expressing E6 and E7 prevented tumour development in some animals challenged with a tumour rendered malignant by co-transformation with HPV-16 and activated ras oncogene, indicating that HPV proteins may function as tumour rejection antigens. Immunity to a further tumour challenge lasted for 3 months after tumour rejection. The immunological basis of tumour rejection was not determined in this model, nor was the HPV specificity of rejection confirmed using a non-HPV tumour challenge.

B7 is a cell surface molecule normally found only on professional antigen-presenting cells, including B cells, Langerhans cells, activated T cells and activated macrophages. This molecule is a natural ligand for the T cell molecule CD28, and stimulation of CD28 on resting T cells at the same time as these cells are presented with antigen is now established as critical to the induction of effector T cells (NABAVI et al. 1992). In a murine model where HPV proteins were the target antigen, it was shown that HPV-16 E7-transfected murine melanoma cells grew progressively in immunocompetent murine hosts, but transfection of this E7-expressing tumour with B7 produced a tumour which regressed in immunocompetent hosts and induced the simultaneous rejection by such hosts of a $B7^-$, $E7^+$ melanoma clone, but not the $B7^-$, $E7^-$ parent melanoma. Rejection was mediated by $CD8^+$, $CD28^+$ CTL (CHEN et al. 1992a). Other groups have demonstrated a similar effect of B7 expression on induction of tumour-specific CTL, including induction of CTL specific for melanomas not expressing E7, where presumably melanoma-specific antigens are the target of the immune response (TOWNSEND and ALLISON 1993). By analogy, T cell stimulation by cells co-expressing tumour or, presumably, viral antigen should enhance the induction of specific cytotoxity in vivo in humans.

Mouse keratinocytes transfected with most of the E7 genome can be grafted onto a syngeneic mouse using a transplantation technique permitting the formation of a differentiated epithelium. This model allows viral antigens to be presented to the immune system in a manner comparable to natural infection. Animals grafted with E7 expressing keratinocyte-derived epithelia developed $CD4^+$ T cell-mediated delayed-type hypersensitivity (DTH) responses, which could be elicited by ear challenge with vaccinia recombinant for E7, but not by challenge with non-recombinant vaccinia (McLEAN et al. 1993).

6 Virus-Mediated Evasion of the Host Immune Response

A number of strategies have evolved in viruses to allow them to escape from immune recognition. Mechanisms to evade antibody-mediated immunity include escape from viral neutralisation by genetic drift, as used by orthomyxoviruses, but the genome of the HPV, as with all double-stranded DNA viruses, is relatively stable (HO et al. 1991). Secretion by viruses of proteins which interfere with the immune response to virally infected cells is also described; these include complement-inhibiting and interleukin (IL1) binding proteins from pox viruses, IL10 and BCL-2-like molecules from Epstein-Barr virus (EBV), and early region proteins which render infected cells less susceptible to tumour necrosis factor or lymphotoxin-mediated cytolysis, by undefined mechanisms, from adenoviruses. No similar function has been ascribed to any HPV proteins. There is no evidence that HPV, like cytomegalovirus, can prevent antigen presentation by blocking the transport of peptide-loaded MHC class I molecules into the Golgi compartment (DEL VAL et al. 1992), and there are no reports of HPV modifying transporter gene function and thereby affecting antigen presentation (SPIES et al. 1992), though both of these mechanisms should be examined as part of the investigation of the relative lack of immunogenicity of HPV infection.

If HPV infection were to effect immune evasion by altering the local cytokine environment, one strategy might be to down-regulate interferon expression by PV-infected cells. Interferon gamma is probably not produced in response to HPV infection as there is little induction of MHC class 2 on PV-infected epithelial cells. Data for other interferons are unknown. Interferons have been tried as therapy for HPV infection with mixed results (KIRBY et al. 1988; HOGAN et al. 1991; GROSS 1988). Local and, to a much lesser extent, parenteral administration of interferon alpha and gamma species reduces visible genital wart disease and the lesions of JLP (AUBORN and STEINBERG 1990). The disease frequently recurs when therapy is discontinued, and the effect of interferon is probably anti-proliferative rather than anti-viral or immunostimulatory. Interferons, locally or systemically, have little or no effect on the clearance of the apparently more immunogenic skin warts (ANDROPHY 1986), whereas wart-specific immunotherapy with autologous wart vaccines has some effect (BIBERSTEIN 1925). Thus it seems unlikely that failure of PV infection to induce a local interferon response is relevant to the failure of host defences to control HPV infection.

Viral infection commonly interferes with antigen processing and presentation by MHC in the infected cell; this has been observed with herpesviruses, adenoviruses and rhabdoviruses. Loss of MHC class 1 expression in cervical carcinomas has been reported (CONNOR and STERN 1990), though no direct correlation with HPV was found, and the level of MHC class 1 and class 2 expression on cervical cancers and CIN appeared to relate more to the degree of differentiation of the epithelial cells than to HPV infection. Normal cervical epithelium expresses no MHC class 2, and only moderate amounts of MHC class 1 basally, with diminution of expression towards the superficial layers of the epithelium. HPV DNA-positive and HPV DNA-negative CIN lesions may show up-regulation of MHC class 1 and class 2 expression, and tumours with all patterns of MHC expression are described.

Establishment and maintenance of peripheral (i.e. non-thymic) tolerance to self and other antigens is multifactorial (ARNOLD et al. 1993) and depends on the level of antigen presentation in the periphery, the immunodominance of the expressed epitopes, and the degree to which accessory molecules for T cell adherence and activation, and cytokines, are locally expressed. The expression of most HPV proteins, particularly the products of the ORFs, in suprabasal epithelial stem cells is likely to be low, leading to a low "antigen + MHC" density at the cell surface. As HPV infection induces no local inflammation and no cell death, there will be no concentration of antigen by professional antigen-presenting cells, including skin Langerhans cells, and local cytokine and "second signal" molecules will be absent. Presentation of antigens by keratinocytes without such local inflammation or second signal molecules appears to tolerogenic rather than immunogenic (BAL et al. 1990).

Virally induced immunosuppression is well documented but classically occurs in viruses causing systemic infection, such as measles, HIV-1, and lymphocytic choriomeningitis virus, particularly if the cells of the lymphoreticular system are directly infected. A number of early reports of depressed immune function among patients with HPV infection but no other concomitant disease or therapy (CARSON et al. 1986; JABLONSKA et al. 1978) invited the interpretation that HPV infection per se causes immunosuppression. A decreased peripheral blood leucocyte-associated NK cell activity has been reported in patients with HPV infection (MALEJCZYK et al. 1989), but host immune function is generally normal in conventional assays; randomised (VARDY et al. 1990) and other controlled studies (ANDROPHY et al. 1984; MOHANTY et al. 1987) have failed to demonstrate significant immunosuppression in patients with chronic HPV infection.

The increased RR for HPV-associated disease, including cancer in immunosuppressed transplant recipients, argues that some effective immune response to HPV occurs in immunocompetent individuals; the RR for cancer of the cervix is about 5 in transplant recipients, and for vulval and penile cancer >30 (SHIEL 1991), whereas the RR of developing most tumours for which viral infection is not thought to play a part is between 1 and 2. The median time to onset of HPV-associated tumour after onset of immune suppression is quite short, on average 5 years. The difference in RR for cervical cancer compared with the other HPV-associated cancers could be used to argue that immune surveillance is less effective in the cervix than elsewhere; the cervix, as with the rest of the uterus, may to some extent be an immunologically privileged site. Alternatively, the differing RR may reflect the greater role of cytotoxic drug-induced destabilisation of metastable epithelium at the squamo-columnar junction in the cervix; the short time from onset of immunosuppression to onset of cancer suggests that the effect of immunosuppressive/cytotoxic drugs is to accelerate development of pre-existing HPV-related cervical and other epithelial dysplasia, rather than to permit new HPV infection to occur, particularly as the exposure rates amongst transplant recipients to possible new sources of genital HPV infection is very low (I. Frazer, unpublished data).

7 Vaccines Against HPV Infection

7.1 Vaccine Strategies

A goal of studies of HPV immunology is to develop a vaccine for prophylactic and therapeutic use since destructive elimination of all viral infection is often not technically feasible and no specific antiviral agent is currently available. A vaccine could also be targeted at HPV-related tumours since these continue to express HPV proteins, particularly the E6 and E7 proteins. The natural immune response to HPV is slow and inadequate, and lesions may persist for months or years. HPV infection demonstrates few of the characteristics of viral infections against which vaccines have been successful in preventing disease (ADA 1990; KARZON et al. 1992). These characteristics are:

- A pathogenesis which includes an obligatory viraemia before infection of the target organ.
- Persistent circulating antibody protects against infection and disease, whether that antibody is acquired naturally or by vaccination; i.e. circulating neutralising antibody is a marker of protection.
- Latency is not present.
- Typically a self-limiting illness follows infection, though death may intervene.

- Following recovery and elimination of virus, immune protection is complete and often life-long.
- An animal model is available which reproduces the human disease.

A vaccine needs to achieve four goals (ADA 1990):

- Activation of antigen-presenting cells.
- Overcoming genetic variability in T cell responses.
- Generation of high levels of T and B memory cells.
- Persistence of antigen.

In particular, a vaccine must be designed to elicit the appropriate immune response. If a predominantly cytotoxic T cell response will be needed for protection, as with HPV-associated tumours, then antigen dose, adjuvant and route of delivery must reflect the need to activate the Th1 subset of helper T cells, which secrete cytokines, including interferon gamma, which bias the immune response towards cytotoxic T cell responses and DTH. Conversely, if antibody responses are thought to be protective, as may be the case for prevention of genital HPV infection, then activation of the Th2 subset of T cells, with predominant IL-4, and IL-5 secretion, will be required.

7.2 HPV Vaccines

There are three stages in the biology of HPV infection at which immune intervention may be attempted:

- Prior to HPV infection of epithelium.
- During viral replication in basal epithelial cells.
- For oncogenic genotypes: after viral integration, cell transformation and sub-epidermal invasion.

7.2.1 Prophylactic Vaccine Design

Although HPV DNA has been reported in peripheral blood mononuclear cells of patients with urogenital HPV infection (PAO et al. 1991), HPV infection is not generally thought to have a systemic phase before colonising epithelium. A putative vaccine must therefore seek to erect an immunological barrier at the portal of entry, and presumably this would be secretory IgA directed against L1 and L2 viral capsid proteins (CASON et al. 1993). Although there are cross-reactive epitopes between the various PV genotypes, these appear to be masked in complete virions and are unlikely to be targets for neutralising antibodies; genotype-specific vaccines will probably be required in humans as in cattle (JARRETT et al. 1990b). Infection of human foreskin keratinocytes, transplanted under the renal capsules of athymic nude mice, by HPV-11 particles could be blocked by anti-virion antibodies raised in rabbits, i.e. viral neutralisation occurred (CHRISTENSEN and KREIDER 1990). To exploit this observation in a vaccine it would be necessary to administer PV antigen by a route which would favour mucosal IgA secretion. It would also be desirable to characterise the cellular HPV receptor in order that vaccine-induced antibody response could be tailored to the receptor ligand on the infecting virions.

7.2.2 Therapeutic Vaccines Aimed at Elimination of Infection

To eliminate virus replicating in epithelial cells, a vaccine should aim to prime for CTLs directed to cytotoxic epitopes derived from endogenously processed viral proteins presented in the context of any of the MHC class 1 alleles displayed by each host.

During the period of viral latency, the E1 and E2 proteins would be the targets of choice as these are the only PV proteins likely to be expressed in the self-renewing suprabasal epithelial stem cell population. However, expression of these proteins in the stem cell population is likely to be below the critical level at which a primed cytotoxic T cell can see its target as this requires 100–200 epitopes presented by MHC class 1 molecules per cell. It will probably therefore be necessary to find means of increasing E1 and E2 expression in infected tissues prior to successful therapeutic vaccination.

While vegetative HPV reproduction is occurring, E6 and E7 are also expressed in suprabasal epithelial cells and can be additional targets for a vaccine. The most abundant proteins during vegetative reproduction are probably only expressed in the terminally differentiated keratinocyte; these include the L1 and L2 capsid proteins and the E4 protein, which probably supports the cytoskeletal architecture in the virally infected differentiating keratinocyte. Targeting a cytotoxic T cell response to these differentiated cells would therefore only be therapeutic if the adjacent self-renewing stem cells were to be eliminated by some bystander effect.

7.2.3 Therapeutic Vaccines to Control HPV-Induced Tumours

The HPV proteins consistently expressed in HPV-associated tumours include only E6 and E7. In tumours, as in virally infected cells, extensive selection pressure forces modification of tumour or viral antigen presentation in an immunocompetent host. Elimination of E6 and E7 expression from HPV-associated tumours is, however, unlikely because this would, on current understanding of the mechanisms of HPV-induced tumorigenesis, require simultaneous acquisition by the HPV-infected tumour cell of three mutations, comprising two RB gene knockouts (MÜNGER et al. 1989) and a dominant p53 mutation (SCHEFFNER et al. 1990). However, by analogy with other tumours, other mechanisms of immune evasion will occur in HPV-induced tumours including loss of expression of MHC alleles, down-regulation of endogenous peptide processing for MHC presentation and loss of cell surface adhesion molecules recognised by immune effector cells. Thus it may be necessary to use additional mechanisms to enhance tumour-specific immunity including induction of antibody to tumour antigens to confer lymphocyte-activated killer (LAK) sensitivity (TAKAHASHI et al. 1993), introduction into ex vivo induced tumour-specific cytotoxic T cells of cytokine-encoding genes such as IL2 (GREENBERG et al. 1993), vaccination with tumour cells engineered to secrete cytokines (DRANOFF et al. 1993) or immunisation with antigen-pulsed dendritic cells (INABA et al. 1990).

7.3 With What to Vaccinate

7.3.1 Peptides/Proteins as Immunogens

Lack of availability of adequate quantities of HPV antigens from in vivo or in vitro sources is overcome by bacterial or yeast expression of cloned HPV ORF genes. Should post-translational modification of peptides, e.g. glycosylation, be shown to be necessary for immunogenicity, high-quantity expression of tertiary-modified protein from HPV genes transfected into insect cells via baculovirus vectors may be preferable. It remains to be determined whether denatured HPV capsid proteins will be effective in humans at inducing host-protective neutralising antibodies, though data from animal models suggest that native capsid proteins are more effective than denatured recombinant protein. Recent recombinant approaches to capsid production have yielded particles (ZHOU et al. 1991; KIRNBAUER et al. 1992; HAGENSEE et al. 1993) which appear to elicit neutralising antibodies and whose parenteral administration may circumvent the constraints on immunogenicity seen in natural infection.

For the induction of CTL for therapeutic immunity, the drawback that exogenous peptide/protein may not enter the cytosolic MHC class I antigen presentation pathway necessary for CTL induction has been overcome by the development of membrane-fusogenic delivery vehicles. It should be possible to elicit specific cellular cytotoxicity as well as antibody by immunisation with appropriate HPV-immunogenic peptides bound to or enveloped by liposomes (NAIR et al. 1992; MILLER et al. 1992), membranous vehicles (ZHOU et al 1992), multiple antigen peptide systems (MAPS) (CHAI et al. 1992), P3CSS (DERES et al. 1989) or ISCOMs (MOREIN 1988). Protein administered and combined with aluminium hydroxide, the only adjuvant licensed for use in humans, has also been reported to induce CTL (DILLON et al. 1992).

7.3.2 Live Vaccines

"Attenuated" HPV alone is unlikely to prove successful as a potential vaccine because of its extremely restricted host cell range; primary HPV infection is not host protective, and an attenuated vaccine is therefore highly unlikely to do better. While parenterally administered killed virus may overcome the restriction of inadequate antigen presentation in natural infection, inability to grow adequate amounts of virions in vitro or in vivo limit this approach. One of the potential advantages in using a live virus vaccine is that it will allow infected cells to express HPV genes endogenously, generally with correct post-translational modification, and should thus stimulate CTL responses. However, live recombinant vaccines do not generally provide a chronic immunogenic challenge which may be necessary for the maintenance of long-term antigen-specific CTL memory (OEHEN et al. 1992), although there are experimental models where this does not apply (DOHERTY et al. 1992).

Vaccinia virus is an attractive candidate expression vector because it has a large capacity for additional foreign DNA and could contain the entire HPV genome. Attenuated vaccinia has an established history as a human vaccine, and both humoral and cell-mediated immune responses are made to the expressed recombinant protein. Vaccinia virus recombinant for HPV-16 E6/E7 and double recombinants containing in addition cytokine genes, especially those designed to give IL2 or GM-CSF expression, show particular promise.

Sabin type 1 polio virus recombinant in antigenic site 1 for L1 protein of HPV-16 is a potent immunogen for antibody production (JENKINS et al. 1990). Sabin type 1 is amongst the safest of human vaccines and has the advantages that it can be administered orally and induces mucosal immunity. Evidence that CTLs to the inserted antigen are generated is lacking since experimental animals lack the polio virus receptor, and intracellular infection, necessary for CTL generation, cannot therefore occur. Mice transgenic for the human polio virus receptor are now available (REN and RACANIELLO 1992) which should circumvent this experimental problem.

Recombinant human adenoviruses are one of the more promising vaccine vectors because live adenovirus vaccines have been given orally to large numbers of military recruits without adverse reactions. Immunising with adenovirus recombinant for immunogenic fragments of a number of viruses (PREVEC et al. 1989; HANKE et al. 1991) has been shown to elicit neutralising antibodies which protected mice against infection, and adenovirus infection in rodents elicits a strong virus-specific CTL response (HANKE et al. 1991).

If constant immune stimulation by HPV proteins proves essential to sustain immunity, which may be the case since natural infection, even when chronic or repeated, seems not to be immunogenic, then EBV could be a vaccine vehicle; however, the constraints on immunising humans with a transforming human pathogenic virus recombinant for the transforming proteins of another human pathogenic virus are considerable!

BCG has an established history as a human vaccine and generates long-lived immunity in the majority of individuals with a single dose. It is generally associated with Th1-type responses. BCG containing HPV sequences driven by the heat-shock protein-70 promoter (ALDOVINI and YOUNG 1991) should generate antibody and cell-mediated responses without the need for further adjuvant.

Several recombinant salmonella systems merit attention as potential HPV delivery vehicles (CARDENAS and CLEMENTS 1992; HACKETT 1990). Mice immunised with recombinant salmonella expressing E7 or E6/E7 made good antibody responses to the insert protein and there was some evidence of growth control of E7 tumour metastases (M. Krul, personal communication). A major problem with recombinant salmonella is that they "cure" themselves of the recombinant plasmid; this problem can be overcome, in the murine model, by use of the NirB promoter which is active only in the low oxygen tension environment of the infected macrophage—the recombinant gene is only expressed in the host antigen presenting cell, where a good Th1 response to the recombinant protein is induced (CHATFIELD et al. 1992). Salmonella transformed by a plasmid-containing foreign viral or parasite genes cloned into the hepatitis B core antigen gene have been shown to generate specific CTL when administered orally to experimental animals. The oral recombinant cholera subunit B–whole cell vaccine approach (HUSBAND 1993) for HPV would, like salmonella, have the potential advantage of inducing mucosal immunity.

7.4 Adjuvants

Progress in the development of contemporary adjuvants furnishes an alternative approach to live recombinant vaccines. Administration of peptide(s) in a pluronic polyol or in muramyl dipeptide will elicit both cellular (CTL) and humoral immunity. These adjuvants preferentially stimulate IgG2a production which is the subclass with maximal complement fixing ability and efficacy in antibody-dependent cell-mediated cytotoxicity (ADCMC), and this bias to a Th2-type response may be desirable for a prophylactic vaccine. In contrast, a therapeutic vaccine will require the ability to invoke a strong Th1 response and induce good DTH and CTL responses to the immunising protein; some newer synthetic adjuvants have been shown to have this property.

7.5 MHC Restriction

Since generation of CTL to a specific antigenic determinant and CTL effector function are both MHC-restricted through class I, the need to overcome MHC restriction is a major consideration for peptide or subunit vaccine design in out-bred populations such as humans. In most ethnic communities, 90% of the population will express at least one of the six most common MHC alleles of that community, and a vaccine based on HPV CTL epitopes recognised in context of these allelic forms of class I would therefore cover most individuals. Another approach would be to seek to identify promiscuous

CTL epitopes within HPV proteins which would function in the context of most MHC class I haplotypes (OLDSTONE et al. 1992).

There is very limited evidence that immune response to particular infections, as opposed to particular antigenic epitopes, is significantly MHC restricted. It is possible, for HPV-16-associated cervical cancer, that the limited range of HPV protein-derived epitopes available from the two small eORF proteins (E6, E7) consistently expressed in tumour cells might result in functional MHC restriction of a host protective tumour immune response. In one study (WANK and THOMSSEN 1991) has shown an association of cervical cancer with DQW3, an MHC class 2 association implying a T helper-dependent mechanism for host protection, although this association of cervical cancer with MHC class 2 alleles has yet to be confirmed.

8 Conclusions

Three conclusions can be drawn from the reported data:

- Infection with most HPV genotypes is generally followed by an immune response to viral capsid proteins and such immune responses are in large part HPV genotype specific. As might be expected, the relative abundance of HPV protein in the clinical lesion appears to be a major determinant of the magnitude of the host immune response. Responses are smaller than might be seen with viruses giving rise to systemic infection.
- It is unlikely that serology will be a robust marker for the epidemiology of HPV infection as it is for the majority of systemic viral infections.
- HPV-associated tumours engender a host response to virally encoded tumour-specific antigen E7. While such responses are clearly not host protective in patients presenting with clinical disease, similar immune responses may have controlled at least some tumours that would otherwise have presented clinically.

The next few years will show the extent to which HPV-specific immune responses are able to clear existing lesions and prevent future infections. Even if the host immune response to HPV infection is inadequate to control the infection because presentation of HPV antigens to the immune system after natural infection lacks accessory signals or the quantity of antigen necessary to invoke host-protective immunity, vaccines to cure HPV-associated disease may be feasible. The requirement for a successful therapeutic vaccine will be that HPV-encoded proteins are presented in sufficient density on self-renewing stem cell populations or can be made, by appropriate local or systemic manipulation, to be suitably presented. Current evidence suggests that only the E1 and E2 ORFs of HPV are likely to be expressed in these cells, unless neoplastic change occurs, when the stem cell population appears to express E6 and E7 ORF products also.

A vaccine to prevent mucosal HPV infections will need to generate neutralising antibodies to viral capsid proteins at mucosal surfaces: further work is needed to define the relevant neutralising epitopes and the virus receptor. Whatever viral protein is targeted in HPV vaccines, the challenge for immunologists will be to develop vaccine delivery systems that can invoke the correct type of immune response in the correct site for long enough to be clinically useful. As this problem is shared by all newer generation vaccines for "difficult" viruses, including HIV-1, hepatitis C and herpesviruses, progress towards solving this problem over the next few years is likely to be rapid.

Acknowledgments. Work from the authors' laboratory was in part funded from NIH grant RO1-CA-57789-01, grants from the NHMRC of Australia, the Queensland Cancer Fund, the Lions Kidney and Medical Research Foundation, and the Mayne bequest. The authors are grateful to Ms. Kerrie Webb for assistance with preparation of the manuscript and to Dr. Paul Lambert for helpful critique.

References

Abcarian H, Sharon N (1977) The effectiveness of immunotherapy in the treatment of anal condyloma acuminatum. J Surg Res 22: 231–236
Ada GL (1990) The immunological principles of vaccination. Lancet 335: 523–526
Aiba S, Rokugo M, Tagami H (1986) Immunohistologic analysis of the phenomenon of spontaneous regression of numerous flat warts. Cancer 58: 1246–1251
Aldovini A, Young RA (1991) Humoral and cell-mediated immune responses to live recombinant BCG-HIV vaccines. Nature 351: 479–482
Almeida JD, Goffe AP (1965) Antibody to wart virus in human sera demonstrated by electron microscopy and precipitin tests. Lancet ii: 1205–1207
Altmann A, Jochmus-Kudielka I, Frank R, Gausepohl H, Moebius U, Gissmann L, Meuer SC (1992) Definition of immunogenic determinants of the human papillomavirus type 16 nucleoprotein E7. Eur J Cancer 28: 326–333
Androphy EJ (1986) Papillomaviruses and interferon. Ciba Found Symp 120: 221–234
Androphy EJ, Dvoretzky I, Maluish AE, Wallace HJ, Lowy DR (1984) Response of warts in epidermodysplasia verruciformis to treatment with systemic and intralesional alpha interferon. J Am Acad Dermatol 11: 197–202
Anisimová E, Barták P, Vlcek D, Hirsch I, Brichácek B, Vonka V (1990) Presence and type specificity of papillomavirus antibodies demonstrable by immunoelectron microscopy tests in samples from patients with warts. J Gen Virol 71: 419–422
Arnold B, Schönrich G, Hämmerling GJ (1993) Multiple levels of peripheral tolerance. Immunol Today 14: 12–14
Auborn KJ, Steinberg BM (1990) Therapy of papillomavirus induced lesions. In: Pfister H(ed) Papillomaviruses and human cancer. CRC Press, Boca Raton, Florida, pp 203–223
Bal V, McIndoe A, Denton G, Hudson D, Lombardi G, Lamb J, Lechler R (1990) Antigen presentation by keratinocytes induces tolerance in human T cells. Eur J Immunol 20: 1893–1897
Banks L, Moreau F, Vousden K, Pim D, Matlashewski G (1991) Expression of the human papillomavirus E7 oncogene during cell transformation is sufficient to induce susceptibility to lysis by activated macrophages. J Immunol 146: 2037–2042
Barber SR, Werdel J, Symbula M, Williams J, Burkett BA, Taylor PT, Roche JK, Crum CP (1992) Seroreactivity to HPV-16 proteins in women with early cervical neoplasia. Cancer Immunol Immunother 35: 33–38
Bender ME (1986) Concepts of wart regression. Arch Dermatol 122: 644–647
Biberstein H (1925) Versuche uber immunotherapie der warzen und kondylome. Klin Wochenschr 14: 638–639

Bleul C, Müller M, Frank R, Gausephol H, Koldovsky U, Mgaya HN, Luande J, Pawlita M, Ter Meulen J, Viscidi R, Gissmann L (1991) Human papillomavirus type 18 E6 and E7 antibodies in human sera: increased anti-E7 prevalence in cervical cancer patients. J Clin Microbiol 29: 1579–1588
Bonnez W, Da Rin C, Rose RC, Reichman RC (1991) Use of human papillomavirus type 11 virions in an ELISA to detect specific antibodies in humans with condylomata acuminata. J Gen Virol 72: 1343–1347
Bonnez W, Kashima HK, Leventhal B, Mounts P, Rose RC, Reichman RC, Shah KV (1992) Antibody response to human papillomavirus (HPV) type 11 in children with juvenile-onset recurrent respiratory papillomatosis (RRP). Virology 188: 384–387
Bunney MH, Barr BB, McLaren K, Smith IW, Benton EC, Anderton JL, Hunter JAA (1987) Human papillomavirus type 5 and skin cancer in renal allograft recipients: letter. Lancet 2:151–152
Campo MS (1991) Vaccination against papillomavirus. Cancer Cells 3: 421–426
Cardenas L, Clements D (1992) Oral immunisation using live attenuated salmonella species as carriers of foreign antigens. Clin Microbiol Rev 5: 328–342
Carson LF, Twiggs LB, Fokushima M, Ostrow RS, Faras AJ (1986) Human genital papilloma infections: an evaluation of immunologic competence in the genital neoplasia-papilloma syndrome. Am J Obstet Gynecol 155: 784–789
Carter JJ, Yaegashi N, Jenison SA, Galloway DA (1991) Expression of human papillomavirus proteins in yeast Saccharomyces cerevisiae. Virology 182: 513–521
Cason J, Kambo PK, Best JM, McCance DJ (1992) Detection of antibodies to a linear epitope on the major coat protein (L1) of human papillomavirus type-16 (HPV-16) in sera from patients with cervical intraepithelial neoplasia and children. Int J Cancer 50: 349–355
Cason J, Khan SA, Best JM (1993) Towards vaccines against human papillomavirus type 16 genital infections. Vaccine 11, 603–611
Chai SK, Clavijo P, Tam JP, Zavala F (1992) Immunogenic properties of multiple antigen peptide systems containing defined T and B epitopes. J Immunol 149: 2385–2390
Chardonnet Y, Beauve P, Viac J, Schmitt D (1983) T-cell subsets and Langerhans cells in wart lesions. Immunol Lett 6: 191–196
Chatfield SN, Charles IG, Makoff AJ, Oxer MD, Dougan G, Pickard D, Slater D, Fairweather NF (1992) Use of the nirB promoter to direct the stable expression of the heterologous antigens in salmonella oral vaccine strains: development of a single-dose oral tetanus vaccine. BioTechnology, 10: 888–892
Chen H-D, Zhao Y, Sun G, Yang C (1989) Occurrence of Langerhans cells and expression of class II antigens on keratinocytes in malignant and benign epithelial tumors of the skin: an immunohistopathologic study with monoclonal antibodies. J Am Acad Dermatol 20: 1007–1014
Chen L, Thomas EK, Hu S-L, Hellstrom I, Hellstrom KE (1991) Human papillomavirus type 16 nucleoprotein E7 is a tumour rejection antigen. Proc Natl Acad Sci USA 88: 110–114
Chen L, Ashe S, Brady WA, Hellström I, Hellström KE, Ledbetter JA, McGowan P, Linsley PS (1992a) Costimulation of antitumor immunity by the B7 counterreceptor for the T lymphocyte molecules CD28 and CTLA-4. Cell 71: 1093–1102
Chen L, Mizuno MT, Singhal MC, Hu S-L, Galloway DA, Hellström I, Hellström KE (1992b) Induction of cytotoxic T lymphocytes specific for a syngeneic tumor expressing the E6 oncoprotein of human papillomavirus type 16. J Immunol 148: 2617–2621
Christensen ND, Kreider JW (1990) Antibody-mediated neutralization in vivo of infectious papillomaviruses. J Virol 64: 3151–3156
Christensen ND, Kreider JW, Shah KV, Rando RF (1992) Detection of human serum antibodies that neutralize infectious human papillomavirus type 11 virions. J Gen Virol 73: 1261–1267
Comerford SA, McCance DJ, Dougan G, Tite JP (1991) Identification of T- and B-cell epitopes of the E7 protein of human papillomavirus type 16. J Virol 65: 4681–4690
Connor ME, Stern PL (1990) Loss of MHC class 1 expression in cervical carcinomas. Int J Cancer 46: 1029–1034
Cromme FV, Meijer CJLM, Snijders PJF, Uyterlinde A, Kenemans P, Helmerhorst T, Stern PL, Van den Brule AJC, Walboomers JMM (1993) Analysis of MHC-class I and II expression in relation to presence of HPV genotypes in premalignant and malignant cervical lesions. Br J Cancer 67: 1372–1380
Davies DH, McIndoe GAJ, Chain BM (1991) Cancer of the cervix: prospects for immunological control. Int J Exp Pathol 72: 239–251

Del Val M, Hengel H, Häcker H, Hartlaub U, Ruppert T, Lucin P, Koszinowski UH (1992) Cytomegalovirus prevents antigen presentation by blocking the transport of peptide-loaded major histocompatibility complex class I molecules into the medial-Golgi compartment. J Exp Med 176: 729–738
Deres K, Schild H, Weismuller KH, Jung G, Rammensee HG (1989) In vitro priming of virus specific cytotoxic T lymphocytes with synthetic lipopeptide vaccine. Nature 342: 561–564
Dillner J (1990) Mapping of linear epitopes of human papillomavirus type 16: the E1, E2, E4, E5, E6 and E7 open reading frames. Int J Cancer 46: 703–711
Dillner J, Dillner L, Robb J, Willems J, Jones I, Lancaster W, Smith R, Lerner R (1989) A synthetic peptide defines a serologic IgA response to a human papillomavirus-encoded nuclear antigen expressed in virus-carrying cervical neoplasia. Proc Natl Acad Sci USA 86: 3838–3841
Dillner J, Dillner L, Utter G, Eklund C, Rotola A, Costa S, DiLuca D (1990) Mapping of linear epitopes of human papillomavirus type 16: the L1 and L2 open reading frames. Int J Cancer 45: 529–535
Dillner L, Bekassy Z, Jonsson N, Moreno-Lopez J, Blomberg J (1989) Detection of IgA antibodies against human papillomavirus in cervical secretions from patients with cervical intraepithelial neoplasia. Int J Cancer 43: 36–40
Dillner L, Moreno-Lopez J, Dillner J (1990) Serological responses to papillomavirus group-specific antigens in women with neoplasia of the cervix uteri. J Clin Microbiol 28: 624–627
Dillner L, Fredriksson A, Persson E, Forslund O, Hansson B-G, Dillner J (1993) Antibodies against papillomavirus antigens in cervical secretions from condyloma patients. J Clin Microbiol 31: 192–197
Dillon SB, Demuth SG, Schneider MA, Weston CB, Jones CS, Young JF, Scott M, Bhatnaghar PK, LoCastro S, Hanna N (1992) Induction of protective class I MHC-restricted CTL in mice by a recombinant influenza vaccine in aluminium hydroxide adjuvant. Vaccine 10: 309–318
Doherty PC, Allan W, Eichelberger M, Carding SR (1992) Roles of $\alpha\beta$ and $\gamma\delta$ T cell subsets in viral immunity. Annu Rev Immunol 10: 123–151
Dranoff G, Jaffee E, Lazenby A, Golumbe KP, Levitsky H, Brose K, Jackson V, Hamada H, Pardoll D, Mulligan RC (1993) Vaccination with irradiated tumour cells engineered to secrete murine granulocyte-macrophage colony-stimulating factor stimulates potent, specific and long-lasting anti-tumour immunity. Proc Natl Acad Sci USA 90: 3593–3543
Eisinger M, Lee AK (1976) Cell mediated immunity (CMI) to human wart virus and wart-associated tissue antigens. Clin Exp Immunol 26: 419–424
Feltkamp MCW, Smits HL, Vierboom MPM, Minnaar RP, De Jongh BM, Drijfhout J, Ter Scheyget J, Melief CJM, Kast WM (1993) Vaccination with cytotoxic T lymphocyte epitope-containing peptide protects against a tumour induced by human papillomavirus type 16-transformed cells. Eur J Immunol 23: 2242–2249
Frazer IH, Tindle RW (1992) Cell mediated immunity to papillomaviruses. Papillomavirus Report 3: 53–58
Frazer IH, Medley G, Crapper RM, Brown TC, Mackay IR (1986) Association between anorectal dysplasia, human papillomavirus, and human immunodeficiency virus infection in homosexual men. Lancet 2: 657–660
Ghim S-J, Jenson AB, Schlegel R (1992) HPV-1 L1 protein expressed in cos cells displays conformational epitopes found on intact virions. Virology 190: 548–552
Gissmann L, Jochmus-Kudielka I (1989) Immune reponse to papillomavirus infections. In: Fritsch P, Schuler G, Hintner H (eds) Current problems in dermatology, vol 18. Karger, Basel, pp 162–167
Greenberg PL, Watanabe K, Gilbert M, Nelson B, Riddell S (1993) Reconstitution of viral immunity in immunocompromised humans by the adoptive transfer of T cell clones. J Cell Biochem [Suppl] 17D: 99
Gross G (1988) Interferon and genital warts. JAMA 260: 2066
Hackett J (1990) Salmonella-based vaccines. Vaccine 8:5–11
Hagensee ME, Yaegashi N, Galloway DA (1993) Self-assembly of human papillomavirus type 1 capsids by expression of the L1 protein alone or by coexpression of the L1 and L2 capsid proteins. J Virol 67: 315–322
Hanke T, Graham FL, Rosenthal KL, Johnson DC (1991) Identification of an immunodominant cytotoxic T-lymphocyte recognition site in glycoprotein B of herpes simplex virus by using recombinant adenovirus vectors and synthetic peptides. J Virol 65: 1177–1186

Hashido M, Kanda T, Zanma S, Watanabe S, Komiyama N, Yoshikawa H, Yamaguchi N, Kawana T, Yoshiike K (1991) Detection of human antibody against the human papillomavirus type 16 E7 protein. Jpn J Cancer Res 82: 1406–1412
Ho L, Chan S-Y, Chow V, Chong T, Tay S-K, Villa LL, Bernard H-U (1991) Sequence variants of human papillomavirus type 16 in clinical samples permit verification and extension of epidemiological studies and construction of a phylogenetic tree. J Clin Microbiol 29: 1765–1772
Hogan PG, Frazer IH, International study group (1991) A randomised placebo controlled study of Interferon alpha2a as treatment for recurrent genital warts. JAMA 265: 2684–2687
Husband AJ (1993) Novel vaccination strategies for the control of mucosal infection. Vacccine 11: 107–112
Inaba K, Metlay JP, Crowley MT, Steinman RM (1990) Dendritic cells pulsed with protein antigens in vitro can prime antigen-specific MHC restricted T cells in situ. J Exp Med 172: 631–640
Jablonska S, Orth G, Jarzabek-Chorzelska M, Rzesa G, Obalek S (1978) Immunological studies in epidermodysplasia verruciformis. Bull Cancer (Paris) 65: 183–190
Jarrett WFH, O'Neil BW, Gaukroger JM, Laird HM, Smith KT, Campo MS (1990a) Studies on vaccination against papillomaviruses: a comparison of purified virus, tumour extract and transformed cells in prophylactic vaccination. Vet Rec 126: 449–452
Jarrett WFH, O'Neil BW, Gaukroger JM, Smith KT, Laird HM, Campo MS (1990b) Studies on vaccination against papillomaviruses: the immunity after infection and vaccination with bovine papillomaviruses of different types. Vet Rec 126: 473–475
Jarrett WFH, Smith KT, O'Neil BW, Gaukroger JM, Chandrachud LM, Grindlay GJ, McGarvie GM, Campo MS (1991) Studies on vaccination against papillomaviruses: prophylactic and therapeutic vaccination with recombinant structural proteins. Virology 184: 33–42
Jenison SA, Firzlaff JM, Langenberg A, Galloway DA (1988) Identification of immunoreactive antigens of human papillomavirus type 6b using Escherichia coli-expressed fusion proteins. J Virol 62: 2115–2123
Jenison SA, Yu X-P, Valentine JM, Galloway DA (1989) Human antibodies react with an epitope of the human papillomavirus type 6b L1 open reading frame which is distinct from the type-common epitope. J Virol 63: 809–818
Jenison SA, Yu X, Valentine JM, Koutsky LA, Christiansen AE, Beckmann AM, Galloway DA (1990) Evidence of prevalent genital-type human papillomavirus infections in adults and children. J infect Dis 162: 60–69
Jenison SA, Yu X-P, Valentine JM, Galloway DA (1991) Characterization of human antibody-reactive epitopes encoded by human papillomavirus types 16 and 18. J Virol 65: 1208–1218
Jenkins O, Cason J, Burke KL, Lunney D, Gillen A, Patel D, McCance DJ, Almond JW (1990) An antigen chimera of poliovirus induces antibodies against human papillomavirus type 16. J Virol 64: 1201–1206
Jochmus-Kudielka I, Schneider A, Braun R, Kimmig R, Koldovsky U, Schneweis KE, Seedorf K, Gissmann L (1989) Antibodies against the human papillomavirus type 16 early proteins in human sera: correlation of anti-E7 reactivity with cervical cancer. JNCI 81: 1698–1704
Kanda T, Onda T, Zanma S, Yasugi T, Furuno A, Watanabe S, Kawana T, Sugase M, Ueda K, Sonoda T, Suzuki S, Yamashiro T, Yoshikawa H, Yoshiike K (1992) Independent association of antibodies against human papillomavirus type 16 E1/E4 and E7 proteins with cervical cancer. Virology 190: 724–732
Karzon DT, Bolognesi DP, Koff WC (1992) Development of a vaccine for the prevention of AIDS. A critical appraisal. Vaccine 10: 1039–1052
Kast WM, Brandt RHP, Drijfhout JW, Melief CJM (1993) Human leucocyte antigen-A2.1 restricted candidate cytotoxic T lymphocyte epitopes of human papillomavirus type 16 E6 and E7 proteins identified by using the processing-defective human cell line T2. J Immunother 14(2): 115–120
Kienzler JL (1985) Humoral immunity to human papillomaviruses. Clinics in Dermatology 3: 144–155
Kirby PK, Kiviat N, Beckman A, Wells D, Sherwin S, Corey L (1988) Tolerance and efficacy of recombinant human interferon gamma in the treatment of refractory genital warts. Am J Med 85: 183–188
Kirnbauer R, Booy F, Cheng N, Lowy DR, Schiller JT (1992) Papillomavirus L1 major capsid protein self-assembles into virus-like particles that are highly immunogenic. Proc Natl Acad Sci USA 89: 12180–12184

Kirschner M (1986) Immunology of human papillomavirus infection. Prog Med Virol 33: 1–41
Kiviat NB, Critchlow CW, Holmes KK, Kuypers J, Sayer J, Dunphy C, Surawicz C, Kirby P, Wood R, Daling JR (1993) Association of anal dysplasia and human papillomavirus with immunosuppression and HIV infection among homosexual men. AIDS 7: 43–49
Köchel HG, Monazahian M, Sievert K, Höhne M, Thomssen C, Teichmann A, Arendt P, Thomssen R (1991) Occurrence of antibodies to L1, L2, E4 and E7 gene products of human papillomavirus types 6b, 16 and 18 among cervical cancer patients and controls. Int J Cancer 48: 682–688
Krchnák V, Vágner J, Suchánková A, Krcmár M, Ritterová L, Vonka V (1990) Synthetic peptides derived from E7 region of human papillomavirus type 16 used as antigens in ELISA. J Gen Virol 71: 2719–2724
Kreider JW, Howett MK, Leure-Dupree AE, Zaino RJ, Weber JA (1987) Laboratory production in vivo of infectious human papillomavirus type 11. J Virol 61: 590–593
Lehtinen M, Leminen A, Kuoppala T, Tiikkainen M, Lehtinen T, Lehtovirta P, Punnonen R, Vesterinen E, Paavonen J (1992a) Pre- and posttreatment serum antibody responses to HPV 16 E2 and HSV 2 ICP8 proteins in women with cervical carcinoma . J Med Virol 37: 180–186
Lehtinen M, Leminen A, Paavonen J, Lehtovirta P, Hyöty H, Vesterinen E, Dillner J (1992b) Predominance of serum antibodies to synthetic peptide stemming from HPV 18 open reading frame E2 in cervical adenocarcinoma. J Clin Pathol 45: 494–497
Li CC, Shah KV, Seth A, Gilden RV (1987) Identification of the human papillomavirus type 6b L1 open reading frame protein in condylomas and corresponding antibodies in human area. J Virol 61: 2684–2690
Lin Y-L, Borenstein LA, Selvakumar R, Ahmed R, Wettstein FO (1992) Effective vaccination against papilloma development by immunization with L1 or L2 structural protein of cottontail rabbit papillomavirus. Virology 187: 612–619
Lin Y-L, Borenstein LA, Selvakumar R, Ahmed R, Wettstein FO (1993) Progression from papilloma to carcinoma is accompanied by changes in antibody response to papillomavirus proteins. J. Virol 67: 382–389
Malejczyk J, Majewski S, Jablonska S, Rogozinski TT, Orth G (1989) Abrogated NK-cell lysis of human papillomavirus (HPV)-16-bearing keratinocytes in patients with pre-cancerous and cancerous HPV-induced anogenital lesions. Int J Cancer 43: 209–214
Mann VM, De Lao SL, Brenes M, Brinton LA, Rawls JA, Green M, Reeves WC, Rawls WE (1990) Occurrence of IgA and IgG antibodies to select peptides representing human papillomavirus type 16 among cervical cancer cases and controls. Cancer Res 50: 7815–7819
McGlennen RC, Ostrow RS, Carson LF, Stanley MS, Faras AJ (1991) Expression of cytokine receptors and markers of differentiation in human papillomavirus-infected cervical tissues. Am J Obstet Gynecol 165: 696–705
McKenzie J, King A, Hare J, Fulford T, Wilson B, Stanley M (1991) Immunocytochemical characterization of large granular lymphocytes in normal cervix and HPV associated disease. J Pathol 165: 75–80
McLean CS, Sterling JS, Mowat J, Nash AA, Stanley MA (1993) Delayed-type hypersensitivity response to the human papillomavirus type 16 E7 protein in a mouse model. J Gen Virol 74: 239–245
Meneguzzi G, Cerni C, Kieny MP, Lathe R (1991) Immunization against human papillomavirus type 16 tumor cells with recombinant vaccinia viruses expressing E6 and E7. Virology 181: 62–69
Meyers C, Frattini MG, Hudson JB, Laimins LA (1992) Biosynthesis of human papillomavirus from a continuous cell line upon epithelial differentiation. Science 257: 971–973
Miller MD, Gould-Fogerite S, Shen L, Woods RM, Koenig S, Mannino RJ, Letvin NL (1992) Vaccination of rhesus monkeys with synthetic peptide in a fusogenic proteoliposome elicits simian immunodeficiency virus-specific $CD8^+$ cytotoxic T lymphocytes. J Exp Med 176: 1739–1744
Mohanty KC, Scott CS, Limbert HJ, Master PS (1987) Circulating B and T lymphocyte subpopulations in patients with genital warts. Br J Clin Pract 41: 601–603
Morein B (1988) The iscom antigen-presenting system. Nature 332: 287–288
Morris HH, Gatter KC, Sykes G, Casemore V, Mason DY (1983) Langerhans' cells in human cervical epithelium: effects of wart virus infection and intraepithelial neoplasia. Br J Obstet Gynaecol 90: 412–420
Müller M, Viscidi RP, Sun Y, Guerrero E, Hill PM, Shah F, Bosch FX, Muñoz N, Gissmann L, Shah

KV (1992) Antibodies to HPV-16 E6 and E7 proteins as markers for HPV- 16-associated invasive cervical cancer. Virology 187: 508–514

Münger K, Werness BA, Dyson N, Phelps WC, Harlow E, Howley PM (1989) Complex formation of human papillomavirus E7 proteins with the retinoblastoma tumor suppressor gene product. EMBO J 8: 4099–4105

Nabavi N, Freeman GJ, Gault A, Godfrey D, Nadler LM, Glimcher LH (1992) Signalling through the MHC class II cytoplasmic domain is required for antigen presentation and induces B7 expression. Nature 360: 266–268

Nair S, Zhou F, Reddy R, Huang L, Rouse BT (1992) Soluble proteins delivered to dendritic cells via pH-sensitive liposomes induce primary cytotoxic T lymphocyte responses in vitro. J Exp Med 175: 609–612

Oehen S, Waldner H, Kündig TM, Hengartner H, Zinkernagel RM (1992) Antivirally protective cytotoxic T cell memory to lymphocytic choriomeningitis virus is governed by persisting antigen. J Exp Med 176: 1273–1281

Oldstone MB, Tishon A, Geckeler R, Lewicki H, Whitton JL (1992) A common antiviral cytotoxic T-lymphocyte epitope for diverse major histocompatibility complex haplotypes; implications for vaccination. Proc Natl Acad Sci USA 89: 2752–2755

Pao CC, Lin S-S, Lin C-Y, Maa J-S, Lai C-H, Hsieh T-T (1991) Identification of human papillomavirus DNA sequences in peripheral blood mononuclear cells. Am J Clin Pathol 95: 540–546

Pass F, Reissig M, Shah KV, Eisinger M, Orth G (1977) Identification of an immunologically distinct papillomavirus from lesions of epidermodysplasia verruciformis. J Natl Cancer Inst 59: 1107–1112

Pfister H (1990) Immunobiology of papillomaviruses and prospects for vaccination. In: Pfister H (ed) Papillomaviruses and human cancer. CRC Press, Boca Raton, Florida, pp 239–251

Pfister H, Zur Hausen H (1978) Characterization of proteins of human papilloma viruses (HPV) and antibody response to HPV 1. Med Microbiol Immunol (Berl) 166: 13–19

Prevec L, Schneider M, Rosenthal KL, Belbeck LW, Derbyshire JB, Graham FL (1989) Use of human adenovirus based vectors for antigen expression in animals. J Gen Virol 70: 429–434

Pyrhonen S, Penttinen K (1972) Wart virus antibodies and the prognosis of wart disease. Lancet ii: 1330–1332

Reeves WC, Rawls JA, Green M, Rawls WE (1990) Antibodies to human papillomavirus type 16 in patients with cervical neoplasia. Lancet 335: 551–552

Ren R, Racaniello VR (1992) Human poliovirus receptor gene expression and poliovirus tissue tropism in transgenic mice. J Virol 66: 296–304

Rogozinski TT, Jablonska S, Jarzabek-Chorzelska M (1988) Role of cell-mediated immunity in spontaneous regression of plane warts. Int J Dermatol 27: 322–326

Rose RC, Bonnez W, Reichman RC, Garcea RL (1993) Expression of human papillomavirus type 11 L1 protein in insect cells: in vivo and in vitro assembly of viruslike particles. J Virol 67: 1936–1944

Rudlinger R, Smith IW, Bunney MH, Hunter JA (1986) Human papillomavirus infections in a group of renal transplant recipients. Br J Dermatol 115: 681–692

Scheffner M, Werness BA, Huibregtse JM, Levine AJ, Howley PM (1990) The E6 oncoprotein encoded by human papillomavirus types 16 and 18 promotes the degradation of p53. Cell 63: 1129–1136

Shepherd PS, Tran TTT, Rowe AJ, Cridland JC, Comerford SA, Chapman MG, Rayfield LS (1992) T cell responses to the human papillomavirus type 16 E7 protein in mice of different haplotypes. J Gen Virol 73: 1269–1274

Shiel AGR (1991) Cancer report. In: Disney APS (ed) ANZDATA report 1991, Australia and New Zealand dialysis and transplant registry, Adelaide, South Australia, pp 100–108

Smillie AE, Tindle RW, O'Connor DT, Kennedy L, Frazer IH (1990) Humoral response to human papillomavirus type 16 and 18 open reading frame antigens in subjects with anogenital papillomavirus infections and controls. Immunol Infect Dis 1: 13–17

Spies T, Cerundolo V, Colonna M, Cresswell P, Townsend A, DeMars R (1992) Presentation of viral antigen by MHC class I molecules is dependent on a putative peptide transporter heterodimer. Nature 355: 644–646

Spradbrow PB (1988) Immune responses to papillomavirus infection. In: Syrjanen K, Gissmann L, Koss LG (eds) Papillomaviruses and human disease. Springer, Berlin Heidelberg New York, pp 371–392

Stacey SN, Bartholomew JS, Ghosh A, Stern PL, Mackett M, Arrand JR (1992) Expression of human papillomavirus type 16 E6 protein by recombinant baculovirus and use for detection of anti-E6 antibodies in human sera. J Gen Virol 73: 2337–2345
Stauss HJ, Davies H, Sadovnikova E, Chain B, Horowitz N, Sinclair C (1992) Induction of cytotoxic T lymphocytes with peptides in vitro: identification of candidate T-cell epitopes in human papilloma virus. Proc Natl Acad Sci USA 89: 7871–7875
Steele JC, Gallimore PH (1990) Humoral assays of human sera to disrupted and nondisrupted epitopes of human papillomavirus type 1. Virology 174: 388–398
Steger G, Olszewsky M, Stockfleth E, Pfister H (1990) Prevalence of antibodies to human papillomavirus type 8 in human sera. J Virol 64: 4399–4406
Strang G, Hickling JK, McIndoe GAJ, Howland K, Wilkinson D, Ikeda H, Rothbard JB (1990) Human T cell responses to human papillomavirus type 16 L1 and E6 synthetic peptides: identification of T cell determinants, HLA-DR restriction and virus type specificity. J Gen Virol 71: 423–431
Suchánková A, Ritter O, Hirsch I, Krchnák V, Kalos Z, Hamsíková E, Brichácek B, Vonka V (1990) Presence of antibody reactive with synthetic peptide derived from L2 open reading frame of human papillomavirus types 6b and 11 in human sera. Acta Virol (Praha) 34: 433–442
Suchánková A, Ritterová L, Krcmár M, Krchnák V, Vágner J, Jochmus I, Gissmann L, Kanka J, Vonka V (1991) Comparison of ELISA and Western blotting for human papillomavirus type 16 E7 antibody determination. J Gen Virol 72: 2577–2581
Suchánková A, Krcmár M, Krchnák V, Hamsíková E, Kanka J, Vágner J, Vonka V (1992) Range of HPV 16 E7 antibodies in cervical cancer patients and healthy subjects. Int J Cancer 51: 837–838
Syrjänen K, Väyrynen M, Mantyjarvi R, Castren O (1986) Natural killer (NK) cells with HNK-1 phenotype in the cervical biopsies of women followed-up for human papillomavirus (HPV) lesions. Acta Obstet Gynecol Scand 65: 139–145
Takahashi H, Nakada T, Puisieux I (1993) Inhibition of human colon cancer cell growth by antibody-directed human LAK cells in SCID mice. Science 259: 1460–1463
Tay SK, Jenkins D, Maddox P, Campion M, Singer A (1987a) Subpopulations of Langerhans' cells in cervical neoplasia. Br J Obstet Gynaecol 94: 10–15
Tay SK, Jenkins D, Maddox P, Singer A (1987b) Lymphocyte phenotypes in cervical intraepithelial neoplasia and human papillomavirus infection. Br J Obstet Gynaecol 94: 16–21
Tay SK, Jenkins D, Singer A (1987c) Natural killer cells in cervical intraepithelial neoplasia and human papillomavirus infection. Br J Obstet Gynaecol 94: 901–906
Tindle RW, Fernando GJP, Sterling JC, Frazer IH (1991) A "public" T-helper epitope of the E7 transforming protein of human papillomavirus 16 provides cognate help for several E7 B-cell epitopes from cervical cancer-associated human papillomavirus genotypes. Proc Natl Acad Sci USA 88: 5887–5891
Tindle RW, Murray B, Herd K, Kennedy L, O'Connor DT, Fernando GJP, Frazer IH (1992) Humoral immune response to the E7 open reading frame protein of human papillomavirus genotype 16 in subjects with anogenital papillomavirus infection and controls. Immunol Infect Dis 2: 223–228
Townsend SE, Allison JP (1993) Tumor rejection after direct costimulation of $CD8^+$ T cells by B7-transfected melanoma cells. Science 259: 368–370
Vardy DA, Baadsgaard O, Hansen ER, Lisby S, Vejlsgaard GL (1990) The cellular immune response to human papillomavirus infection. Int J Dermatol 29: 603–610
Viac J, Thivolet J, Hegazy MR, Chardonnet Y, Dambuyant C (1977) Comparative study of delayed hypersensitivity skin reactions and antibodies to human papillomavirus (HPV). Clin Exp Immunol 29: 240–246
Viac J, Staquet MJ, Miguet M, Chabanon M, Thivolet J (1978) Specific immunity to human papilloma virus (HPV) in patients with genital warts. Br J Vener Dis 54: 172–175
Viac J, Chomel J-J, Chardonnet Y, Aymard M (1990a) Incidence of antibodies to human papillomavirus type 1 in patients with cutaneous and mucosal papillomas. J Med Virol 32: 18–21
Viac J, Guérin-Reverchon I, Chardonnet Y, Brémond A (1990b) Langerhans cells and epithelial cell modifications in cervical intraepithelial neoplasia: Correlation with human papillomavirus infection. Immunobiology 180: 328–338
Wank R, Thomssen C (1991) High risk of squamous cell carcinoma of the cervix for women with HLA-DOw3. Nature 352: 723–725

Yaegashi N, Jenison SA, Batra M, Galloway DA (1992) Human antibodies recognise multiple distinct type-specific and cross-reactive regions of the minor capsid proteins of human papillomavirus types 6 and 11. J Virol 66: 2008–2019

Zhou F, Rouse BT, Huang L (1992) Induction of cytotoxic T lymphocytes in vivo with protein antigen entrapped in membranous vehicles. J Immunol 149: 1599–1604

Zhou J, Sun XY, Stenzel DJ, Frazer IH (1991) Expression of vaccinia recombinant HPV 16 L1 and L2 ORF proteins in epithelial cells is sufficient for assembly of HPV virion-like particles. Virology 185: 251–257

Zhou J, Sun X-Y, Davies H, Crawford L, Park D, Frazer IH (1992) Definition of linear antigenic regions of the HPV 16 L1 capsid protein using synthetic viron-like particles. Virology 189: 592–599

Vaccination Against Papillomavirus in Cattle

M.S. CAMPO

1 Introduction

Papillomaviruses are oncogenic viruses, involved in the induction of squamous cell carcinomas. The evidence for this is incontrovertible in animals and extremely strong in the human subject. Experimental reproduction of cancer induced by papillomavirus has been achieved in cattle and rabbits (CAMPO and JARRETT 1986; CAMPO et al. 1992; WETTSTEIN 1987), and whilst similar experiments cannot, for obvious reasons, be performed in humans, the epidemiological link between papillomavirus infection and anogenital cancer, in conjunction with a large body of laboratory work on the cell transformation capabilities of the virus (see Delius and Hofmann, ter Schegget and van der Noordaa, this volume), leaves little room for doubting the papillomaviral origin of a number of human cancers, first and foremost squamous cell carcinoma of the uterine cervix.

In addition to their involvement in cancer, papillomaviruses cause a number of diseases which, even if presenting a low risk of malignant progression, are debilitating and life threatening, like laryngeal papillomas in children (STEINBERG 1987), or greatly reduce the quality of life as is the case for anogenital condylomas (ORIEL 1971). In animals, bovine papillomavirus (BPV) infection of the teats and udders of cows or of the penis of bulls poses

The Beatson Institute for Cancer Research, Garscube Estate, Bearsden, Glasgow, G61 1BD, Scotland

Current Topics in Microbiology and Immunology, Vol. 186

serious agricultural problems (JARRETT 1985), and papillomavirus-induced sarcomas in horses are a great financial burden to breeders and racers (SUNDBERG et al. 1977).

Therefore it would be desirable to arm both the clinician and the veterinary doctor with effective prophylaxis and therapy capable of preventing or eradicating papillomavirus infection, respectively. Also, once the role of the virus in carcinogenesis is accepted, it follows that any action directed at preventing or curing viral infection will ultimately result in a decrease in the incidence of cancer. This reasoning has already been applied to hepatitis B virus (HPV), one of the main factors of hepatocellular carcinoma in eastern countries, and global vaccination programmes against the virus are already in operation (STEPHENNE 1988; KNISKERN et al. 1989).

This chapter reviews the vaccination programmes against papillomavirus in animal models, with particular emphasis on the cattle system, and deals with the prospect of similar vaccination schemes against human papillomavirus (HPV).

2 Animal Models

The same ethical considerations which restrict the experimental reproduction of papillomavirus-associated cancer in animals apply to the development and testing of antiviral vaccines. Thus the cattle and the rabbit systems have proved invaluable in this respect, and, as will be discussed later, successful vaccination against BPV and cottontail rabbit papillomavirus (CRPV) has shown that the twin goals of preventing and curing viral infection are achievable and indicates the way forward for the immunological management of papillomavirus infection in humans.

The suitability of cattle and rabbits as models for the human subject is not confined to the possibility of direct experimentation. Indeed these two animal models present other benefits: (a) as they are the natural hosts to BPVs and CRPV, respectively, the responses to experimental viral infection, be they immunological or cellular, are meaningful and reflect biological processes that occur naturally; (b) the natural history of viral infection in both systems is reasonably well known (JARRETT 1985; KREIDER and BARTLETT 1981), and this knowledge can be exploited in the design of new experiments; (c) the animals can be infected in controlled conditions and longitudinal studies can be pursued, both in the field of immunology (JARRETT et al. 1991; OKABAYASHI et al. 1991; HAN et al. 1992; CAMPO et al. 1993) and in that of carcinogenesis (CAMPO and JARRETT 1986; CAMPO et al. 1992; BRANDSMA et al. 1991; PENG et al. 1993); (d) last, but not least, both cattle and rabbits are outbred, an obvious advantage when, for instance, the immune response to papillomavirus infection and the immunological control of it have to be compared or used as guidance in the human system. In this context, both cattle and rabbits score higher than mice. Mice have proved invaluable in showing that the immune response to particular epitopes of HPV-16 is MHC restricted (SHEPHERD et al. 1992; TINDLE et al. 1990, 1991), but these results have to be carefully evaluated in the light of the demonstration that the so-called "public epitope" of HPV-16 E7, first identified in mice (TINDLE et al. 1991), is not recognized by infected human subjects (FRAZER and TINDLE 1992) and may be specific to mice only. Mice are not naturally infected with HPV, indeed no papillomavirus has yet been isolated from mice in the wild, and, contrary to the situation encountered with cattle and rabbits, the immune system of laboratory mice has no knowledge of HPV until experimental infection.

3 Papillomaviruses and Cancer in Cattle

Cattle, like humans, are infected by several different BPV types which display different degrees of molecular and immunological homology (JARRETT et al. 1984a). Of the six types of BPV characterized in detail so far, two are associated with cancer. BPV-2 commonly infects the skin and induces warts (CAMPO et al. 1981), but it can also infect the rumen mucosa (JARRETT et al. 1984b) where it causes non-productive lesions, and the urinary bladder, where evidence strongly suggests it has a causative role in cancer (CAMPO et al. 1992). BPV-4 infects the mucosa of the upper alimentary canal and causes papillomas (CAMPO et al. 1980). These are benign and self-limiting tumours but progress to cancer in animals eating bracken fern (JARRETT et al. 1978). Bracken contains immunosuppressants (W.C. EVANS et al. 1982) which cause a dramatic and sustained drop in number of lymphocytes and other white blood cells (CAMPO et al. 1992), thus abrograting immunosurveillance and allowing the papillomas to spread and persist (CAMPO and JARRETT 1986), and mutagens (I.A. EVANS et al. 1982) which powerfully contribute to cell transformation by BPV-4 (PENNIE and CAMPO 1992). It is apposite to compare the natural history of the BPV-4 papilloma–carcinoma complex with that of the cervical lesions induced by HPV-16. HPV-16 infection of the cervical mucosa induces lesions which become malignant in the presence of other cofactors. These are multiple and some still ill defined (BRINTON 1992), but smoking stands out as a critical factor. Tobacco smoke induces immunosuppression, with diminished IgG, IgA and IgM concentrations, suppression of T-cell functions and of natural killer cells (HOLT 1987), and it contains mutagens which concentrate in the cervical mucus of smokers, itself mutagenic (HOLLY et al. 1986); DNA from cervical biopsies of smokers contains smoking-related adducts (PHILLIPS and NI SHE 1993), further evidence of smoking-induced mutations. The comparison between the action of BPV-4 and bracken in cancer of the alimentary mucosa in cattle, and that of HPV-16 and smoking in cancer of the genital mucosa in humans springs easily to mind.

4 Papillomavirus Vaccination in Cattle

The first vaccines against papillomavirus in cattle were based on crude wart extracts (OLSON and SKIDMORE 1959; OLSON et al. 1960), similar to those already employed by SHOPE (1937) in rabbits. These vaccines had a rather unpredictable outcome, imparting complete, partial or no protection towards further challenge in different experiments. The reason for this variability was not fully understood, but it is likely to be due to the immunological heterogeneity of BPVs (JARRETT et al. 1984a): a vaccine raised against a particular virus type would not be effective against a different pathogen. This was experimentally demonstrated by using purified virus vaccines (JARRETT et al. 1990a); it was clear that animals vaccinated, say, against BPV-2 were immune to further challenge with this virus but not to challenge with any other BPV type, and the same held true for all the other virus types. Only the very closely related BPV-1 and BPV-2 afforded cross-immunity. In the same series of experiments, it was shown that protection is due to the presence in the vaccinated animals of high-titre neutralizing antibodies (JARRETT et al. 1990b), and that the neutralizing response must be due to the structural components of the virus as vaccines based on transformed cells which express only the early viral proteins but not the virion proteins (MOAR et al. 1981) are ineffective. This was subsequently confirmed by the use of subunit vaccines against BPV.

Recent studies on BPV vaccines have focussed on BPV-2 and BPV-4 which affect the skin and the mucous epithelium of the alimentary canal, respectively (JARRETT et al. 1984a)

4.1 BPV-2 Vaccines

The structural proteins L1 and L2 of BPV-2, produced in bacteria as β galactosidase fusion products (Fig. 1), were used in vaccination experiments (JARRETT et al. 1991). Cattle vaccinated with L1 were resistant to further challenge by BPV-2; protection was accompanied by production of virus-neutralizing antibodies, confirming that neutralizing epitopes are encoded by a structural protein. Similar results have been obtained with BPV-1 L1 (PILACHINSKI et al. 1986) and CRPV L1 (LIN et al. 1992), indicating conservation of immunological functions across papillomavirus types. Vaccination with BPV-2 L2 produced surprising results. Vaccinated animals were not

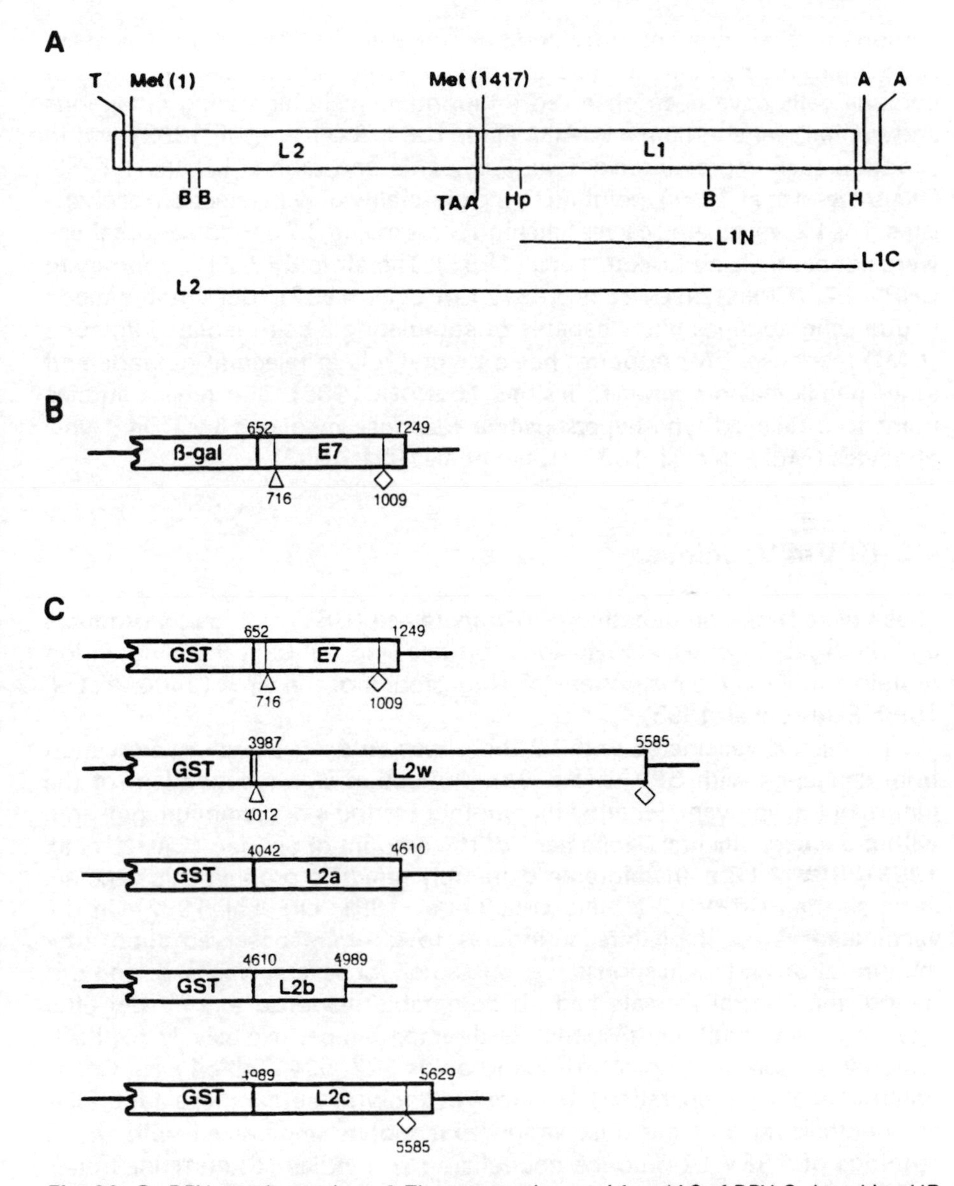

Fig. 1A–C. BPV protein vaccines. **A** The structural genes L1 and L2 of BPV-2 cloned in pUR (RUTHER and MULLER-HILL 1983) as β-gal fusions. *T*, TATA box; *A*, polyadenylation site; *Met*, translation initiation codon; *TAA*, translational termination codon; *B*, BamHI site; *Hp*, HpaI site; H, HindIII site. The DNA fragments cloned in pUR are indicated *L1N* (HpaI-BamHI), *L1C* (BamHI-HindIII) and *L2* (BamHI-BamHI). For details see JARRETT et al. (1991). **B** The E7 gene of BPV-4 cloned in pUR. Nucleotide positions are shown. The *triangle* points to the ATG codon and the *diamond* to the termination codon. **C** The E7 and L2 genes of BPV-4 cloned in pGEX (SMITH and JOHNSON 1988) as glutathione-S-transferase fusions. The L2 open reading frame was cloned in its entirety (L2w), and as three fragments: a 5′ end fragment (*L2a*), a middle fragment (*L2b*) and a 3′ end fragment (*L2c*). Nucleotide positions are shown. The *triangles* point to the ATG codon and the *diamonds* to the termination codons. For details see CAMPO et al. (1993)

immune to challenge and developed skin warts; the warts, however, were soon infiltrated by immune cells and rapidly regressed. Similar infiltrates of immune cells have been observed in spontaneously regressing cutaneous and genital warts in humans (AIBA et al. 1986; BENTON et al. 1992) and in spontaneously regressing skin warts in cattle and rabbits (JARRETT 1985; OKABAYASHI et al. 1991), pointing to the generality of wart rejection mechanisms. The L2-vaccinated calves had high titres of anti-L2 antibodies but these were not neutralizing (JARRETT et al. 1991). Therefore BPV-2 L2, contrary to CRPV L2 (CHRISTENSEN et al. 1991; LIN et al. 1992), does not encode neutralizing epitopes but is capable of stimulating a cell-mediated immune (CMI) response. CMI response has a pivotal role in rejection of warts and other papillomavirus-induced lesions (BENDER 1986), and recent studies point to a delayed type hypersensitive response mediated by $CD4^+$ lymphocytes (MCLEAN et al. 1993; HOPFL et al. 1991, 1993).

4.2 BPV-4 Vaccines

These were based on glutathione-S-transferase (GST)–L2 fusion products and on β-gal–E7 or GST–E7 fusion products (Fig. 1). L2 is the minor virion protein and E7 is the major transforming protein of the virus (JAGGAR et al. 1990; PENNIE et al. 1993).

In animals vaccinated with L2 there was virtually complete protection from challenge with BPV-4 (Fig. 2a). Protection was independent of the nature of the adjuvant, whether incomplete Freund's or aluminium gel, and, within the experimental parameters, of the amount of antigen (CAMPO et al. 1993). BPV-4 L2 is therefore an extremely effective prophylactic vaccine, more so than CRPV L2 (CHRISTENSEN et al. 1991; LIN et al. 1992). In the vaccinated calves, high-titre antibodies to L2 were observed soon after immunization and the response was sustained for several weeks; during this period, the control animals had no detectable response to L2 even after challenge. The antibody response is directed almost exclusively to the C terminal portion of the protein (amino acids 327–524; Table 1) (L. Chandrachud et al., in preparation); it is not yet known whether these antibodies are neutralizing, but the observation that rabbits immunized with the C terminus of CRPV L2 produce neutralizing antibodies (CHRISTENSEN et al. 1991) suggests that the same may be the case for BPV-4 L2.

Animals vaccinated with BPV-4 E7 were not protected from challenge and developed alimentary canal papillomas. The papillomas, however, never achieved their full size and rapidly regressed (Fig. 2b) (CAMPO et al. 1993). E7 therefore proved to be a successful therapeutic vaccine. Vaccination with E7 was followed by vigorous humoral and cellular immune response to the vaccine. Both responses appeared much earlier and had a greater amplitude in the vaccinated animals than in the control calves. Indeed, some of the control animals never developed detectable antibodies to E7 throughout the

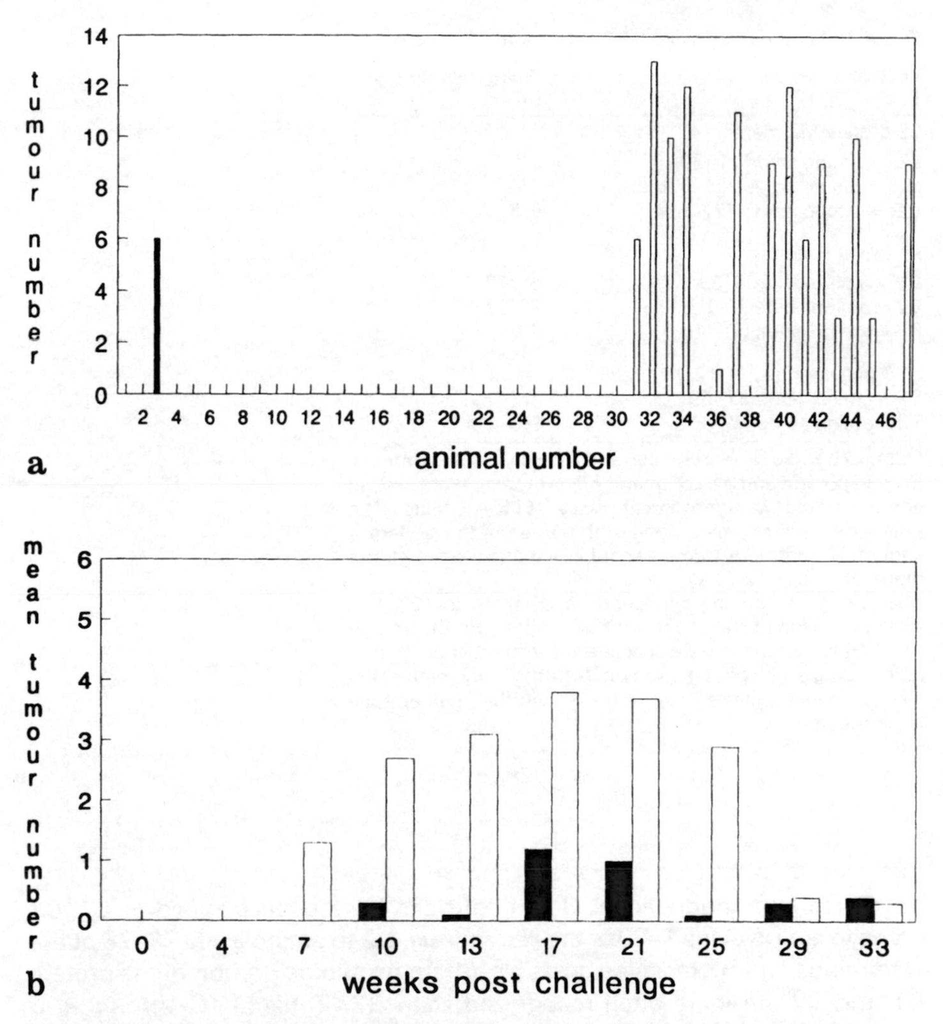

Fig. 2a, b. Prophylactic and therapeutic vaccination against BPV-4. **a** L2 vaccination. Tumour numbers in vaccinated and control animals 11 weeks post-challenge; animals 1–15 were inoculated with 1 mg/animal of L2 fusion peptides and animals 16–30 with 100 μg/animal. Adjuvant was aluminium gel. ***Solid column***, 1–30, L2 vaccine; ***open columns***, 31–47, control. For details see CAMPO et al. (1993). **b** E7 vaccination. Mean number of stage 3 tumours per animal in vaccinated (*solid columns*) and control animals (*open columns*). Vaccinated animals were inoculated with 1 mg/animal of E7 fusion peptide in incomplete Freund's adjuvant. For details see CAMPO et al. (1993)

course of the experiment (CHANDRACHUD et al. 1994). The T cell response was particularly strong especially when compared to control animals (G. McGarvie et al., in preparation) and may explain the efficacy of the vaccine, although the possible contribution of anti-E7 antibodies cannot be discounted. Vaccine E7 is therefore successfully presented to both effector arms of the immune system while viral E7 is poorly presented.

Table 1. Immune response to BPV-4 L2 and E7 vaccines

Vaccine	Responder animals (*n*)
L2 protein fragments[a]	
L2a = amino acids 11–200	2/6
L2b = amino acids 201–326	0/6
L2c = amino acids 327–524	6/6
E7 B cell epitopes[b]	
B1 = amino acids 1–30	21/27
B2 = amino acids 79–98	22/27
B3 = amino acids 50–69	6/27
E7 T cell epitopes[b]	
T1 = amino acids 31–59	2/3
T2 = amino acids 70–88	1/3

[a] L2a, L2b and L2c, representing, respectively, the N terminus, the mid portion and the C terminus of the protein, were used in enzyme-linked immunosorbent assay (ELISA) tests. The number of animals recognizing each portion of the protein is shown. The antibody titres against L2c are three to nine times higher than against L2a.
[b] Synthetic overlapping oligopeptides spanning the E7 proteins were used to map the B and T cell epitopes in ELISA and T cell proliferation assays, respectively. The number of animals recognizing each epitope is shown. Note that only three of the E7-vaccinated animals have been analysed for T cell epitope recognition.

Three immunodominant B cell epitopes have been mapped in E7. B1 maps to amino acids 1–30 at the N terminus, B2 to amino acids 79–98 at the C terminus and B3 to amino acids 51–69 in the middle portion of the protein B1 and B2 are more often recognized than B3 (Table 1) (CHANDRACHUD et al. 1994). Two T cell epitopes have also been mapped in E7: T1 maps to amino acids 31–59 and T2 to amino acids 70–88; T1 appears to be recognized more often than T2 (Table 1), although this analysis is still in progress (G. McGarvie et al., unpublished observations). An interesting observation derives from these results. The regions of BPV-4 E7 where the epitopes have been mapped have amino acid and functional homology to HPV-16 E7 (Rb-binding domain I and II and Zn-binding domain) (JAGGAR et al. 1990) and the same areas in HPV-16 E7 have been shown to contain B and T cell epitopes (KRCHNAK et al. 1990; TINDLE et al. 1990, 1991; COMERFORD et al. 1991). The immunological homology between conserved areas of BPV-4 and HPV-16 E7 proteins gives weight to the suggestion that HPV-16 E7 can be a successful therapeutic vaccine in humans as BPV-4 E7 is in cattle.

5 Prospects for Human Vaccines

The results obtained in the cattle and rabbit models encourage the optimistic prediction that vaccination will soon be possible also in human subjects against the most problematic or dangerous forms of HPVs.

Of course, the vaccination protocols used for animals will have to be changed and adapted for human use. Thus it is unlikely that a bacterial fusion product will be accepted for treatment of humans; it remains to be seen if viral proteins free of bacterial constituents are as effective vaccines as their fusion counterparts, although the experiments done with CRPV L1 expressed in recombinant vaccinia virus (LIN et al. 1992) provide strong indication that this is the case. Also, bacterially expressed proteins are unlikely to be accepted for human vaccines because of the possible presence of endotoxins in the preparation. On the other hand, yeast is already in use as a source of HBV surface antigen for vaccination against HBV (STEPHENNE 1988; KNISKERN et al. 1989) and it therefore provides the best candidate for HPV protein production. Concerning the formulation of the vaccine, the observation that, in the case of BPV-4 L2, aluminium gel is as good an adjuvant as incomplete Freund's is an important one as alum is an accepted adjuvant for human vaccines.

Probably the most important and most difficult questions regarding HPV vaccination concern which groups will need vaccination and when, and whether to vaccinate prophylactically or therapeutically.

In the opinion of this writer, the vaccine of choice would be prophylactic and directed against the HPVs that pose the highest risk for genital cancer, and would be administered to the whole population at a young age. Of course, a programme of prophylactic vaccination would take decades to make whole populations immune, and the population sectors currently at risk, or already infected, would not be affected. These could be targeted with either prophylactic or therapeutic vaccines. While this opinion is shared by several experts in each of the different aspects of the field, others disagree primarily on the basis that the link between papillomavirus infection and genital cancer is not yet firmly established, and that it would not be necessary to vaccinate males as the mode of spread of the virus may not be only be sexual contact. Concerning the former point, although it may never be possible to completely dispel doubts about the role of viruses in human cancer, vaccination could indeed be the final proof of viral involvement. If prevention of viral infection leads to prevention of cancer, then the link between cause and effect will have been established. This argument holds true not only for HPV but also for HBV, Epstein-Barr virus (EBV), human T cell leukaemia virus (HTLV) and any other virus which may be found associated with cancer. The current extensive vaccination programme against HBV is expected to lead to a marked decrease in the incidence of hepatocellular carcinoma. Concerning the latter point, while it is true that the genital HPVs may spread by non-sexual routes, sexual contact is the primary one. Males are infected just like females, and it would not make sense to vaccinate half of the population only to leave a reservoir of infectious virus in the other half.

Surely all these points must be debated and resolved, but the signs are that it will not be long before HPV vaccines enter clinical trials.

Acknowledgments. I wish to thank all my colleagues and collaborators who have helped with this review. My own work is supported by the Cancer Research Campaign.

References

Aiba S, Rokego M, Tagaini H (1986) Immunohistological analysis of the phenomenon of spontaneous regression of numerous flat warts. Cancer 58: 1246–1252

Bender ME (1986) Concepts of wart regression. Arch Dermatol 122: 644–648

Benton C, Shahidullah H, Hunter JAA (1992) Human papillomavirus in the immunosuppressed. Papillomavirus Rep 3: 23–26

Brandsma JL, Yang Z-H, Barthold SW, Johnson EA (1991) Use of a rapid efficient inoculation method to induce papillomas by cottontail rabbit papillomavirus DNA shows that the E7 gene is required. Proc Natl Acad Sci USA 88: 4816–4820

Brinton LA (1992) Epidemiology of cervical cancer-overview. In: Munoz N, Bosch FX, Shah KV, Meheus A (eds) The epidemiology of human papillomavirus and cervical cancer. IARC Sci Publ 119: 3–23

Campo MS, Jarrett WFH (1986) Papillomavirus infection in cattle: viral and chemical cofactors in naturally occurring and experimentally induced tumours. Ciba Found Symb 120: 117–135

Campo MS, Moar MH, Jarrett WFH, Laird HM (1980) A new papillomavirus associated with alimentary cancer in cattle. Nature 286: 180–182

Campo MS, Moar MH, Laird HM, Jarrett WFH (1981) Molecular heterogeneity and lesion site specificity of cutaneous bovine papillomaviruses. Virology 113: 323–335

Campo MS, Jarrett WFH, Barron R, O'Neil BW, Smith KT (1992) Association of bovine papillomavirus type 2 and bracken fern with bladder cancer in cattle. Cancer Res 53: 1–7

Campo MS, Grindlay GJ, O'Neil BW, Chandrachud LM, McGarvie GM, Jarrett WFH (1993) Prophylactic and therapeutic vaccination against a mucosal papillomavirus. J Gen Virol 74: 945–953

Chandrachud LM, O'Neil BW, Jarrett WFH, Grindlay GJ, McGarvie GM, Campo MS (1994) Humoral immune response to the E7 protein of bovine papillomavirus type 4 and identification of B-cell epitopes. Virology (in press)

Christensen ND, Kreider JW, Kan NC, DiAngelo SL (1991) The open reading frame L2 of cottontail rabbit papillomavirus contains antibody-inducing neutralizing epitopes. Virology 181: 582–589

Comerford SA, McCance DJ, Dougan G, Title JP (1991) Identification of T- and B-cell epitopes in the E7 protein of human papillomavirus type 16. J Virol 65: 4681–4690

Evans IA, Prorok JH, Cole RC, Al-Samani MH, Al-Samarrai AM, Patel MC, Smith RMN (1982) The carcinogenic, mutagenic and teratogenic toxicity of bracken. Proc R Soc Edinb 81: 65–77

Evans WC, Patel MC, Koohy Y (1982) Acute bracken poisoning in homogastric and ruminant animals. Proc R Soc Edinb 81: 29–64

Frazer IH, Tindle RW (1992) Cell-mediated immunity to papillomaviruses. Papillomavirus Report 3: 53–58

Han R, Breitburd F, Marche PN, Orth G (1992) Linkage of regression and malignant conversion of rabbit viral papillomas to MHC class II genes. Nature 357: 66–68

Holly EA, Petrakis NL, Friend NF, Sarles DL, Lee RE, Flander LB (1986) Mutagenic mucus in the cervix of smokers. J Natl Cancer Inst 76: 983–986

Holt PG (1987) Immune and inflammatory function in cigarette smokers. Thorax 42: 241–250
Hopfl R, Sandbichler M, Sepp N, Heim K, Muller-Holzner E, Wartusch B, Dapunt O, Jochmus-Kudielka I, ter Meulen J, Gissman L, Fritsch P (1991) Skin test for HPV type 16 proteins in cervical intraepithelial neoplasia. Lancet 337: 373–374
Hopfl R, Christensen ND, Angell MG, Kreider JW (1993) Skin test to assess immunity against cottontail rabbit papillomavirus antigens in rabbits with progressing papillomas or after papilloma regression. J Invest Dermatol 101: 227–231
Jaggar RT, Pennie WD, Smith KT, Jackson ME, Campo MS (1990) Cooperation between bovine papillomavirus type 4 and ras in the morphological transformation of primary bovine fibroblasts. J Gen Virol 71: 3041–3046
Jarret WFH (1985) The natural history of bovine papillomavirus infection. Adv Viral Oncol 5: 83–102
Jarrett WFH, McNeal PE, Grimshaw TR, Selman IE, McIntyre WIM (1978) High incidence area of cattle cancer with a possible interaction between an environmental carcinogen and a papillomavirus. Nature 274: 215–217
Jarrett WFH, Campo MS, O'Neil BW, Laird HM, Coggins LW (1984a) A novel bovine papillomavirus (BPV-6) causing true epithelial papillomas of the mammary gland skin: a member of a proposed new subgroup. Virology 136: 256–264
Jarrett WFH, Campo MS, Blaxter ML, O'Neil BW, Laird HM, Moar MH, Sartirana ML (1984b) Alimentary fibropapilloma in cattle: a spontaneous tumour, nonpermissive for papillomavirus replication. J Natl Cancer Inst 73: 499–504
Jarrett WFH, O'Neil BW, Gaukroger JM, Smith KT, Laird HM, Campo MS (1990a) Studies on vaccination against papillomaviruses: the immunity after infection and vaccination with bovine papillomaviruses of different types. Vet Rec 126: 483–485
Jarrett WFH, O'Neil BW, Gaukroger JM, Laird HM, Smith KT, Campo MS (1990b) Studies on vaccination against papillomaviruses: a comparison of purified virus, tumour extract and transformed cells in prophylactic vaccination. Vet Rec 126: 450–453
Jarrett WFH, Smith KT, O'Neil BW, Gaukroger JM, Chandrachud LM, Grindlay GJ, McGarvie GM, Campo MS (1991) Studies on vaccination against papillomaviruses: prophylactic and therapeutic vaccination with recombinant structural proteins. Virology 184: 33–42
Kniskern PJ, Hagopian A, Burke P, Dunn N, Montgomery DL, Schultz LD, Schulman CA, Carty CE, Maitgetter RZ, Wampler DE, Lehman ED, Yamazaki S, Kubek DJ, Emini EA, Miller WJ, Hurni WM, Ellis RW (1989) The application of molecular biology to the development of novel vaccines. Adv Exp Med Biol 252: 83–98
Krchnak V, Vagner J, Suchankova A, Krcmar M, Ritterova L, Vonka V (1990) Synthetic peptides derived from E7 region of human papillomavirus type 16 used as antigens in ELISA. J Gen Virol 71: 2719–2724
Kreider JW, Bartlett GL (1981) The Shope papilloma-carcinoma complex of rabbits: a model system of neoplastic progression and spontaneous rejection. Adv Cancer Res 35: 81–110
Lin Y-L, Borenstein LA, Selvakumar R, Ahmed R, Wettstein FO (1992) Effective vaccination against papilloma development by immunization with L1 or L2 structural proteins of cottontail rabbit papillomavirus. Virology 187: 612–619
McLean CS, Sterling JS, Mowat J, Nash AA, Stanley MA (1993) Delayed type hypersensitivity response to the human papillomavirus type 16 E7 protein in a mouse model. J Gen Virol 74: 239–245
Moar MH, Campo MS, Laird HM, Jarrett WFH (1981) Persistence of non-integrated viral DNA in bovine cells transformed in vitro by bovine papillomavirus type 2. Nature 293: 750–752
Okabayashi M, Angell MG, Christensen ND, Kreider JW (1991) Morphometric analysis and identification of infiltrating leukocytes in regressing and progressing Shope rabbit papillomas. Int J Cancer 50: 919–923
Olson C, Skidmore LV (1959) Therapy of experimentally produced bovine cutaneous papillomatosis with vaccines and excision. J Am Vet Med Assoc 135: 339–343
Olson C, Segre D, Skidmore LV (1960) Further observations on immunity to bovine cutaneous papillomatosis. Am J Vet Res 21: 233–242
Oriel JD (1971) Natural history of genital warts. Br J Vener Dis 47: 1–13
Peng X, Olson RO, Christan CB, Lang CM, Kreider JW (1993) Papillomas and carcinomas in transgenic rabbits carrying EJ-ras DNA and cottontail rabbit papillomavirus. J Virol 67: 1698–1701
Pennie WD, Campo MS (1992) Synergism between bovine papillomavirus type 4 and the flavonoid quercetin in cell transformation in vitro. Virology 190: 861–865

Pennie WD, Grindlay GJ, Cairney M, Campo MS (1993) Analysis of the transforming functions of bovine papillomavirus type 4. Virology 193: 614–620
Phillips DH, Ni She M (1993) Smoking-related DNA adducts in human cervical biopsies. In: Phillips DH, Castegtnaro M, Bartsch H (eds) Postlabelling methods for the detection of DNA damage. IARC Sci Publ 124: 327–330
Pilachinski WP, Glassman DL, Glassman KF, Reed DE, Lum MA, Marshall RF, Muscoplat CC, Faras AJ (1986) Immunization against bovine papillomavirus infection. Ciba Found Symp 120: 136–149
Ruther U, Muller-Hill B (1983) Easy identification of cDNA clones. EMBO J 2: 1791–1794
Shepherd PS, Tran TTT, Rowe AJ, Cridland JC, Comerford SA, Chapman MG, Rayfield LS (1992) T-cell responses to the human papillomavirus type 16 E7 protein in mice of different aplotypes. J Gen Virol 73: 1269–1274
Shope RE (1937) Immunization of rabbits to infectious papillomatosis. J Exp Med 65: 219–231
Smith BD, Johnson KS (1988) Single step purification of polypeptides expressed in Escherichia coli as fusions with glutathione-S-transferase. Gene 67: 31–40
Steinberg BM (1987) Laryngeal papillomas. Clinical aspects and in vitro studies. In: Salzman NP, Howley PM (eds) The papovaviridae, vol 2. Plenum, New York, pp 265–292
Stephenne J (1988) Recombinant versus plasma-derived hepatitis B vaccines: issues of safety, immunogenicity and cost-effectiveness. Vaccine 6: 299–303
Sundberg JP, Burnstein T, Page EH, Kirkham WW, Robinson FR (1977) Neoplasms in equidae. J Am Vet Med Assoc 170: 150–152
Tindle RW, Smith JA, Geysen HM, Selvey LA, Frazer IH (1990) Identification of B-epitopes in human papillomavirus type 16 E7 open reading frame protein. J Gen Virol 71: 1347–1354
Tindle RW, Fernando GJ, Sterling JC, Frazer IH (1991) A "public" T-helper epitope of the E7 transforming protein of human papillomavirus 16 provides cognate help for several E7 B-cell epitopes from cervical cancer-associated human papillomavirus genotypes. Proc Natl Acad Sci USA 88: 5887–5891
Wettstein FO (1987) Papillomaviruses and carcinogenic progression: cottontail rabbit papillomavirus. In: Salzman NP, Howley PM (eds) The papovaviridae, vol 2. Plenum, New York, pp 167–186

Subject Index

Current Topics in Microbiology and Immunology

Volumes published since 1989 (and still available)

Vol. 142: **Schüpach, Jörg:** Human Retrovirology. Facts and Concepts. 1989. 24 figs. 115 pp. ISBN 3-540-50455-9

Vol. 143: **Haase, Ashley T.; Oldstone, Michael B. A. (Ed.):** In Situ Hybridization. 1989. 22 figs. XII, 90 pp. ISBN 3-540-50761-2

Vol. 144: **Knippers, Rolf; Levine, A. J. (Ed.):** Transforming. Proteins of DNA Tumor Viruses. 1989. 85 figs. XIV, 300 pp. ISBN 3-540-50909-7

Vol. 145: **Oldstone, Michael B. A. (Ed.):** Molecular Mimicry. Cross-Reactivity between Microbes and Host Proteins as a Cause of Autoimmunity. 1989. 28 figs. VII, 141 pp. ISBN 3-540-50929-1

Vol. 146: **Mestecky, Jiri; McGhee, Jerry (Ed.):** New Strategies for Oral Immunization. International Symposium at the University of Alabama at Birmingham and Molecular Engineering Associates, Inc. Birmingham, AL, USA, March 21–22, 1988. 1989. 22 figs. IX, 237 pp. ISBN 3-540-50841-4

Vol. 147: **Vogt, Peter K. (Ed.):** Oncogenes. Selected Reviews. 1989. 8 figs. VII, 172 pp. ISBN 3-540-51050-8

Vol. 148: **Vogt, Peter K. (Ed.):** Oncogenes and Retroviruses. Selected Reviews. 1989. XII, 134 pp. ISBN 3-540-51051-6

Vol. 149: **Shen-Ong, Grace L. C.; Potter, Michael; Copeland, Neal G. (Ed.):** Mechanisms in Myeloid Tumorigenesis. Workshop at the National Cancer Institute, National Institutes of Health, Bethesda, MD, USA, March 22, 1988. 1989. 42 figs. X, 172 pp. ISBN 3-540-50968-2

Vol. 150: **Jann, Klaus; Jann, Barbara (Ed.):** Bacterial Capsules. 1989. 33 figs. XII, 176 pp. ISBN 3-540-51049-4

Vol. 151: **Jann, Klaus; Jann, Barbara (Ed.):** Bacterial Adhesins. 1990. 23 figs. XII, 192 pp. ISBN 3-540-51052-4

Vol. 152: **Bosma, Melvin J.; Phillips, Robert A.; Schuler, Walter (Ed.):** The Scid Mouse. Characterization and Potential Uses. EMBO Workshop held at the Basel Institute for Immunology, Basel, Switzerland, February 20–22, 1989. 1989. 72 figs. XII, 263 pp. ISBN 3-540-51512-7

Vol. 153: **Lambris, John D. (Ed.):** The Third Component of Complement. Chemistry and Biology. 1989. 38 figs. X, 251 pp. ISBN 3-540-51513-5

Vol. 154: **McDougall, James K. (Ed.):** Cytomegaloviruses. 1990. 58 figs. IX, 286 pp. ISBN 3-540-51514-3

Vol. 155: **Kaufmann, Stefan H. E. (Ed.):** T-Cell Paradigms in Parasitic and Bacterial Infections. 1990. 24 figs. IX, 162 pp. ISBN 3-540-51515-1

Vol. 156: **Dyrberg, Thomas (Ed.):** The Role of Viruses and the Immune System in Diabetes Mellitus. 1990. 15 figs. XI, 142 pp. ISBN 3-540-51918-1

Vol. 157: **Swanstrom, Ronald; Vogt, Peter K. (Ed.):** Retroviruses. Strategies of Replication. 1990. 40 figs. XII, 260 pp. ISBN 3-540-51895-9

Vol. 158: **Muzyczka, Nicholas (Ed.):** Viral Expression Vectors. 1992. 20 figs. IX, 176 pp. ISBN 3-540-52431-2

Vol. 159: **Gray, David; Sprent, Jonathan (Ed.):** Immunological Memory. 1990. 38 figs. XII, 156 pp. ISBN 3-540-51921-1

Vol. 160: **Oldstone, Michael B. A.; Koprowski, Hilary (Eds.):** Retrovirus Infections of the Nervous System. 1990. 16 figs. XII, 176 pp. ISBN 3-540-51939-4

Vol. 161: **Racaniello, Vincent R. (Ed.):** Picornaviruses. 1990. 12 figs. X, 194 pp. ISBN 3-540-52429-0

Vol. 162: **Roy, Polly; Gorman, Barry M. (Eds.):** Bluetongue Viruses. 1990. 37 figs. X, 200 pp. ISBN 3-540-51922-X

Vol. 163: **Turner, Peter C.; Moyer, Richard W. (Eds.):** Poxviruses. 1990. 23 figs. X, 210 pp. ISBN 3-540-52430-4

Vol. 164: **Bækkeskov, Steinnun; Hansen, Bruno (Eds.):** Human Diabetes. 1990. 9 figs. X, 198 pp. ISBN 3-540-52652-8

Vol. 165: **Bothwell, Mark (Ed.):** Neuronal Growth Factors. 1991. 14 figs. IX, 173 pp. ISBN 3-540-52654-4

Vol. 166: **Potter, Michael; Melchers, Fritz (Eds.):** Mechanisms in B-Cell Neoplasia 1990. 143 figs. XIX, 380 pp. ISBN 3-540-52886-5

Vol. 167: **Kaufmann, Stefan H. E. (Ed.):** Heat Shock Proteins and Immune Response. 1991. 18 figs. IX, 214 pp. ISBN 3-540-52857-1

Vol. 168: **Mason, William S.; Seeger, Christoph (Eds.):** Hepadnaviruses. Molecular Biology and Pathogenesis. 1991. 21 figs. X, 206 pp. ISBN 3-540-53060-6

Vol. 169: **Kolakofsky, Daniel (Ed.):** Bunyaviridae. 1991. 34 figs. X, 256 pp. ISBN 3-540-53061-4

Vol. 170: **Compans, Richard W. (Ed.):** Protein Traffic in Eukaryotic Cells. Selected Reviews. 1991. 14 figs. X, 186 pp. ISBN 3-540-53631-0

Vol. 171: **Kung, Hsing-Jien; Vogt, Peter K. (Eds.):** Retroviral Insertion and Oncogene Activation. 1991. 18 figs. X, 179 pp. ISBN 3-540-53857-7

Vol. 172: **Chesebro, Bruce W. (Ed.):** Transmissible Spongiform Encephalopathies. 1991. 48 figs. X, 288 pp. ISBN 3-540-53883-6

Vol. 173: **Pfeffer, Klaus; Heeg, Klaus; Wagner, Hermann; Riethmüller, Gert (Eds.):** Function and Specificity of γ/δ T Cells. 1991. 41 figs. XII, 296 pp. ISBN 3-540-53781-3

Vol. 174: **Fleischer, Bernhard; Sjögren, Hans Olov (Eds.):** Superantigens. 1991. 13 figs. IX, 137 pp. ISBN 3-540-54205-1

Vol. 175: **Aktories, Klaus (Ed.):** ADP-Ribosylating Toxins. 1992. 23 figs. IX, 148 pp. ISBN 3-540-54598-0

Vol. 176: **Holland, John J. (Ed.):** Genetic Diversity of RNA Viruses. 1992. 34 figs. IX, 226 pp. ISBN 3-540-54652-9

Vol. 177: **Müller-Sieburg, Christa; Torok-Storb, Beverly; Visser, Jan; Storb, Rainer (Eds.):** Hematopoietic Stem Cells. 1992. 18 figs. XIII, 143 pp. ISBN 3-540-54531-X

Vol. 178: **Parker, Charles J. (Ed.):** Membrane Defenses Against Attack by Complement and Perforins. 1992. 26 figs. VIII, 188 pp. ISBN 3-540-54653-7

Vol. 179: **Rouse, Barry T. (Ed.):** Herpes Simplex Virus. 1992. 9 figs. X, 180 pp. ISBN 3-540-55066-6

Vol. 180: **Sansonetti, P. J. (Ed.):** Pathogenesis of Shigellosis. 1992. 15 figs. X, 143 pp. ISBN 3-540-55058-5

Vol. 181: **Russell, Stephen W.; Gordon, Siamon (Eds.):** Macrophage Biology and Activation. 1992. 42 figs. IX, 299 pp. ISBN 3-540-55293-6

Vol. 182: **Potter, Michael; Melchers, Fritz (Eds.):** Mechanisms in B-Cell Neoplasia. 1992. 188 figs. XX, 499 pp. ISBN 3-540-55658-3

Vol. 183: **Dimmock, Nigel J.:** Neutralization of Animal Viruses. 1993. 10 figs. VII, 149 pp. ISBN 3-540-56030-0

Vol. 184: **Dunon, Dominique; Mackay, Charles R.; Imhof, Beat A. (Eds.):** Adhesion in Leukocyte Homing and Differentiation. 1993. 37 figs. IX, 260 pp. ISBN 3-540-56756-9

Vol. 185: **Ramig, Robert F. (Ed.):** Rotaviruses. 1994. Approx. IX, 375 pp. ISBN 3-540-56761-5

Springer-Verlag and the Environment

We at Springer-Verlag firmly believe that an international science publisher has a special obligation to the environment, and our corporate policies consistently reflect this conviction.

We also expect our business partners – paper mills, printers, packaging manufacturers, etc. – to commit themselves to using environmentally friendly materials and production processes.

The paper in this book is made from low- or no-chlorine pulp and is acid free, in conformance with international standards for paper permanency.

Lightning Source UK Ltd.
Milton Keynes UK
UKOW05f0131200913

217523UK00001B/11/P